OXFORD STATISTICAL SCIENCE SERIES

SERIES EDITORS

A. C. ATKINSON   J. B. COPAS
D. A. PIERCE   M. J. SCHERVISH
D. M. TITTERINGTON

OXFORD STATISTICAL SCIENCE SERIES

A. C. Atkinson: *Plots, Transformations, and Regression*
M. Stone: *Coordinate-Free Multivariable Statistics*
W. J. Krzanowski: *Principles of Multivariate Analysis: A User's Perspective*
M. Aitkin, D. Anderson, B. Francis, and J. Hinde: *Statistical Modelling in GLIM*
Peter J. Diggle: *Time Series: A Biostatistical Introduction*
Howell Tong: *Non-linear Time Series: A Dynamical Approach*
V. P. Godambe: *Estimating Functions*
A. C. Atkinson and A. N. Donev: *Optimum Experimental Designs*
U. N. Bhat and I. V. Basawa: *Queuing and Related Models*
J. K. Lindsey: *Models for Repeated Measurements*
N. T. Longford: *Random Coefficient Models*
P. J. Brown: *Measurement, Regression, and Calibration*
Peter J. Diggle, Kung-Yee Liang, and Scott L. Zeger: *Analysis of Longitudinal Data*
J. I. Ansell and M. J. Phillips: *Practical Methods for Reliability Data Analysis*
J. K. Lindsey: *Modelling Frequency and Count Data*
J. L. Jensen: *Saddlepoint Approximations*
Steffen L. Lauritzen: *Graphical Models*
A. W. Bowman and A. Azzalini: *Applied Smoothing Methods for Data Analysis*

Graphical Models

STEFFEN L. LAURITZEN

Department of Mathematics and Computer Science
Aalborg University

CLARENDON PRESS · OXFORD

This book has been printed digitally and produced in a standard specification in order to ensure its continuing availability

OXFORD
UNIVERSITY PRESS

Great Clarendon Street, Oxford OX2 6DP

Oxford University Press is a department of the University of Oxford.
It furthers the University's objective of excellence in research, scholarship,
and education by publishing worldwide in

Oxford New York

Auckland Bangkok Buenos Aires Cape Town Chennai
Dar es Salaam Delhi Hong Kong Istanbul Karachi Kolkata
Kuala Lumpur Madrid Melbourne Mexico City Mumbai Nairobi
São Paulo Shanghai Taipei Tokyo Toronto

Oxford is a registered trade mark of Oxford University Press
in the UK and in certain other countries

Published in the United States
by Oxford University Press Inc., New York

© Steffen L. Lauritzen 1996

The moral rights of the author have been asserted

Database right Oxford University Press (maker)

Reprinted 2004 with corrections

All rights reserved. No part of this publication may be reproduced,
stored in a retrieval system, or transmitted, in any form or by any means,
without the prior permission in writing of Oxford University Press,
or as expressly permitted by law, or under terms agreed with the appropriate
reprographics rights organization. Enquiries concerning reproduction
outside the scope of the above should be sent to the Rights Department,
Oxford University Press, at the address above

You must not circulate this book in any other binding or cover
And you must impose this same condition on any acquirer

ISBN 0-19-852219-3

Preface

In 1976 Terry Speed invited me to Perth, Australia where he conducted a research seminar exploring relations between statistics and statistical physics. Among other things we studied the relation between the notion of interaction as used in contingency table analysis and in thermodynamics. To our delight they were formally the same and one of the most inspiring periods in my life as a researcher was initiated. In the next couple of months we worked day and night and laid essentially the foundations for the papers Darroch *et al.* (1980) and Lauritzen *et al.* (1984).

In 1979 I was invited to lecture at the Swedish summer school in statistics. Here I met Nanny Wermuth and with her as the main source of inspiration we set out to investigate possibilities for making graphical models that simultaneously dealt with discrete and continuous variables.

In 1985 David Spiegelhalter contacted me with the purpose of discussing possibilities for using graphical models in artificial intelligence.

These three meetings have had a profound influence on my life as researcher in general and in particular on the development of the material described in the present book. I am deeply indebted to the inspiration and ideas that Terry, Nanny, David, and others have provided.

Over the years I have enjoyed being part of many research groups. Here I will particularly mention three. The Danes that have been enthusiastically interested in graphical models from their early beginning: Jens Henrik Badsberg, David Edwards, Morten Frydenberg, Svend Kreiner, and a bit later also Poul Svante Eriksen. The BAIES group working on probabilistic expert systems, consisting of Robert Cowell, Phil Dawid, and David Spiegelhalter. The ODIN group in Aalborg involving far too many to mention, but Stig Andersen, Finn Jensen, Frank Jensen, Uffe Kjærulff, and Kristian Olesen have been there all the time. The enthusiasm from members of these groups has kept up my spirits.

Countless colleagues in Denmark and throughout the world have encouraged me every time I was losing hope and energy. It is plainly impossible to mention them all, but it is clear that without these the book would never have existed. It is a privilege to belong to a scientific community with so many fine people.

Heidi Andersen, David Edwards, Jinglong Wang and Dorte Sørensen have read parts of the book in some detail and I am extremely grateful for their comments and criticism. This also applies to the students at Aalborg

University who were exposed to the somewhat compact material in the spring of 1993 and survived.

Various institutions deserve thanks for hosting me while I was writing parts of the book. This includes the Statistical Laboratory in Cambridge, where I began the writing during my sabbatical leave in 1987. Chapter 6 was largely written during a wonderful stay at the institution of San Cataldo, a former nunnery beautifully situated in Southern Italy and run as a study home for Danish researchers and artists.

The Danish Research Councils and the Carlsberg Foundation have in various ways contributed financially to the research.

Oxford University Press and my colleague Frank Jensen have been of invaluable assistance with typographical matters.

Aalborg   S.L.L.
December 1995

Contents

1 Introduction   **1**
    1.1 Graphical models 1
    1.2 Outline of book 2

2 Graphs and hypergraphs   **4**
    2.1 Graphs 4
        2.1.1 Notation and terminology 4
        2.1.2 Decompositions of marked graphs 7
        2.1.3 Simplicial subsets and perfect sequences 13
        2.1.4 Subgraphs of decomposable graphs 19
    2.2 Hypergraphs 21
        2.2.1 Basic concepts 21
        2.2.2 Graphs and hypergraphs 22
        2.2.3 Junction trees and forests 24
    2.3 Notes 26

3 Conditional independence and Markov properties   **28**
    3.1 Conditional independence 28
    3.2 Markov properties 32
        3.2.1 Markov properties on undirected graphs 32
        3.2.2 Markov properties on directed acyclic graphs 46
        3.2.3 Markov properties on chain graphs 53
    3.3 Notes 60

4 Contingency tables   **62**
    4.1 Examples 62
    4.2 Basic facts and concepts 67
        4.2.1 Notation and terminology 67
        4.2.2 Saturated models 70
        4.2.3 Log–affine and log–linear models 71
    4.3 Hierarchical models 81
        4.3.1 Estimation in hierarchical log–affine models 82
        4.3.2 Test in hierarchical models 85
        4.3.3 Interaction graphs and graphical models 88
    4.4 Decomposable models 90

	4.4.1 Basic factorizations	90
	4.4.2 Maximum likelihood estimation	91
	4.4.3 Exact tests in decomposable models	98
	4.4.4 Asymptotic tests in decomposable models	105
4.5	Recursive models	106
	4.5.1 Recursive graphical models	107
	4.5.2 Recursive hierarchical models	112
4.6	Block-recursive models	113
	4.6.1 Chain graph models	114
	4.6.2 Block-recursive hierarchical models	118
	4.6.3 Decomposable block-recursive models	119
4.7	Notes	121
	4.7.1 Collapsibility	121
	4.7.2 Bibliographical notes	121

5 Multivariate normal models 123

5.1	Basic facts and concepts	123
	5.1.1 Notation	123
	5.1.2 The saturated model	124
	5.1.3 Conditional independence	129
	5.1.4 Interaction	131
5.2	Covariance selection models	131
	5.2.1 Maximum likelihood estimation	132
	5.2.2 Deviance tests	142
5.3	Decomposable models	144
	5.3.1 Basic factorizations	144
	5.3.2 Maximum likelihood estimation	145
	5.3.3 Exact tests in decomposable models	149
5.4	Notes	153
	5.4.1 Chain graph models	153
	5.4.2 Lattice models	156
	5.4.3 Collapsibility	156
	5.4.4 Bibliographical notes	156

6 Models for mixed data 158

6.1	Basic facts and concepts	158
	6.1.1 CG distributions	158
	6.1.2 The saturated models	168
6.2	Graphical interaction models	173
	6.2.1 CG interactions	173
	6.2.2 Maximum likelihood estimation	175
6.3	Decomposable models	187
	6.3.1 Basic factorizations	187
	6.3.2 Maximum likelihood estimation	188

		6.3.3	Exact tests in decomposable models	191

- 6.4 Hierarchical interaction models 199
  - 6.4.1 General properties 199
  - 6.4.2 Generators and canonical statistics 201
  - 6.4.3 Maximum likelihood estimation 205
  - 6.4.4 Mixed hierarchical model subspaces 213
- 6.5 Chain graph models 216
  - 6.5.1 CG regressions 217
  - 6.5.2 Estimation in chain graph models 218
- 6.6 Notes 219
  - 6.6.1 Collapsibility 219
  - 6.6.2 Bibliographical notes 220

7 Further topics 221
- 7.1 Probabilistic expert systems 221
  - 7.1.1 Specification of the joint distribution 223
  - 7.1.2 Local computation algorithm 226
  - 7.1.3 Extensions 228
- 7.2 Model selection 229
- 7.3 Modelling complexity 230
  - 7.3.1 Markov chain Monte Carlo methods 231
  - 7.3.2 Applications 232
- 7.4 Missing-data problems 233
  - 7.4.1 The EM algorithm 233
  - 7.4.2 Hierarchical log–linear models 234
  - 7.4.3 Recursive models 235

Appendices

A Various prerequisites 237
- A.1 Inequalities 237
- A.2 Kullback–Leibler divergence 238
- A.3 Möbius inversion 239
- A.4 Iterative partial maximization 239
- A.5 Sufficiency 241

B Linear algebra and random vectors 243
- B.1 Matrix results 243
- B.2 Factor subspaces and interactions 246
- B.3 Random vectors 250

C The multivariate normal distribution 254
- C.1 Basic properties 254
- C.2 The Wishart distribution 258
- C.3 Other derived distributions 262
  - C.3.1 Box-type distributions 262
  - C.3.2 Wilks's distribution 263
  - C.3.3 Test for identical covariances 264

D Exponential models 266
- D.1 Regular exponential models 266
  - D.1.1 Basic terminology 266
  - D.1.2 Analytic properties 267
  - D.1.3 Maximum likelihood estimation 268
  - D.1.4 Affine hypotheses 268
  - D.1.5 Iterative computational methods 269
- D.2 Curved exponential models 272
  - D.2.1 The non-singular case 272
  - D.2.2 The singular case 276

Bibliography 278

Index 295

1
Introduction

1.1 Graphical models

Graphical models have their origin in several scientific areas. One of these is statistical physics. Here the ideas can be traced back to Gibbs (1902). The object of interest is a large system of particles, possibly the atoms of a gas or solid. Each particle is occupying a site where it can be in different states. The total energy of the system is composed by an external potential plus a potential due to *interaction* of groups of particles. It is usually assumed that only particles at sites close to each other interact. Sites that are close to each other are termed *neighbours* and an undirected graph is determined by the neighbour relationship. The total energy of the system is determining the behaviour of the system through the so-called Gibbsian distribution

$$p(x) = Z^{-1} \exp\{-E(x)\},$$

where $E(x)$ is the total energy of the system when this is in configuration x, and Z is a normalizing constant.

Another origin is in the subject of genetics, where the graphical models go back to Wright (1921, 1923, 1934), who founded the so-called path analysis. Here one is studying heritable properties of natural species and the graph relations are directed, with arrows moving from parent to child. Ideas of path analysis were later taken up in economics and social sciences (Wold 1954; Blalock 1971).

The third source has less obvious relation to graphs. Bartlett (1935) studied the notion of interaction in a three-way contingency table. It turns out that, apart from differences due to notational conventions, this notion of interaction is formally identical to the notion used in statistical physics (Darroch *et al.* 1980).

With this understanding, the graphical models are ready at hand for use in statistics. Their fundamental and universal applicability is due to a number of factors. Firstly, the graphs can visually represent the scientific content of a given model and facilitate communication between researcher and statistician. Secondly, the models are naturally modular so that complex problems can be described and handled by careful combination of simple elements. Thirdly, the graphs are natural data structures for modern digital computers. Thus models can be efficiently communicated to these and the road is paved for exploiting their computational power.

Over the years the theory and methodology have developed and been extended in a multitude of directions. This is true to such an extent that it has not been possible to cover the area properly in the present book. Luckily a number of other books have been written that take care of some of the omissions in the present. Whittaker (1990b) focuses on the ideas behind graphical modelling and has many examples. The same holds for the book by Edwards (1995), just that here the emphasis is on models for mixed data and special attention is paid to problems that are suitable for analysis using the program MIM. Cox and Wermuth (1996) is concentrating attention on the interpretation and application of multivariate systems, in particular graphical models. Neapolitan (1990) deals with probabilistic expert systems based on graphical models, and the collection Oliver and Smith (1990) is primarily concerned with graphical models as influence diagrams, i.e. used as tools in decision analysis.

The present book is primarily concerned with the fundamental mathematical and statistical theory of graphical models. The book is mostly based on a traditional statistical approach, discussing aspects of maximum likehood methods and significance testing in the different variety of models. It is believed that these results are basic and therefore hopefully will have interest for anyone who wants to enter the world of graphical models, independently of statistical paradigms.

The book is a research monograph and it has been written primarily for researchers and graduate students in mathematical statistics. However, I have tried to keep the mathematics at a reasonable level of abstraction and exactness. No doubt, certain sections are more difficult to read than others, but it is my hope that a relatively wide audience will be able to take advantage of the central parts of the book.

1.2 Outline of book

There are five main chapters. In Chapter 2 the basic graph theory is developed. It may not be advisable to read this chapter in detail before continuing to chapters with a more statistical content, but some familiarity with the notions is necessary to understand even the basic ideas in the subsequent developments. Some of the heavier parts can be omitted at first reading and then returned to when necessary.

Chapter 3 is concerned with conditional independence and Markov properties and these are also essential for later chapters. If desired, the reader can skip the description of directed and chain graph Markov properties at first reading, and then return to these later.

The next three chapters of the book form the core of the statistical theory. Chapter 4 is concerned with discrete variables, Chapter 5 with variables that are normally distributed, and Chapter 6 describes models for systems that contain both discrete and continuous variables. These three chapters are largely written so that they can be read independently of each other. However, as the results of Chapter 6 unify and generalize the results in Chapters 4 and 5, some readers may want to read them in the given order. On the other hand, Chapter 6 is the chapter which contains most original material that has not appeared elsewhere in the literature and others may want to attack this chapter directly.

Chapter 7 is largely a guide to the literature on topics that have not been treated in the book but constitute important and natural extensions.

Finally the book contains four appendices with some material that is used at various places and may not be sufficiently familiar to the reader. Appendix A is a collection of various little topics. Appendix B has some results from linear algebra. Appendix C treats the multivariate normal distribution in some detail, and Appendix D describes basic elements of the theory of exponential families. By having these ready at hand I hope to help readers who do not have all of this present in their minds. However, it is not advisable to learn the material in all four appendices from the given expositions alone.

2
Graphs and hypergraphs

2.1 Graphs

2.1.1 Notation and terminology

A *graph*, as we use it throughout this book, is a pair $\mathcal{G} = (V, E)$, where V is a finite set of *vertices* and the set of *edges* E is a subset of the set $V \times V$ of ordered pairs of distinct vertices. Thus our graphs are *simple*, i.e. there are no multiple edges and they have no loops.

Edges $(\alpha, \beta) \in E$ with both (α, β) and (β, α) in E are called *undirected*, whereas an edge (α, β) with its *opposite* (β, α) not in E is called *directed*.

In a large part of the book we deal with graphs where the vertices are *marked* in the sense that they are partitioned into groups. We then use the term *marked graph*. Throughout the book, vertices in marked graphs are partitioned into two types, such that the vertex set has the structure

$$V = \Delta \cup \Gamma \text{ with } \Delta \cap \Gamma = \emptyset.$$

The vertices in the set Δ represent qualitative variables and those in Γ quantitative variables. Therefore we say that the vertices in Δ are *discrete* and the vertices in Γ are *continuous*. A graph is *pure* if it has only one kind of vertex.

A basic feature of the notion of a graph is that it is a visual object. It is conveniently represented by a picture, where a *dot* is used for a discrete vertex and a *circle* for a continuous. Further, a *line* joining α to β represents an undirected edge, whereas an *arrow* from α pointing towards β is used for a directed edge (α, β) with $(\beta, \alpha) \notin E$. Figure 2.1 contains an illustration of these conventions. Correspondingly we use the notation $\alpha \to \beta$, $\alpha \sim \beta$ to signify that

$$(\alpha, \beta) \in E \wedge (\beta, \alpha) \notin E, \quad (\alpha, \beta) \in E \wedge (\beta, \alpha) \in E$$

and $\alpha \not\to \beta$, $\alpha \not\sim \beta$ for

$$(\alpha, \beta) \notin E, \quad (\alpha, \beta) \notin E \wedge (\beta, \alpha) \notin E.$$

Note that if $\alpha \not\sim \beta$ then there is neither an arrow nor a line between the vertices α and β.

Fig. 2.1. A marked graph and its undirected version. Discrete vertices are represented by dots and continuous vertices by circles. Directed edges are represented by arrows and undirected edges by lines.

If the graph has only undirected edges it is an *undirected* graph and if all edges are directed, the graph is said to be *directed*. In an undirected graph it is often more convenient to represent the edges as unordered pairs $\{\alpha, \beta\}$.

The *undirected version* $\mathcal{G}^\sim$ of a graph $\mathcal{G}$ is the undirected graph obtained from $\mathcal{G}$ by substituting lines for arrows. Conversely, suppose that an irreflexive order relation $\prec$ on the vertex set V of a graph $\mathcal{G}$ is given. Then the corresponding *ordered* graph $\mathcal{G}^\prec$ has edges between exactly the same vertices α and β as the original graph $\mathcal{G}$, but the edge is directed if $\alpha \prec \beta$, and undirected otherwise. If $\mathcal{G}^\prec$ is a directed graph, we say that it is a *directed version* of $\mathcal{G}$.

If $A \subseteq V$ is a subset of the vertex set, it induces a subgraph $\mathcal{G}_A = (A, E_A)$, where the edge set $E_A = E \cap (A \times A)$ is obtained from $\mathcal{G}$ by keeping edges with both endpoints in A.

A graph is *complete* if all vertices are joined by an arrow or a line. A subset is *complete* if it induces a complete subgraph. A complete subset that is maximal (with respect to $\subseteq$) is called a *clique*.

If there is an arrow from α pointing towards β, α is said to be a *parent* of β and β a *child* of α. The set of parents of β is denoted as pa(β) and the set of children of α as ch(α).

If there is a line between α and β, α and β are said to be *adjacent* or *neighbours*. If there is a neither a line nor an arrow between α and β, i.e. $\alpha \not\sim \beta$, then α and β are said to be *non-adjacent*. The set of neighbours of a vertex α is denoted as ne(α).

The expressions pa(A), ch(A), and ne(A) denote the collection of parents, children, and neighbours of vertices in A that are not themselves elements of A:

$$\text{pa}(A) = \cup_{\alpha \in A} \text{pa}(\alpha) \setminus A$$
$$\text{ch}(A) = \cup_{\alpha \in A} \text{ch}(\alpha) \setminus A$$
$$\text{ne}(A) = \cup_{\alpha \in A} \text{ne}(\alpha) \setminus A.$$

The *boundary* bd(A) of a subset A of vertices is the set of vertices in $V \setminus A$ that are parents or neighbours to vertices in A. In symbols we then

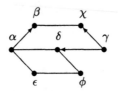

Fig. 2.2. Illustration of graph theoretic concepts. We have $\alpha \to \beta$ and also $\gamma \to \delta$ but $\delta \not\to \gamma, \alpha \not\sim \chi$ whereas, for example, $\epsilon \sim \phi$. Also $\mathrm{pa}(\chi) = \{\gamma\}$ and $\mathrm{ch}(\gamma) = \{\delta, \chi\}$ as well as $\mathrm{bd}(\delta) = \{\alpha, \phi, \gamma\}$.

have $\mathrm{bd}(A) = \mathrm{pa}(A) \cup \mathrm{ne}(A)$. The *closure* of A is $\mathrm{cl}(A) = A \cup \mathrm{bd}(A)$. See Fig. 2.2 for further illustration.

A *path* of length n from α to β is a sequence $\alpha = \alpha_0, \ldots, \alpha_n = \beta$ of distinct vertices such that $(\alpha_{i-1}, \alpha_i) \in E$ for all $i = 1, \ldots, n$. If there is a path from α to β we say that α *leads to* β and write $\alpha \mapsto \beta$. If both $\alpha \mapsto \beta$ and $\beta \mapsto \alpha$ we say that α and β *connect* and write $\alpha \rightleftharpoons \beta$. Clearly $\rightleftharpoons$ is an equivalence relation and the corresponding equivalence classes $[\alpha]$, where

$$\beta \in [\alpha] \iff \alpha \rightleftharpoons \beta,$$

are the *connectivity components* of $\mathcal{G}$. If $\alpha \in A \subseteq V$, the symbol $[\alpha]_A$ denotes the connectivity component of α in $\mathcal{G}_A$.

A subset $C \subseteq V$ is said to be an (α, β)-*separator* if all paths from α to β intersect C. Thus, in an undirected graph, C is an (α, β)-separator if and only if

$$[\alpha]_{V \setminus C} \neq [\beta]_{V \setminus C}.$$

The subset C is said to *separate A from B* if it is an (α, β)-separator for every $\alpha \in A, \beta \in B$.

The vertices α such that $\alpha \mapsto \beta$ and $\beta \not\mapsto \alpha$ are the *ancestors* $\mathrm{an}(\beta)$ of β, and the *descendants* $\mathrm{de}(\alpha)$ of α are the vertices β such that $\alpha \mapsto \beta$ and $\beta \not\mapsto \alpha$. The *non-descendants* are $\mathrm{nd}(\alpha) = V \setminus (\mathrm{de}(\alpha) \cup \{\alpha\})$.

If $\mathrm{bd}(\alpha) \subseteq A$ for all $\alpha \in A$ we say that A is an *ancestral* set. In a directed graph the set A is ancestral if and only if $\mathrm{an}(\alpha) \subseteq A$ for all $\alpha \in A$. In an undirected graph, the ancestral sets are unions of connectivity components. The intersection of a collection of ancestral sets is again ancestral. Hence, for any subset A of vertices there is a smallest ancestral set containing A which is denoted by $\mathrm{An}(A)$.

A *chain* of length n from α to β is a sequence $\alpha = \alpha_0, \ldots, \alpha_n = \beta$ of distinct vertices such that $\alpha_{i-1} \to \alpha_i$ or $\alpha_i \to \alpha_{i-1}$ for all $i = 1, \ldots, n$.

An *n-cycle* is a path of length n with the modification that $\alpha = \beta$, i.e. it begins and ends in the same point. The cycle is said to be *directed* if it contains an arrow.

A *tree* is a connected, undirected graph without cycles. It has a unique path between any two vertices. A *rooted tree* is the directed acyclic graph obtained from a tree by choosing a vertex as root and directing all edges away from this root. A *forest* is an undirected graph where all connectivity components are trees.

A class of graphs of special interest to us is the class of *chain graphs*. These are the graphs where the vertex set V can be partitioned into numbered subsets, forming a so-called *dependence chain* $V = V(1) \cup \cdots \cup V(T)$ such that all edges between vertices in the same subset are undirected and all edges between different subsets are directed, pointing from the set with lower number to the one with higher number. Such graphs are characterized by *having no directed cycles* and the connectivity components form a partitioning of the graph into *chain components*. A graph is a chain graph if and only if its connectivity components induce undirected subgraphs. The graph in Fig. 2.2 is a chain graph. Its chain components are $\{\alpha, \delta, \epsilon, \phi\}, \{\gamma\}, \{\beta, \chi\}$. The chain components are most easily found by removing all arrows before taking connectivity components. An undirected graph is a special case of a chain graph. A directed, acyclic graph is a chain graph with all chain components consisting of one vertex.

For a chain graph $\mathcal{G}$ we define its *moral graph* $\mathcal{G}^m$ as the undirected graph with the same vertex set but with α and β adjacent in $\mathcal{G}^m$ if and only if either $\alpha \to \beta$ or $\beta \to \alpha$ or if there are γ_1, γ_2 in the same chain component such that $\alpha \to \gamma_1$ and $\beta \to \gamma_2$. In the graph of Fig. 2.2, the moral graph is obtained by adding an edge between α and γ that both have children in the same chain component $\{\beta, \chi\}$, and then ignoring directions. If no edges have to be added to form the moral graph, the chain graph is said to be *perfect*. We warn the reader that the notion of a perfect graph in most graph theory literature refers to something quite different.

In the special case of a directed, acyclic graph the moral graph is obtained from the original graph by 'marrying parents' with a common child and subsequently deleting directions on all arrows.

A chain component C is said to be *terminal* if none of the vertices in C have children. A chain graph has always at least one terminal chain component. A terminal component with only one vertex is a *terminal vertex*.

2.1.2 Decompositions of marked graphs

In this subsection we study decompositions and decomposable graphs. Since the notion is fundamental, we state formally

Definition 2.1 A triple (A, B, C) of disjoint subsets of the vertex set V of an undirected, marked graph $\mathcal{G}$ is said to form a (strong) *decomposition* of $\mathcal{G}$ if $V = A \cup B \cup C$ and the three conditions below all hold:

(i) C separates A from B;

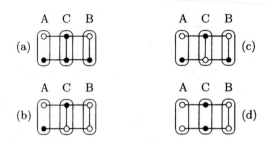

Fig. 2.3. Illustration of the notions of (strong) and weak decompositions. In (a) we see a decomposition with $C \subseteq \Delta$ and in (b) with $B \subset \Gamma$. In (c) the decomposition is only weak, because neither of these two conditions is fulfilled. In (d) we do not have a decomposition because the separator C is not complete.

(ii) C is a complete subset of V;

(iii) $C \subseteq \Delta \lor B \subseteq \Gamma$.

When this is the case we say that (A, B, C) *decomposes* $\mathcal{G}$ into the *components* $\mathcal{G}_{A \cup C}$ and $\mathcal{G}_{B \cup C}$. Occasionally, to avoid misunderstanding, we shall use the qualifier *strong* to distinguish from a *weak decomposition*, which is defined below:

Definition 2.2 A triple (A, B, C) of disjoint subsets of the vertex set V of an undirected, marked graph $\mathcal{G}$ is said to form a *weak decomposition* of $\mathcal{G}$ if $V = A \cup B \cup C$ and the two conditions below both hold:

(i) C separates A from B;

(ii) C is a complete subset of V.

Thus a weak decomposition satisfies the first two conditions of Definition 2.1, but not necessarily (iii). In the pure cases (iii) holds automatically and all weak decompositions are also decompositions. Note that we allow some of the sets in (A, B, C) to be empty. If the sets A and B in (A, B, C) are both non-empty, we say that the decomposition is *proper*. Figure 2.3 illustrates the notions of (strong) and weak decompositions.

A decomposable graph is one that can be successively decomposed into its cliques. Again we choose to state this formally through a recursive definition as

Definition 2.3 An undirected, marked graph is said to be *decomposable* if it is complete, or if there exists a proper decomposition (A, B, C) into decomposable subgraphs $\mathcal{G}_{A \cup C}$ and $\mathcal{G}_{B \cup C}$.

$a \quad c \quad b$

Fig. 2.4. The smallest non-decomposable graph. The graph is weakly decomposable.

Note that the definition makes sense because the decomposition is assumed to be proper, such that both subgraphs $\mathcal{G}_{A\cup C}$ and $\mathcal{G}_{B\cup C}$ have fewer vertices than the original graph $\mathcal{G}$.

Analogously, a *weakly decomposable* graph is one that can be decomposed into cliques by weak decompositions. It is then obvious that *any decomposable graph is weakly decomposable*. The converse is in general false. The smallest graph that is not decomposable is shown in Fig. 2.4. The only candidate for a separating set C is the vertex c but since this is continuous there are not allowed to be discrete variables in both sets A and B, for then condition (iii) of Definition 2.1 would be violated. But the graph is certainly weakly decomposable because $(\{a\}, \{b\}, \{c\})$ forms a weak decomposition.

A *triangulated* graph is an undirected graph with the property that every cycle of length $n \geq 4$ possesses a *chord*, i.e. two non-consecutive vertices that are neighbours. A definition such as this is a so-called 'forbidden path' definition. It follows immediately that the property must be stable under the operation of taking subgraphs, stated formally below.

Proposition 2.4 *If $\mathcal{G}$ is triangulated and $A \subset V$, then $\mathcal{G}_A$ is triangulated.*

A classical result states

Proposition 2.5 *The following conditions are equivalent for an undirected, marked graph $\mathcal{G}$:*

(i) $\mathcal{G}$ *is weakly decomposable;*

(ii) $\mathcal{G}$ *is triangulated;*

(iii) *every minimal (α, β)-separator is complete.*

Proof: We show this partly by induction on the number of vertices $|V|$ of $\mathcal{G}$. The result is trivial for a graph with no more than three vertices since the three conditions then are automatically fulfilled. So assume the result holds for all graphs with $|V| \leq n$ and consider a graph $\mathcal{G}$ with $n+1$ vertices. We then argue cyclically as (i) $\Longrightarrow$ (ii) $\Longrightarrow$ (iii) $\Longrightarrow$ (i).

First we show (i) $\Longrightarrow$ (ii). Suppose that $\mathcal{G}$ is weakly decomposable. If it is complete, it is obviously triangulated. Otherwise it can be weakly decomposed into decomposable subgraphs $\mathcal{G}_{A\cup C}$ and $\mathcal{G}_{B\cup C}$, both with fewer vertices. The inductive assumption implies that these are triangulated.

Fig. 2.5. The smallest graph that is not weakly decomposable.

Thus the only possibility for a chordless cycle is one that intersects both A and B. But, because C separates A from B, such a cycle must intersect C at least twice. But then it contains a chord because C is complete.

Then (ii) $\implies$ (iii). Let C be a minimal (α, β)-separator. If C has only one vertex, it is complete. If not it contains at least two, γ_1 and γ_2, say. Since C is a minimal separator, there will be paths from α to β via γ_1 and back via γ_2. The sequence

$$(\alpha, \ldots, \gamma_1, \ldots, \beta, \ldots, \gamma_2, \ldots, \alpha)$$

forms a cycle, with the modification that it can have repeated points. These, and chords other than a link between γ_1 and γ_2, can be used to shorten the cycle, still leaving at least one vertex in the component $[\alpha]_{V \setminus C}$ and one in $[\beta]_{V \setminus C}$. This produces a cycle of length at least 4, which must have a chord. Hence we get $\gamma_1 \sim \gamma_2$. Repeating the argument for all pairs of vertices in C gives that C is complete.

And finally that (iii) $\implies$ (i). Suppose that every minimal (α, β)-separator is complete. If $\mathcal{G}$ is complete there is nothing to show. Else it has at least two non-adjacent vertices α and β. Let C be a minimal (α, β)-separator and partition the vertex set into $[\alpha]_{V \setminus C}$, $[\beta]_{V \setminus C}$, C and the set of remaining vertices D. Then, since C is complete, the triple (A, B, C), where $A = [\alpha]_{V \setminus C} \cup D$, and $B = [\beta]_{V \setminus C}$, form a weak decomposition of $\mathcal{G}$. But each of the subgraphs $\mathcal{G}_{A \cup C}$ and $\mathcal{G}_{B \cup C}$ must be decomposable. For if C_1 is a minimal (α_1, β_1)-separator in $\mathcal{G}_{A \cup C}$, it is contained in a minimal (α_1, β_1)-separator in $\mathcal{G}$ which is complete by assumption and C_1 is therefore itself complete. The inductive assumption implies then that $\mathcal{G}_{A \cup C}$ is weakly decomposable, and similarly with $\mathcal{G}_{B \cup C}$. Thus we have weakly decomposed $\mathcal{G}$ into weakly decomposable subgraphs and the proof is complete. $\square$

The smallest graph that is not weakly decomposable is therefore a 4-cycle and shown in Fig. 2.5.

An elegant construction, due to Leimer (1989), makes it possible to take full advantage of the wealth of knowledge pertaining to triangulated graphs when discussing decomposable graphs. If $\mathcal{G}$ is an undirected, marked graph, its *star graph* is denoted as $\mathcal{G}^\star = (V^\star, E^\star)$ and constructed as follows: The vertex set V is to be extended with an extra vertex, denoted $\star$, such that $V^\star = V \cup \{\star\}$. This extra vertex, the *star*, is then connected to all discrete vertices by a line, such that $E^* = E \cup \{(\star, \delta), (\delta, \star), \delta \in \Delta\}$. The fact which makes this valuable is the following

Proposition 2.6 (Leimer) *An undirected, marked graph $\mathcal{G}$ is decomposable if and only if $\mathcal{G}^*$ is triangulated.*

Proof: We first observe that for any subgraph of $\mathcal{G}$, we have

$$(\mathcal{G}_A)^* = (\mathcal{G}^*)_{A \cup \{\star\}}. \tag{2.1}$$

The main part of the proof is by induction as usual. So let $\mathcal{G}$ be decomposable. If it is complete, its star graph is obviously triangulated. Else it can be decomposed into decomposable subgraphs $\mathcal{G}_{A \cup C}$ and $\mathcal{G}_{B \cup C}$, where $B \subseteq \Gamma$ or $C \subseteq \Delta$.

In the case $B \subseteq \Gamma$, the triple $(A \cup \{\star\}, B, C)$ is a weak decomposition of $\mathcal{G}^*$. The first component is, by (2.1) and the inductive assumption, triangulated, and the other component is decomposable, and therefore triangulated.

In the case $C \subseteq \Delta$, the triple $(A, B, C \cup \{\star\})$ is a weak decomposition of $\mathcal{G}^*$, and (2.1) and the inductive assumption again give that both these are triangulated. Thus $\mathcal{G}^*$ has in both cases been weakly decomposed into triangulated subgraphs, implying that it must itself be triangulated.

Conversely let us assume that $\mathcal{G}^*$ is triangulated. If $\mathcal{G}$ is not complete it contains two non-adjacent vertices α and β. But these are also non-adjacent in $\mathcal{G}^*$ and a minimal (α, β)-separator C will define a decomposition of $\mathcal{G}^*$ with $\alpha \in A$ and $\beta \in B$ as in the proof of Proposition 2.5. By symmetry we can assume $\star \notin B$ and have two cases.

First assume $\star \in A$. Then, since C separates A from B, we must have $B \subseteq \Gamma$ and the triple $(A \setminus \{\star\}, B, C)$ is a proper decomposition of $\mathcal{G}$. Each part has triangulated star graphs and (2.1) and the inductive assumption yield that the parts are decomposable.

Then let $\star \in C$. Because C is complete, $C \setminus \{\star\} \subseteq \Delta$, such that the triple $(A, B, C \setminus \{\star\})$ is a proper decomposition of $\mathcal{G}$. And the graph has been decomposed into decomposable subgraphs by (2.1) and the inductive assumption. $\square$

The construction of star graphs is illustrated in Fig. 2.6.

It follows from Proposition 2.6 that a forbidden path characterization exists for decomposable graphs.

Corollary 2.7 *An undirected, marked graph is decomposable if and only if it is triangulated and does not contain any path $(\delta_1 = \alpha_0, \ldots, \alpha_n = \delta_2)$ between two non-adjacent discrete vertices passing through only continuous vertices, i.e. with $\delta_1 \not\sim \delta_2$ and $\alpha_i \in \Gamma$ for $0 < i < n$.*

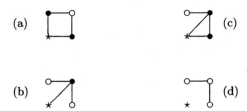

Fig. 2.6. The construction of star graphs. In (a), which is the star graph of the non-decomposable graph from Fig. 2.4, the star creates a 4-cycle while the other three decomposable graphs have triangulated star graphs.

Fig. 2.7. Path which is forbidden in a decomposable graph.

Proof: The forbidden paths are exactly those that give rise to chordless cycles in $\mathcal{G}^\star$. □

We have illustrated the typical forbidden path in Fig. 2.7. See also Figs 2.4 and 2.6, the latter indicating clearly how the forbidden path creates a 4-cycle in the star graph.

Since we now have established a forbidden path characterization of decomposable graphs, we have

Corollary 2.8 *If $\mathcal{G}$ is decomposable and $A \subseteq V$, $\mathcal{G}_A$ is decomposable.*

Figures 2.8 and 2.9 show all connected, marked graphs with no more than four vertices that are not decomposable.

Fig. 2.8. Marked graphs with no more than four vertices that are neither decomposable nor triangulated.

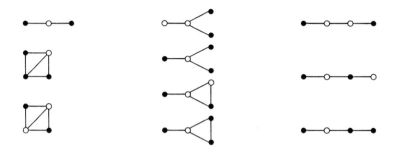

Fig. 2.9. Triangulated, marked graphs with no more than four vertices that are not decomposable.

2.1.3 Simplicial subsets and perfect sequences

Closely related to the notion of a decomposition is the notion of a *simplicial subset*, which is a subset B that satisfies the following two conditions:

(i) $\mathrm{bd}(B)$ is complete;

(ii) $B \subseteq \Gamma \vee \mathrm{bd}(B) \subseteq \Delta$.

A subset is said to be *weakly* simplicial if the first of these conditions is satisfied. Clearly, when a subset is (weakly) simplicial the triple $(V \setminus \mathrm{cl}(B), B, \mathrm{bd}(B))$ is a (weak) decomposition of $\mathcal{G}$. The notion is illustrated in Fig. 2.10. A vertex α is said to be (weakly) simplicial if the subset $\{\alpha\}$ is. The following lemma, due to Dirac (1961), plays a central role.

Lemma 2.9 (Dirac) *Let $\mathcal{G}$ be a triangulated graph with at least two vertices. Then $\mathcal{G}$ has at least two weakly simplicial vertices. If $\mathcal{G}$ is not complete these can be chosen to be non-adjacent.*

Proof: Induction on $|V|$. If $|V| = 2$ the lemma is obviously true. Assume that the lemma holds for all graphs with $|V| \leq n$ and let $|V| = n + 1$. If $\mathcal{G}$ is complete the statement is obvious. Otherwise there exists a proper weak decomposition (A, B, C) of $\mathcal{G}$. The induction assumption used on $\mathcal{G}_{A \cup C}$ yields a pair (α_1, α_2) of non-adjacent vertices that are weakly simplicial in $\mathcal{G}_{A \cup C}$. At least one of these, α_1 say, must then be in A, because C is complete. By symmetry there is a vertex β in B that is weakly simplicial in $\mathcal{G}_{B \cup C}$. Because C separates A from B, (α_1, β) must be a pair of non-adjacent vertices that are weakly simplicial in $\mathcal{G}$. □

Note that it is not true that any decomposable, marked graph has at least two simplicial vertices. A counterexample is given in Fig. 2.11. But it has at least one, which is stated in the following

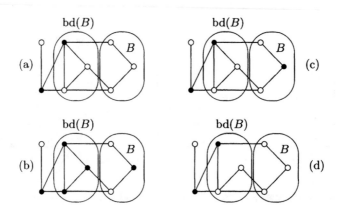

Fig. 2.10. Simplicial subsets. In (a), B is simplicial because $B \subseteq \Gamma$ and $\mathrm{bd}(B)$ is complete. In (b), B is simplicial because $\mathrm{bd}(B) \subseteq \Delta$ and $\mathrm{bd}(B)$ is complete. In (c), B is only weakly simplicial because $B \cap \Delta \neq \emptyset$ and $\mathrm{bd}(B) \cap \Gamma \neq \emptyset$. In (d), B is not even weakly simplicial because $\mathrm{bd}(B)$ is not complete.

Fig. 2.11. A decomposable graph with only one simplicial vertex.

Corollary 2.10 *A decomposable graph has at least one simplicial vertex. If $\Gamma \neq \emptyset$, this can be chosen continuous.*

Proof: Using Lemma 2.9 on the star graph gives at least one simplicial vertex which is not the star. This is simplicial also in $\mathcal{G}$, which gives the first part of the statement. Assume now that $\Gamma \neq \emptyset$. Let $\mathcal{G}'$ be obtained from $\mathcal{G}$ by adding edges between any discrete pair of vertices. This must also be decomposable as no forbidden paths can be created this way. The star graph of $\mathcal{G}'$ is not complete and has therefore two non-adjacent simplicial vertices. One of these, say γ, must be continuous. The vertex γ is also simplicial in $\mathcal{G}$. For otherwise it would have two non-adjacent discrete neighbours δ_1 and δ_2 and $(\delta_1, \gamma, \delta_2)$ would be a forbidden path. □

Let $B_1, \cdots, B_k$ be a sequence of subsets of the vertex set V of an undirected, marked graph $\mathcal{G}$. Let

$$H_j = B_1 \cup \cdots \cup B_j, \quad R_j = B_j \setminus H_{j-1}, \quad S_j = H_{j-1} \cap B_j.$$

The sequence is said to be *perfect* if the following conditions are fulfilled:

(i) for all $i > 1$ there is a $j < i$ such that $S_i \subseteq B_j$;

(ii) the sets S_i are complete for all i;

(iii) for all $i > 1$ we have $R_i \subseteq \Gamma$ or $S_i \subseteq \Delta$.

The condition (i) is known as the *running intersection property*. A sequence of subsets that satisfies (i) and (ii) is said to be *weakly perfect*.

We term the sets H_j the *histories*, R_j the *residuals*, and S_j the *separators* of the sequence. The justification for the use of the term *separator* is based on Lemma 2.11 below.

A *perfect numbering* of the vertices V of $\mathcal{G}$ is a numbering $\alpha_1, \ldots, \alpha_k$ such that
$$B_j = \mathrm{cl}(\alpha_j) \cap \{\alpha_1, \ldots, \alpha_j\}, \quad j \geq 1$$
is a perfect sequence of sets. Note that this implies that the sets B_j are all complete.

Similarly we define *weakly perfect* numberings as those where the sequence of sets B_j is only weakly perfect.

Perfect sequences and numberings play important roles in the understanding and manipulation of decomposable graphs, partly because, as we shall see in Proposition 2.17, their existence is a characteristic for decomposable graphs, but also because they form the basis for recursive computational procedures. Before we show the characterization results, we need the following lemmas:

Lemma 2.11 *Let $B_1, \ldots, B_k$ be a perfect sequence of sets which contains all cliques of an undirected, marked graph $\mathcal{G}$. Then for every j, S_j separates $H_{j-1} \setminus S_j$ from R_j in $\mathcal{G}_{H_j}$ and hence (H_{j-1}, R_j, S_j) decomposes $\mathcal{G}_{H_j}$.*

Proof: Let p be the highest number such that B_p is a clique. Then $H_p = V$ and hence $R_j = \emptyset$, so the separation is trivial for $j > p$. Next we must show that S_p separates $H_{p-1} \setminus S_p$ from R_p in $\mathcal{G}$. But suppose there were an edge between $\alpha \in R_p$ and $\beta \in B_j \setminus S_p$ for some $j < p$. Then $\{\alpha, \beta\}$ must be subset of some clique of $\mathcal{G}$. But this cannot be B_p, as $\beta \notin B_p$ and not B_i for some $i < p$ as $\alpha \notin H_{p-1}$. Since all cliques are in the sequence, the edge can therefore not exist and S_p must separate $H_{p-1} \setminus S_p$ from R_p.

Now $B_1, \ldots, B_{p-1}$ is a perfect sequence of sets that contains all cliques of $\mathcal{G}_{H_{p-1}}$. For $S_p \subseteq B_i$ for some $i < p$ and hence, if $R_p = \emptyset$ then $B_i = B_p$ is a clique. If $R_p \neq \emptyset$ the subgraph $\mathcal{G}_{H_{p-1}}$ has one clique fewer. We repeat the argument and continue until the sequence is reduced to a single set. □

If a perfect sequence of sets does not contain all cliques, the sets S_j may not separate; see Fig. 2.12.

Lemma 2.12 *Let $C_1, \ldots, C_p$ be the cliques of $\mathcal{G}$ and assume that they form a weakly perfect sequence. Next let the vertices of $\mathcal{G}$ be numbered with first those in C_1, then those in R_2, R_3 and so on. The numbering $\alpha_1, \ldots, \alpha_k$ so*

Fig. 2.12. A perfect sequence of sets that does not decompose the graph.

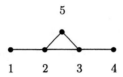

Fig. 2.13. A perfect vertex numbering that does not induce a perfect clique numbering. The numbering of the vertices is perfect, but the cliques, numbered as $(\{1,2\},\{3,4\},\{2,3,5\})$, do not form a perfect sequence of sets.

obtained is weakly perfect. If the sequence $C_1,\ldots,C_p$ is perfect, and within each of the sets $C_1, R_2, R_3,\ldots$ continuous vertices are given higher number than discrete vertices, then the numbering $(\alpha_1,\ldots,\alpha_k)$ is perfect.

Proof: This is immediate. □

The 'converse' to Lemma 2.12 is false in the sense that the sequence of cliques induced by a perfect numbering of the vertices might not be perfect. The induced sequence is here formed by numbering the cliques according to their highest numbered vertex. A counterexample is provided in Fig. 2.13.

Lemma 2.13 Let $C_1,\ldots,C_k$ be a perfect sequence. Assume that $C_t \subseteq C_p$ for some $t \neq p$ and that p is minimal with this property for fixed t. Then

(i) if $p < t$ then $C_1,\ldots,C_{t-1},C_{t+1},\ldots,C_k$ is a perfect sequence;

(ii) if $p > t$ then $C_1,\ldots,C_{t-1},C_p,C_{t+1},\ldots,C_{p-1},C_{p+1},\ldots,C_k$ is a perfect sequence.

Proof: Case (i) is immediate. In case (ii) we first argue that $S_p = C_t$. For we have

$$S_p = C_p \cap (C_1 \cup \cdots \cup C_{p-1}) \supseteq C_p \cap C_t = C_t$$

but also that $S_p \subseteq C_k$ for some $k < p$ from the running intersection property. Hence

$$C_t \subseteq S_p \subseteq C_k.$$

The minimality of p then implies $k = t$. Next

$$S^* = C_p \cap H_{t-1} \subseteq S_p = C_t,$$

whereby also $S^* = S_t$, as then

$$S^* = (C_p \cap H_{t-1}) \cap C_t = S_t.$$

Hence S^* is complete and contained in some C_k for $k < t$. Finally, if $R_p \cap \Delta \neq \emptyset$ then $S_p = C_t \subseteq \Delta$, whereby $S^* \subseteq \Delta$.

For $t < k < p$ we have

$$\begin{aligned} C_k \cap (H_{t-1} \cup C_p \cup C_{t+1} \cup \cdots \cup C_{k-1}) &= C_k \cap (H_{k-1} \cup C_p) \\ &= S_k \cup \{C_k \cap (C_p \setminus C_t)\}. \end{aligned}$$

But as $C_k \cap C_p \subseteq S_p = C_t$, then $\{C_k \cap (C_p \setminus C_t)\} = \emptyset$ and therefore

$$C_k \cap (H_{t-1} \cup C_p \cup C_{t+1} \cup \cdots \cup C_{k-1}) = S_k.$$

For $k > p$ the separators in the new sequence are trivially identical to those in the original sequence. □

Perfect sequences of vertices contain all cliques as stated below.

Lemma 2.14 *Let $\alpha_1, \ldots, \alpha_k$ be a perfect numbering of the vertices of an undirected, marked graph $\mathcal{G}$. Then the sets $B_j = \text{cl}(\alpha_j) \cap \{\alpha_1, \ldots, \alpha_{j-1}\}$ form a perfect sequence that contains all cliques of $\mathcal{G}$.*

Proof: The sequence $B_1, \ldots, B_k$ is perfect by definition. B_k is necessarily a clique. An induction argument now gives the result, as $\alpha_1, \ldots \alpha_{k-1}$ is a perfect numbering of the vertices of $\mathcal{G}_{V \setminus \{\alpha_k\}}$ and the cliques of $\mathcal{G}$ consist of B_k and those cliques of $\mathcal{G}_{V \setminus \{\alpha_k\}}$ that are not subsets of B_k. □

We now have a way of constructing a perfect sequence of cliques from a perfect sequence of vertices by thinning, i.e. simply by using Lemma 2.13 to remove superfluous sets from the sequence constructed in Lemma 2.14. We have further

Lemma 2.15 *If $(\star = \alpha_0, \alpha_1, \ldots, \alpha_k)$ is a weakly perfect numbering of $\mathcal{G}^\star$, then $(\alpha_1, \ldots, \alpha_k)$ is a perfect numbering of $\mathcal{G}$.*

Proof: The construction of the star graph gives $\alpha_q \in \Delta \iff \star \in \text{bd}(\alpha_q)$ Thus if $\text{bd}(\alpha_q) \cap \{\star, \alpha_1, \ldots, \alpha_{q-1}\}$ is complete, and $\alpha_q \in \Delta$, then $\text{bd}(\alpha_q) \cap \{\alpha_1, \ldots, \alpha_{q-1}\} \subseteq \Delta$, which completes the proof. □

And, more generally for sets we have

Lemma 2.16 *Let $(C_1, \ldots, C_p)$ be a weakly perfect sequence of complete sets which contain all cliques in $\mathcal{G}^*$ and assume that $\star \in C_1$. Then the sequence $(C_1 \setminus \{\star\}, \ldots, C_p \setminus \{\star\})$ is perfect in $\mathcal{G}$.*

Proof: As above. □

We can now obtain a recursive characterization of decomposable graphs.

Proposition 2.17 *The following conditions are equivalent for an undirected, marked graph $\mathcal{G}$:*

(i) *the vertices of $\mathcal{G}$ admit a perfect numbering;*

(ii) *the cliques of $\mathcal{G}$ can be numbered to form a perfect sequence;*

(iii) *the graph $\mathcal{G}$ is decomposable.*

Proof: First (i) implies (ii) by using Lemma 2.14 and the thinning procedure described in Lemma 2.13. That (ii) implies (iii) follows by Lemma 2.11 and the definition of decomposability. That (iii) implies (i) is seen by induction on the number of vertices as follows. If $\mathcal{G}$ is decomposable, then by Corollary 2.10 $\mathcal{G}$ has a simplicial vertex and we can label this as α_k. The induction assumption gives us a perfect numbering of the remaining $k-1$ vertices. □

Note that it follows from exploiting Corollary 2.10 fully in the proof that the perfect numbering of the vertices can be chosen such that any discrete vertex has lower number than any continuous vertex.

A perfect numbering of the vertices of $\mathcal{G}$ induces a linear ordering of these and therefore a directed acyclic version $\mathcal{G}^<$ of $\mathcal{G}$ with arrows pointing from vertices with low numbers to vertices with high numbers. The graph $\mathcal{G}^<$ is then called a *perfect directed version* of $\mathcal{G}$.

Proposition 2.17 implies that *an undirected graph is decomposable if and only if it has a perfect directed version.* It follows that the undirected version $\mathcal{G}^\sim$ of a perfect directed acyclic graph is decomposable.

If we consider the pure case, perfect sequences of cliques can be arranged to begin with any clique. More precisely we have the useful lemma below.

Lemma 2.18 *Let C^* be a clique in a triangulated graph $\mathcal{G}$. Then the cliques of $\mathcal{G}$ can be ordered as a weakly perfect sequence $C_1, \ldots, C_k$ with $C_1 = C^*$.*

Proof: We use induction on the number of vertices $n = |V|$ of $\mathcal{G}$. For $n \leq 2$ the statement is obvious. Assume then the lemma to hold for all graphs with $n \leq p$ and let $\mathcal{G}$ have $p+1$ vertices. If $\mathcal{G}$ is complete the lemma is obviously true. Otherwise, by Dirac's Lemma 2.9, $\mathcal{G}$ has at least two non-adjacent weakly simplicial vertices, i.e. one of them, say α, is not in

C^*. This vertex must be a member of exactly one clique C_α. The cliques of $\mathcal{G}'$ are the cliques of $\mathcal{G}$ except C_α, possibly with $C_\alpha \setminus \{\alpha\}$ adjoined. The inductive assumption implies that the cliques of $\mathcal{G}' = \mathcal{G}_{V \setminus \{\alpha\}}$ admit a weakly perfect numbering $C_1, \ldots, C_{k-1}$ or $C_1, \ldots, C_{k-1}, C'_k$ with $C_1 = C^*$. Letting $C_k = C_\alpha$ we obtain a weakly perfect numbering of the cliques of $\mathcal{G}$ with the desired property. □

Also here one should be aware that the strong form of the result is not true for decomposable graphs with both types of vertices; see Fig. 2.11.

2.1.4 Subgraphs of decomposable graphs

Subgraphs of decomposable graphs that are induced by subsets A of V are always decomposable, such as stated in Corollary 2.8. But if a subgraph $\mathcal{G}'$ has the same vertex set, but fewer edges, the situation is more complicated. When removing an edge, one could, for example, be so unlucky as to remove the chord needed to short-cut a 4-cycle. The next lemma characterizes removable edges.

Lemma 2.19 *Let $\mathcal{G} = (V, E)$ be a decomposable, marked graph and let $\mathcal{G}' = (V, E')$ be a subgraph of $\mathcal{G}$ with exactly one edge, e, less. Then $\mathcal{G}'$ is decomposable if and only if both of the conditions below are satisfied:*

(i)   *e is a member of one clique C^* only;*

(ii) *$e \subseteq \Delta \implies C^* \subseteq \Delta$.*

Proof: The condition (i) is necessary because if e were a member of two cliques, $\mathcal{G}'$ would contain a chordless 4-cycle. Similarly (ii) is necessary, because if $e \subseteq \Delta$ and $\gamma \in C^* \cap \Gamma$, the path (α, γ, β) would be a forbidden path as in Corollary 2.7. But the condition is sufficient as well, because no new forbidden paths or chordless cycles are generated if (i) and (ii) hold. □

Apart from knowing which edges are removable, we need to investigate the behaviour of perfect sequences under edge removals. More precisely let e and C^* be as in Lemma 2.19 above and let $C_1, \ldots, C_k$ be the cliques in $\mathcal{G}$ numbered to form a perfect sequence with $C_q = C^*$. Then let $C_{q1} = C^* \setminus \{\beta\}$ and $C_{q2} = C^* \setminus \{\alpha\}$. We then have

Lemma 2.20 *The vertex α can be chosen to fulfil both of the conditions*

(i)   *$\alpha \in \Delta$ if $\{\alpha, \beta\} \cap \Delta \neq \emptyset$;*

(ii) *$\alpha \in S_q$ if $\{\alpha, \beta\} \cap S_q \neq \emptyset$.*

When so chosen, the sequence $C_1, \ldots, C_{q1}, C_{q2}, \ldots, C_k$ is perfect in $\mathcal{G}'$.

Proof: It is immediate that α can be chosen to fulfil (i). But assume then that $\alpha \in \Delta$, $\beta \in \Gamma$ and $\alpha \in R_q$. Since the sequence is perfect, we must have $S_q \subseteq \Delta$, whereby also $\beta \in R_q$ and thus $\{\alpha, \beta\} \cap S_q = \emptyset$.

Note that $\alpha \in S_q$ implies $\beta \in R_q$, since otherwise both α and β are in more than one clique.

The first $q-1$ sets form a perfect sequence since they did in $\mathcal{G}$. Consider then
$$S_{q1} = C_{q1} \cap (C_1 \cup \cdots \cup C_{q-1}), \quad R_{q1} = C_{q1} \setminus S_{q1}.$$

We have
$$S_{q1} = S_q \setminus \{\beta\} = S_q, \quad R_{q1} = R_q \setminus \{\beta\} \subset R_q.$$

Since $R_q \subseteq \Gamma$ or $S_q \subseteq \Delta$, the same holds for R_{q1} and S_{q1}. Similarly, S_{q1} is complete.

The next intersection is given by
$$S_{q2} = C_{q2} \cap (C_1 \cup \cdots \cup C_{q1}), \quad R_{q2} = C_{q2} \setminus S_{q2}.$$

But here we have
$$S_{q2} = C_{q2} \cap C_{q1} = C^* \setminus \{\alpha, \beta\}, \quad R_{q2} = \{\beta\}.$$

Thus S_{q2} is complete in $\mathcal{G}'$ and if $\beta \notin \Gamma$, both α and β are in Δ which, by (ii) of Lemma 2.19 implies that $C^* \subseteq \Delta$.

Finally, for $j > q$ we have exactly the same sets involved. So we need only check that all sets $S_j, j > q$ are complete in $\mathcal{G}'$. But if this were not the case, e would be both in C^* and in C_j which, by Lemma 2.19, contradicts our assumptions. □

Lemma 2.21 *Let $\mathcal{G} = (V, E)$ be a triangulated graph and let $\mathcal{G}' = (V, E')$ be a subgraph of $\mathcal{G}$ that also is triangulated with $|E \setminus E'| = k$. Then there is an increasing sequence $\mathcal{G}' = \mathcal{G}_0 \subset \cdots \subset \mathcal{G}_k = \mathcal{G}$ of triangulated graphs that differ by exactly one edge.*

Proof: It is enough to show that we can find $\mathcal{G}_1$, i.e. an edge e in $E \setminus E'$ such that $(V, E' \cup \{e\})$ is triangulated. We use induction on the number of vertices $|V|$ of V. For $|V| \leq 3$ there is nothing to prove. Assume $n > 3$, $|V| = n+1$ and that the lemma holds for all triangulated graphs with $|V| \leq n$. Then since $\mathcal{G}'$ is not complete it has by Dirac's Lemma 2.9 two non-adjacent vertices α_1 and α_2 with a boundary which is complete in $\mathcal{G}'$. Consider for $i = 1, 2$ the subgraphs $\tilde{\mathcal{G}}_i = \mathcal{G}_{V \setminus \{\alpha_i\}}$ and $\tilde{\mathcal{G}}'_i = \mathcal{G}'_{V \setminus \{\alpha_i\}}$. We now distinguish two cases. Suppose first that $\tilde{\mathcal{G}}_i = \tilde{\mathcal{G}}'_i$ for $i = 1, 2$. Then $\mathcal{G}$ and $\mathcal{G}'$ differ by one edge between α_1 and α_2 only, and the lemma holds. Assume then without loss of generality that $\tilde{\mathcal{G}}_1 \neq \tilde{\mathcal{G}}'_1$. Then, by the induction

Fig. 2.14. A sequence of decomposable graphs that differ by one edge only.

assumption, there must be an edge e, such that $(V \setminus \alpha_1, E'_{V \setminus \alpha_1} \cup \{e\})$ is triangulated. As $\mathrm{bd}(\alpha_1)$ is complete in $\mathcal{G}'$, the triple $(V \setminus \mathrm{bd}(\alpha_1), \{\alpha_1\}, \mathrm{bd}(\alpha_1))$ forms a weak decomposition of $(V, E' \cup \{e\})$ into triangulated subgraphs and hence also $(V, E' \cup \{e\})$ is triangulated. This completes the proof. □

Lemma 2.22 *Let $\mathcal{G} = (V, E)$ be a decomposable, marked graph and let $\mathcal{G}' = (V, E')$ be a subgraph of $\mathcal{G}$ that also is decomposable with $|E \setminus E'| = k$. Then there is an increasing sequence $\mathcal{G}' = \mathcal{G}_0 \subset \cdots \subset \mathcal{G}_k = \mathcal{G}$ of decomposable graphs that differ by exactly one edge. Further, if some of the missing edges connect pairs of discrete vertices, these can be added first, i.e. if $|(E \setminus E') \cap \Delta \times \Delta| = l > 0$, then $\mathcal{G}_i$ can be chosen with $E_i \setminus E_{i-1} \subset \Delta \times \Delta$ for $0 < i \leq l$.*

Proof: We form the two star graphs and essentially repeat the proof of Lemma 2.21. Only, care must be taken if there are discrete missing edges and not all vertices are discrete. Then by Corollary 2.10 we can assume that either α_1 or α_2 is continuous. As in the proof of Lemma 2.21, the star graphs of either $\mathcal{G}_{V \setminus \{\alpha_1\}}$ and $\mathcal{G}'_{V \setminus \{\alpha_1\}}$, or $\mathcal{G}_{V \setminus \{\alpha_2\}}$ and $\mathcal{G}'_{V \setminus \{\alpha_2\}}$, must differ by at least one discrete edge. The proof then continues as before. □

Lemma 2.22 is illustrated in Fig. 2.14.

2.2 Hypergraphs

2.2.1 Basic concepts

A *hypergraph* is a collection $\mathcal{H}$ of subsets of a finite set H, the *base set*. The elements of $\mathcal{H}$ are called *hyperedges*. In most cases of interest to us, the base set will be the union of the hyperedges, i.e. $H = \cup_{h \in \mathcal{H}} h$. This will henceforth be assumed to be the case, when not otherwise explicitly stated.

A typical hypergraph is a set of complete subsets of a graph $\mathcal{G}$, for example the set of cliques $\mathcal{C}(\mathcal{G})$ of the graph, denoted the *clique hypergraph* of $\mathcal{G}$. Another example is the generating class for a hierarchical model subspace, see Section B.2. A hypergraph is *simple* if it has only one hyperedge. A simple hypergraph is the clique hypergraph of a complete graph.

If all hyperedges in $\mathcal{H}$ are pairwise incomparable in the sense that none is a subset of the other, we say that $\mathcal{H}$ is *reduced*. The examples above are reduced hypergraphs. The operation $\text{red}\,\mathcal{H}$ produces a reduced hypergraph from $\mathcal{H}$ by removing all hyperedges that are contained in other hyperedges. If we define join and meet operations for two hypergraphs as

$$\mathcal{H}_1 \vee \mathcal{H}_2 = \text{red}(\mathcal{H}_1 \cup \mathcal{H}_2)$$
$$\mathcal{H}_1 \wedge \mathcal{H}_2 = \text{red}\{h_1 \cap h_2 \mid h_1 \in \mathcal{H}_1, h_2 \in \mathcal{H}_2\},$$

the class of reduced hypergraphs forms a distributive lattice with the partial order

$$\mathcal{H}_1 \preceq \mathcal{H}_2 \iff \text{ for all } h_1 \in \mathcal{H}_1 \text{ there exists an } h_2 \in \mathcal{H}_2 \text{ with } h_1 \subseteq h_2.$$

Two arbitrary hypergraphs are equivalent if their reductions are equal:

$$(\mathcal{H}_1 \preceq \mathcal{H}_2 \text{ and } \mathcal{H}_1 \preceq \mathcal{H}_2) \iff \text{red}(\mathcal{H}_1) = \text{red}(\mathcal{H}_2).$$

The join $\mathcal{H} = \mathcal{H}_1 \vee \mathcal{H}_2$ of two hypergraphs is said to be *direct* if their meet is simple, i.e. if $\mathcal{H}_1 \wedge \mathcal{H}_2 = \{h\}$. Note that then necesssarily $h = H_1 \cap H_2$.

2.2.2 Graphs and hypergraphs

As mentioned above, each undirected graph $\mathcal{G}$ has an associated clique hypergraph $\mathcal{C}(\mathcal{G})$, but conversely with any hypergraph $\mathcal{H}$ we can associate its graph $\mathcal{G}(\mathcal{H}) = (V, E)$, where $V = H$ and

$$(\alpha, \beta) \in E \iff \{\alpha, \beta\} \subseteq h \text{ for some } h \in \mathcal{H}.$$

Clearly we have

$$\mathcal{G}\{\mathcal{C}(\mathcal{G})\} = \mathcal{G} \text{ and } \mathcal{H} \preceq \mathcal{C}\{\mathcal{G}(\mathcal{H})\}.$$

If it also holds that $\mathcal{H}$ contains the cliques of $\mathcal{G}(\mathcal{H})$,

$$\mathcal{C}\{\mathcal{G}(\mathcal{H})\} \preceq \mathcal{H},$$

we say that the hypergraph $\mathcal{H}$ is *graphical*. Then the reduced hypergraph $\text{red}(\mathcal{H})$ consists exactly of the cliques of $\mathcal{G}(\mathcal{H})$. It obviously holds that

$$\mathcal{H}_1 \preceq \mathcal{H}_2 \implies \mathcal{G}(\mathcal{H}_1) \subseteq \mathcal{G}(\mathcal{H}_2)$$
$$\mathcal{G}_1 \subseteq \mathcal{G}_2 \implies \mathcal{C}(\mathcal{G}_1) \preceq \mathcal{C}(\mathcal{G}_2).$$

Further, one readily verifies from the definitions that

$$\mathcal{G}(\mathcal{H}_1 \vee \mathcal{H}_2) = \mathcal{G}(\mathcal{H}_1) \cup \mathcal{G}(\mathcal{H}_2) \qquad (2.2)$$
$$\mathcal{G}(\mathcal{H}_1 \wedge \mathcal{H}_2) = \mathcal{G}(\mathcal{H}_1) \cap \mathcal{G}(\mathcal{H}_2) \qquad (2.3)$$
$$\mathcal{C}(\mathcal{G}_1 \cap \mathcal{G}_2) = \mathcal{C}(\mathcal{G}_1) \wedge \mathcal{C}(\mathcal{G}_2), \qquad (2.4)$$

whereas in general
$$\mathcal{C}(\mathcal{G}_1 \cup \mathcal{G}_2) \succeq \mathcal{C}(\mathcal{G}_1) \vee \mathcal{C}(\mathcal{G}_2). \tag{2.5}$$
In the case of direct joins and weak decompositions we have

Lemma 2.23 *If $\mathcal{H}$ is the direct join of hypergraphs $\mathcal{H}_1$ and $\mathcal{H}_2$, then the triple $(H_1 \setminus H_2, H_2 \setminus H_1, H_1 \cap H_2)$ is a weak decomposition of $\mathcal{G}(\mathcal{H})$. If conversely (A, B, C) is a weak decomposition of the graph $\mathcal{G}$, then*
$$\mathcal{C}(\mathcal{G}) = \mathcal{C}\left(\mathcal{G}_{A\cup C} \cup \mathcal{G}_{B\cup C}\right) = \mathcal{C}\left(\mathcal{G}_{A\cup C}\right) \vee \mathcal{C}\left(\mathcal{G}_{B\cup C}\right) \tag{2.6}$$
and the join is direct.

Proof: It follows from (2.2) that $H_1 \cap H_2$ separates $H_1 \setminus H_2$ from $H_2 \setminus H_1$. Since the join is direct, (2.3) gives that $H_1 \cap H_2$ is complete.

In the case where (A, B, C) forms a decomposition, (2.5) implies that it is enough to show that
$$\mathcal{C}(\mathcal{G}) \preceq \mathcal{C}\left(\mathcal{G}_{A\cup C}\right) \vee \mathcal{C}\left(\mathcal{G}_{B\cup C}\right).$$
But if $c \in \mathcal{C}(\mathcal{G})$, it must be contained in either $A \cup C$ or $B \cup C$ since C separates A from B in $\mathcal{G}$. Assume then $c \subseteq A \cup C$. Because c is a clique it is in $\mathcal{C}(\mathcal{G}_{A\cup C})$. □

An important corollary to this is

Corollary 2.24 *If $\mathcal{H}$ is the direct join of graphical hypergraphs $\mathcal{H}_1$ and $\mathcal{H}_2$, then $\mathcal{H}$ is itself graphical.*

Proof: We must show that $\mathcal{C}\{\mathcal{G}(\mathcal{H})\} \preceq \mathcal{H}$. We find
$$\begin{aligned}\mathcal{C}\{\mathcal{G}(\mathcal{H})\} &= \mathcal{C}\{\mathcal{G}(\mathcal{H}_1) \vee \mathcal{G}(\mathcal{H}_2)\} \\ &= \mathcal{C}\{\mathcal{G}(\mathcal{H}_1)\} \vee \mathcal{C}\{\mathcal{G}(\mathcal{H}_2)\} = \mathcal{H}_1 \vee \mathcal{H}_2 = \mathcal{H},\end{aligned}$$
where we have used (2.6) to obtain the second equality. □

A *decomposable* hypergraph $\mathcal{H}$ is a hypergraph that either is simple or can be obtained by direct joins of hypergraphs that have fewer hyperedges. We then have the following central result.

Theorem 2.25 *A hypergraph $\mathcal{H}$ is decomposable if and only if it is the clique hypergraph of a weakly decomposable graph. In particular all decomposable hypergraphs are graphical.*

Proof: Simple hypergraphs are obviously graphical with complete graphs as their graphs. Corollary 2.24 ensures that this continues to hold when forming direct joins. From Lemma 2.23 we have that direct joins of hypergraphs match weak decompositions of the associated graphs. Thus weakly decomposable graphs must correspond to decomposable hypergraphs and vice versa. □

2.2.3 Junction trees and forests

An important structure associated with computational aspects of decomposable hypergraphs is a tree with a particular property. More precisely, a tree $\mathcal{T}$ with the set $\mathcal{H}$ of hyperedges as vertices of the tree is called a *junction tree* for $\mathcal{H}$ if it holds for any two hyperedges a and b in $\mathcal{H}$ and any h on the unique path in $\mathcal{T}$ between a and b that

$$a \cap b \subseteq h. \tag{2.7}$$

We refer to (2.7) as the *junction property*. It can alternatively be expressed as follows. The subset of hyperedges that contain a given subset $a \subseteq V$ forms a connected subtree $\mathcal{T}_a$ of $\mathcal{T}$ for all a.

A *junction forest* for $\mathcal{H}$ is a collection $\mathcal{F}$ of trees $\mathcal{T}_i$ that are junction trees for $\mathcal{H}_i$, with $\mathcal{H} = \vee_i \mathcal{H}i$ and

$$\mathcal{H}_i \wedge \mathcal{H}_j = \emptyset \quad \text{for } i \neq j.$$

Hence, hyperedges a and b that are in different trees of a junction forest are disjoint, and thus if $a \cap b \neq \emptyset$ there is a path in $\mathcal{F}$ between a and b.

Consider an arbitrary forest $\mathcal{F}$ with the hyperedges $\mathcal{H}$ as vertex set and two hyperedges h_1 and h_2 which are adjacent in $\mathcal{F}$. If the link between h_1 and h_2 is removed, then the tree containing these two hyperedges disconnects. Let $\mathcal{H}_1$ be the set of hyperedges that are still connected to h_1 and let $\mathcal{H}_2$ denote the set of remaining hyperedges in $\mathcal{H}$. The key to the relation between decomposability and junction trees and forests is the following lemma.

Lemma 2.26 *If $\mathcal{F}$ is a junction forest for a reduced hypergraph $\mathcal{H}$ then for any neighbours h_1 and h_2 in $\mathcal{F}$, $\mathcal{H}$ is the direct join of the components $\mathcal{H}_1$ and $\mathcal{H}_2$.*

Proof: Choose two neighbours h_1 and h_2 in $\mathcal{F}$ and define $\mathcal{H}_1$ and $\mathcal{H}_2$ as above. We recall that

$$\mathcal{H}_1 \wedge \mathcal{H}_2 = \text{red}\{a \cap b \mid a \in \mathcal{H}_1, b \in \mathcal{H}_2\}.$$

Assume first that $\mathcal{F}$ is a junction forest for $\mathcal{H}$. For any $a \in \mathcal{H}_1$ and $b \in \mathcal{H}_2$ with $a \cap b \neq \emptyset$, both h_1 and h_2 are on the unique path between a and b in $\mathcal{F}$. Hence, by the junction property (2.7),

$$a \cap b \subseteq h_1 \cap h_2$$

and hence

$$\mathcal{H}_1 \wedge \mathcal{H}_2 = \{h_1 \cap h_2\},$$

whereby $\mathcal{H}$ is the direct join of $\mathcal{H}_1$ and $\mathcal{H}_2$. □

Proposition 2.27 *A reduced hypergraph $\mathcal{H}$ is decomposable if and only if there is a junction forest $\mathcal{F}$ for $\mathcal{H}$.*

Proof: The proof is by induction on the number of hyperedges in $\mathcal{H}$. The statement is trivial for a simple hypergraph. Assume then that the statement holds for any hypergraph with at most n hyperedges and let $\mathcal{H}$ have $n+1$ hyperedges.

First let $\mathcal{H}$ be decomposable. Then it is the direct join of reduced hypergraphs $\mathcal{H}_1$ and $\mathcal{H}_2$ where both of these have fewer hyperedges. As the join is direct we can choose $h_1 \in \mathcal{H}_1$ and $h_2 \in \mathcal{H}_2$ such that

$$\mathcal{H}_1 \wedge \mathcal{H}_2 = \{h_1 \cap h_2\}.$$

The inductive assumption gives two junction forests $\mathcal{F}_1$ and $\mathcal{F}_2$ for $\mathcal{H}_1$ and $\mathcal{H}_2$. Form now $\mathcal{F}$ from $\mathcal{F}_1$ and $\mathcal{F}_2$ by taking their union and adding an edge between h_1 and h_2 if $h_1 \cap h_2 \neq \emptyset$. We must show that $\mathcal{F}$ is a junction forest for $\mathcal{H}$. So let $a, b \in \mathcal{H}$. If both are in $\mathcal{H}_1$ or both in $\mathcal{H}_2$ and h is on the path between a and b, (2.7) follows from the fact that $\mathcal{F}_1$ and $\mathcal{F}_2$ were junction forests. Else we might assume that $a \in \mathcal{H}_1$ and $b \in \mathcal{H}_2$. Since $\mathcal{H}$ is the direct join of $\mathcal{H}_1$ and $\mathcal{H}_2$ we have

$$a \cap b \subseteq h_1 \cap h_2.$$

If $a \cap b \neq \emptyset$ we have also that $a \cap h_1 \neq \emptyset$ and there is therefore a path from a to h_1 in $\mathcal{F}_1$ and similarly a path from b to h_2 in forest $\mathcal{F}_2$, hence a path from a to b in $\mathcal{F}$. If h is on the path between a and b it is either on the path from a to h_1 or from b to h_2. In the former case we find

$$a \cap b \subseteq a \cap h_1 \cap h_2 \subseteq a \cap h_1 \subseteq h,$$

where the junction property of $\mathcal{F}_1$ has been used to give the last inclusion. If h is on the path from b to h_2 we argue analogously. Hence the junction property for $\mathcal{F}$ is established.

Assume conversely that $\mathcal{F}$ is a junction forest for $\mathcal{H}$. By Lemma 2.26, $\mathcal{H}$ is the direct join of hypergraphs $\mathcal{H}_1$ and $\mathcal{H}_2$ that both have fewer hyperedges. Clearly, the induced subgraphs $\mathcal{F}_1 = \mathcal{F}_{\mathcal{H}_1}$ and $\mathcal{F}_2 = \mathcal{F}_{\mathcal{H}_2}$ are junction forests. Hence $\mathcal{H}_1$ and $\mathcal{H}_2$ are decomposable by the inductive assumption. As $\mathcal{H}$ is the direct join of decomposable hypergraphs it is itself decomposable. $\square$

Finally the fundamental relation between junction forests and sequences of sets satisfying the running intersection property is emphasized.

Consider a sequence $B_1, \ldots, B_k$ of finite and distinct sets that satisfies the running intersection property, i.e. for all $i > 1$ there is a $j < i$ such that $S_i \subseteq B_j$ where

$$S_i = B_i \cap (B_1 \cup \cdots \cup B_{i-1}),$$

and define the hypergraph $\mathcal{H} = \{B_1, \ldots, B_k\}$. Construct then an undirected graph $\mathcal{F}$ with these sets as vertices by successively choosing j for each i such that $S_i \subseteq B_j$ and then letting $i \sim j$.

Proposition 2.28 *The graph $\mathcal{F}$ is a junction forest for $\mathcal{H}$.*

Proof: We use induction on the number k of sets in the sequence. The statement is trivial for $k \leq 2$. Assume the statement to hold for sequences of length at most n and assume $k = n + 1$.

Using the construction until B_{k-1} gives a graph $\mathcal{F}_{k-1}$ which is a junction forest by the inductive assumption. Adding the edge from B_k to B^*, where $S_k \subseteq B^*$, produces clearly a forest but the junction property must be checked. It is enough to consider $a \in \mathcal{F}_{k-1}$, $b = B_k$ and h on the path between them. Obviously, then h is also on the path from a to B^*. Using that $a \subseteq B_1 \cup \cdots \cup B_{k-1}$ and the junction property of $\mathcal{F}_{k-1}$ we obtain

$$a \cap B_k \subseteq a \cap (B_1 \cup \cdots \cup B_{k-1}) \cap B_k \subseteq a \cap B^* \subseteq h.$$

Hence $\mathcal{F}$ is a junction forest. $\square$

Conversely, let $\mathcal{F}$ be a junction forest and choose roots arbitrarily in all trees, thereby directing all edges in $\mathcal{F}$. This induces a natural partial order on the vertices of $\mathcal{F}$ by having a before b if there is a directed path from a to b. Assume now that $b_1, \ldots, b_k$ is any numbering of the vertices of $\mathcal{F}$ that is compatible with this ordering.

Proposition 2.29 *The sets $b_1, \ldots, b_k$ have running intersection property.*

Proof: Consider b_i for $i > 1$ and assume this to be part of the tree $\mathcal{T}$ in $\mathcal{F}$ with chosen root R. All sets on the path from R to b_i must be among $b_1, \ldots b_{i-1}$ or the numbering would not be compatible. Let b^* be the hyperedge on this path which is nearest to b_i.

Suppose b_l for $1 < l < i-1$ is in a different tree. Then $b_i \subseteq b_l = \emptyset$. Else b^* is on the path between b_l and b_i. The junction property thus implies $b_l \cap b_j \subseteq b^*$ and therefore

$$b_i \cap (b_1 \cup \cdots \cup b_{i-1}) \subseteq b^*,$$

which shows that we can choose $b_j = b^*$ and have the running intersection property. $\square$

2.3 Notes

The notion of a decomposable graph has deep connections to many areas besides the statistical analysis of association. This includes general graph theory (Diestel 1987), the four-colour problem (Wagner 1937), measure theory (Kellerer 1964a, 1964b; Vorobev 1962, 1963), the solution of systems of linear equations (Parter 1961; Rose 1970), game theory (Vorobev 1967), relational databases (Beeri *et al.* 1981, 1983) and expert systems (Lauritzen and Spiegelhalter 1988). See these as well as Lauritzen *et al.* (1984), Golumbic (1980), Diestel (1990) for further discussion and references.

In the pure case, decomposable graphs are well-studied objects although they usually appear under other names such as, for example, *rigid circuit* (Dirac 1961), *chordal* (Gavril 1972) or *triangulated* (Berge 1973; Rose 1970).

In the case of a marked graph, the notion of a decomposable graph was first introduced by Lauritzen and Wermuth (1984) and studied further by Leimer (1985, 1989).

There is an extensive literature concerned with algorithms for manipulating decomposable graphs in an efficient way. This includes algorithms for checking decomposability of a graph and finding their cliques (Rose *et al.* 1976; Tarjan and Yannakakis 1984), for constructing junction trees for given decomposable graphs (F.V. Jensen and Jensen 1994), for finding optimal decompositions (Tarjan 1985; Leimer 1993), and for finding small decomposable graphs that contain a given graph (Kjærulff 1990; 1992).

3
Conditional independence and Markov properties

3.1 Conditional independence

Throughout this text a central notion is that of conditional independence of random variables, the graphs keeping track of the conditional independence relations.

Formally, if X, Y, Z are random variables with a joint distribution P, we say that *X is conditionally independent of Y given Z under P*, and write $X \perp\!\!\!\perp Y \mid Z\ [P]$, if, for any measurable set A in the sample space of X, there exists a version of the conditional probability $P(A \mid Y, Z)$ which is a function of Z alone. Usually P will be fixed and omitted from the notation. If Z is trivial we say that *X is independent of Y*, and write $X \perp\!\!\!\perp Y$. The notation is due to Dawid (1979) who studied the notion of conditional independence in a systematic fashion. Dawid (1980) gives a formal treatment.

When X, Y, and Z are discrete random variables the condition for $X \perp\!\!\!\perp Y \mid Z$ simplifies as

$$P(X = x, Y = y \mid Z = z) = P(X = x \mid Z = z) P(Y = y \mid Z = z),$$

where the equation holds for all z with $P(Z = z) > 0$. When the three variables admit a joint density with respect to a product measure μ, we have

$$X \perp\!\!\!\perp Y \mid Z \iff f_{XY \mid Z}(x, y \mid z) = f_{X \mid Z}(x \mid z) f_{Y \mid Z}(y \mid z), \qquad (3.1)$$

where this equation is to hold almost surely with respect to P. If all densities are continuous, the equality in (3.1) must hold for all z with $f_Z(z) > 0$. Here it is understood that all functions on a discrete space are considered continuous functions. The condition (3.1) can be rewritten as

$$X \perp\!\!\!\perp Y \mid Z \iff f_{XYZ}(x, y, z) f_Z(z) = f_{XZ}(x, z) f_{YZ}(y, z) \qquad (3.2)$$

and this equality must hold *for all values of* z when the densities are continuous.

The ternary relation $X \perp\!\!\!\perp Y \mid Z$ has the following properties, where h denotes an arbitrary measurable function on the sample space of X:

(C1) if $X \perp\!\!\!\perp Y \mid Z$ then $Y \perp\!\!\!\perp X \mid Z$;

(C2) if $X \perp\!\!\!\perp Y \mid Z$ and $U = h(X)$, then $U \perp\!\!\!\perp Y \mid Z$;

(C3) if $X \perp\!\!\!\perp Y \mid Z$ and $U = h(X)$, then $X \perp\!\!\!\perp Y \mid (Z, U)$;

(C4) if $X \perp\!\!\!\perp Y \mid Z$ and $X \perp\!\!\!\perp W \mid (Y, Z)$, then $X \perp\!\!\!\perp (W, Y) \mid Z$.

We leave the proof of these facts to the reader. Note that the converse to (C4) follows from (C2) and (C3).

If we use f as generic symbol for the probability density of the random variables corresponding to its arguments, the following statements are true:

$$X \perp\!\!\!\perp Y \mid Z \iff f(x,y,z) = f(x,z)f(y,z)/f(z) \quad (3.3)$$
$$X \perp\!\!\!\perp Y \mid Z \iff f(x \mid y, z) = f(x \mid z) \quad (3.4)$$
$$X \perp\!\!\!\perp Y \mid Z \iff f(x, z \mid y) = f(x \mid z)f(z \mid y) \quad (3.5)$$
$$X \perp\!\!\!\perp Y \mid Z \iff f(x,y,z) = h(x,z)k(y,z) \text{ for some } h, k \quad (3.6)$$
$$X \perp\!\!\!\perp Y \mid Z \iff f(x,y,z) = f(x \mid z)f(y,z). \quad (3.7)$$

The equalities above hold apart from a set of triples (x, y, z) with probability zero. If the densities are continuous functions (in particular if the state spaces are discrete), the equations hold whenever the quantitites involved are well defined, i.e. when the densities of all conditioning variables are positive. We also leave the proof of these equivalences to the reader.

Another property of the conditional independence relation is often used:

(C5) if $X \perp\!\!\!\perp Y \mid Z$ and $X \perp\!\!\!\perp Z \mid Y$ then $X \perp\!\!\!\perp (Y, Z)$.

However (C5) does not hold universally, but only under additional conditions — essentially that there be no non-trivial logical relationship between Y and Z. A trivial counterexample appears when $X = Y = Z$ with $P\{X = 1\} = P\{X = 0\} = 1/2$. We have however

Proposition 3.1 *If the joint density of all variables with respect to a product measure is positive and continuous, then the statement* (C5) *will hold true.*

Proof: Assume that the variables have a continuous density $f(x, y, z) > 0$ and that $X \perp\!\!\!\perp Y \mid Z$ as well as $X \perp\!\!\!\perp Z \mid Y$. Then (3.6) gives for all values of (x, y, z) that

$$f(x, y, z) = k(x, z)l(y, z) = g(x, y)h(y, z)$$

for suitable strictly positive functions g, h, k, l. Thus, as the density is assumed continuous, we have that for all z,

$$g(x,y) = \frac{k(x,z)l(y,z)}{h(y,z)}.$$

Choosing a fixed $z = z_0$ we have $g(x,y) = \pi(x)\rho(y)$ where $\pi(x) = k(x, z_0)$ and $\rho(y) = l(y, z_0)/h(y, z_0)$. Thus $f(x,y,z) = \pi(x)\rho(y)h(y,z)$ and hence $X \perp\!\!\!\perp (Y, Z)$ as desired. $\square$

The proposition can be weakened to more general functions than continuous functions, but we abstain from pursuing this here.

It is illuminating to think of the properties (C1)–(C5) as purely formal expressions, with a meaning that is not necessarily tied to probability. If we interpret the symbols used for random variables as abstract symbols for pieces of knowledge obtained from, say, reading books, and further interpret the symbolic expression $X \perp\!\!\!\perp Y \mid Z$ as:

Knowing Z, reading Y is irrelevant for reading X,

the properties (C1)–(C4) translate to the following:

(I1) if, knowing Z, reading Y is irrelevant for reading X, then so is reading X for reading Y;

(I2) if, knowing Z, reading Y is irrelevant for reading the book X, then reading Y is irrelevant for reading any chapter U of X;

(I3) if, knowing Z, reading Y is irrelevant for reading the book X, it remains irrelevant after having read any chapter U of X;

(I4) if, knowing Z, reading the book Y is irrelevant for reading X and even after having also read Y, reading W is irrelevant for reading X, then reading of both Y and W is irrelevant for reading X.

Thus one can view the relations (C1)–(C4) as pure formal properties of the notion of irrelevance. The property (C5) is slightly more subtle. In a certain sense, also the symmetry (C1) is a somewhat special property of probabilistic conditional independence, rather than general irrelevance.

It is thus tempting to use the relations (C1)–(C4) as formal axioms for conditional independence or irrelevance. A *semi-graphoid* is an algebraic structure which satisfies (C1)–(C4) where X, Y, Z are disjoint subsets of a finite set and $U = h(X)$ is replaced by $U \subseteq X$ (Pearl 1988). If also (C5) holds for disjoint subsets, it is called a *graphoid*. Below we give further examples of such structures.

CONDITIONAL INDEPENDENCE

Example 3.2 A very important example of a model for the irrelevance axioms above is that of *graph separation* in undirected graphs. Let A, B, and C be subsets of the vertex set V of a finite undirected graph $\mathcal{G} = (V, E)$. Define
$$A \stackrel{\mathcal{G}}{\perp} B \,|\, C \iff C \text{ separates } A \text{ from } B \text{ in } \mathcal{G}.$$

Then it is not difficult to see that graph separation has the following properties:

(S1) if $A \stackrel{\mathcal{G}}{\perp} B \,|\, C$ then $B \stackrel{\mathcal{G}}{\perp} A \,|\, C$;

(S2) if $A \stackrel{\mathcal{G}}{\perp} B \,|\, C$ and U is a subset of A, then $U \stackrel{\mathcal{G}}{\perp} B \,|\, C$;

(S3) if $A \stackrel{\mathcal{G}}{\perp} B \,|\, C$ and U is a subset of B, then $A \stackrel{\mathcal{G}}{\perp} B \,|\, (C \cup U)$;

(S4) if $A \stackrel{\mathcal{G}}{\perp} B \,|\, C$ and $A \stackrel{\mathcal{G}}{\perp} D \,|\, (B \cup C)$, then $A \stackrel{\mathcal{G}}{\perp} (B \cup D) \,|\, C$.

Even the analogue of (C5) holds when all involved subsets are disjoint. Hence graph separation satisfies the graphoid axioms. □

Example 3.3 As another fundamental example, consider *geometric orthogonality* in Euclidean vector spaces. Let L, M, and N be linear subspaces of a Euclidean space V and define
$$L \perp M \,|\, N \iff (L \ominus N) \perp (M \ominus N), \tag{3.8}$$

where $L \ominus N = L \cap N^\perp$. If (3.8) is satisfied, then L and M are said to *meet orthogonally in N*. Again, it is not hard to see that the orthogonal meet has the following properties:

(O1) if $L \perp M \,|\, N$ then $M \perp L \,|\, N$;

(O2) if $L \perp M \,|\, N$ and U is a linear subspace of L, then $U \perp M \,|\, N$;

(O3) if $L \perp M \,|\, N$ and U is a linear subspace of M, then $L \perp M \,|\, (N+U)$;

(O4) if $L \perp M \,|\, N$ and $L \perp R \,|\, (M+N)$, then $L \perp (M+R) \,|\, N$.

The analogue of (C5) does not hold in general; for example if $M = N$ we may have
$$L \perp M \,|\, N \text{ and } L \perp N \,|\, M,$$
but if L and M are not orthogonal then it is false that $L \perp (M+N)$. □

An abstract theory for conditional independence based on graphoids and semi-graphoids has recently developed (Studený 1989, 1993; Matúš 1992a). It was conjectured (Pearl 1988) that the properties (C1)–(C4) were sound and complete axioms for probabilistic conditional independence. However, the completeness fails. In fact, finite axiomatization of probabilistic conditional independence is not possible (Studený 1992). See also Geiger and Pearl (1993) for a systematic study of the logical implications of conditional independence.

3.2 Markov properties

In this section we consider conditional independence in the special situation where we have a collection of random variables $(X_\alpha)_{\alpha \in V}$ taking values in probability spaces $(\mathcal{X}_\alpha)_{\alpha \in V}$. The probability spaces are either real finite-dimensional vector spaces or finite and discrete sets. For A being a subset of V we let $\mathcal{X}_A = \times_{\alpha \in A} \mathcal{X}_\alpha$ and further $\mathcal{X} = \mathcal{X}_V$. Typical elements of $\mathcal{X}_A$ are denoted as $x_A = (x_\alpha)_{\alpha \in A}$. Similarly $X_A = (X_\alpha)_{\alpha \in A}$. We then use the short notation
$$A \perp\!\!\!\perp B \mid C$$
for
$$X_A \perp\!\!\!\perp X_B \mid X_C$$
and so on. The set V is assumed to be the vertex set of a graph $\mathcal{G} = (V, E)$.

3.2.1 Markov properties on undirected graphs

Associated with an undirected graph $\mathcal{G} = (V, E)$ and a collection of random variables $(X_\alpha)_{\alpha \in V}$ as above there is a range of different Markov properties. A probability measure P on $\mathcal{X}$ is said to obey

(P) *the pairwise Markov property*, relative to $\mathcal{G}$, if for any pair (α, β) of non-adjacent vertices
$$\alpha \perp\!\!\!\perp \beta \mid V \setminus \{\alpha, \beta\};$$

(L) *the local Markov property*, relative to $\mathcal{G}$, if for any vertex $\alpha \in V$
$$\alpha \perp\!\!\!\perp V \setminus \mathrm{cl}(\alpha) \mid \mathrm{bd}(\alpha);$$

(G) *the global Markov property*, relative to $\mathcal{G}$, if for any triple (A, B, S) of disjoint subsets of V such that S separates A from B in $\mathcal{G}$
$$A \perp\!\!\!\perp B \mid S.$$

The Markov properties are related as described in the proposition below.

Proposition 3.4 *For any undirected graph $\mathcal{G}$ and any probability distribution on $\mathcal{X}$ it holds that*

$$(G) \implies (L) \implies (P). \tag{3.9}$$

Proof: Firstly, (G) implies (L) because $\text{bd}(\alpha)$ separates α from $V \setminus \text{cl}(\alpha)$. Assume next that (L) holds. We have $\beta \in V \setminus \text{cl}(\alpha)$ because α and β are non-adjacent. Hence

$$\text{bd}(\alpha) \cup ((V \setminus \text{cl}(\alpha)) \setminus \{\beta\}) = V \setminus \{\alpha, \beta\},$$

and it follows from (L) and (C3) that

$$\alpha \perp\!\!\!\perp V \setminus \text{cl}(\alpha) \mid V \setminus \{\alpha, \beta\}.$$

Application of (C2) then gives $\alpha \perp\!\!\!\perp \beta \mid V \setminus \{\alpha, \beta\}$ which is (P). □

It is worth noting that (3.9) only depends on the properties (C1)–(C4) of conditional independence and hence holds for any semi-graphoid. The various Markov properties are different in general, as the following examples show.

Example 3.5 Define the joint distribution of five binary random variables U, W, X, Y, Z as follows: U and Z are independent with

$$P(U = 1) = P(Z = 1) = P(U = 0) = P(Z = 0) = 1/2,$$

$W = U$, $Y = Z$, and $X = WY$. The joint distribution so defined is easily seen to satisfy (L) but not (G) for the graph below.

$$\underset{U}{\bullet} \quad \underset{W}{\bullet} \quad \underset{X}{\bullet} \quad \underset{Y}{\bullet} \quad \underset{Z}{\bullet}$$

In fact, Matúš (1992b) shows that the global and local Markov properties coincide if and only if the dual graph $\check{\mathcal{G}}$ (defined as $\alpha \tilde{\sim} \beta$ if and only if $\alpha \not\sim \beta$) does not have the 4-cycle as an induced subgraph. □

Example 3.6 A simple example of a probability distribution of (X, Y, Z) that satisfies the pairwise Markov property (P) with respect to the graph

$$\underset{X}{\bullet} \quad \underset{Y}{\bullet} \!\!\!-\!\!\! \underset{Z}{\bullet}$$

but does not satisfy the local Markov property (L) can be constructed by letting $X = Y = Z$ with $P\{X = 1\} = P\{X = 0\} = 1/2$. It can be shown (Matúš 1992b) that the global and pairwise Markov properties coincide if and only if the dual graph $\check{\mathcal{G}}$ does not have a subset of three vertices with two or three edges in its induced subgraph. □

If it holds for all disjoint subsets A, B, C, and D that

if $A \perp\!\!\!\perp B \,|\, (C \cup D)$ and $A \perp\!\!\!\perp C \,|\, (B \cup D)$ then $A \perp\!\!\!\perp (B \cup C) \,|\, D$, (3.10)

then the Markov properties are all equivalent. This condition is analogous to (C5) and holds, for example, if P has a positive and continuous density with respect to a product measure μ. This is seen as in Proposition 3.1. The result is stated in the theorem below, due to Pearl and Paz (1987); see also Pearl (1988).

Theorem 3.7 (Pearl and Paz) *If a probability distribution on $\mathcal{X}$ is such that (3.10) holds for disjoint subsets A, B, C, D then*

$$(G) \iff (L) \iff (P).$$

Proof: We need to show that (P) implies (G), so assume that S separates A from B in $\mathcal{G}$ and that (P) as well as (3.10) hold. Without loss of generality we can also assume that both A and B are non-empty. The proof is then backward induction on the number of vertices $n = |S|$ in S. If $n = |V| - 2$ then both A and B consist of one vertex and the required conditional independence follows from (P).

So assume $|S| = n < |V| - 2$ and that separation implies conditional independence for all separating sets S with more than n elements. We first assume that $V = A \cup B \cup S$, implying that at least one of A and B has more than one element, A, say. If $\alpha \in A$ then $S \cup \{\alpha\}$ separates $A \setminus \{\alpha\}$ from B and also $S \cup A \setminus \{\alpha\}$ separates α from B. Thus by the induction hypothesis

$$A \setminus \{\alpha\} \perp\!\!\!\perp B \,|\, S \cup \{\alpha\} \text{ and } \alpha \perp\!\!\!\perp B \,|\, S \cup A \setminus \{\alpha\}.$$

Now (3.10) gives $A \perp\!\!\!\perp B \,|\, S$.

If $A \cup B \cup S \subset V$ we choose $\alpha \in V \setminus (A \cup B \cup S)$. Then $S \cup \{\alpha\}$ separates A and B, implying $A \perp\!\!\!\perp B \,|\, S \cup \{\alpha\}$. Further, either $A \cup S$ separates B from $\{\alpha\}$ or $B \cup S$ separates A from $\{\alpha\}$. Assuming the former gives $\alpha \perp\!\!\!\perp B \,|\, A \cup S$. Using (3.10) and (C2) we derive that $A \perp\!\!\!\perp B \,|\, S$. The latter case is similar. □

Note that the proof only exploits (C1)–(C4) and (3.10) and therefore applies to any graphoid.

The global Markov property (G) is important because it gives a general criterion for deciding when two groups of variables A and B are conditionally independent given a third group of variables S.

As conditional independence is intimately related to factorization, so are the Markov properties. A probability measure P on $\mathcal{X}$ is said to *factorize* according to $\mathcal{G}$ if for all complete subsets $a \subseteq V$ there exist non-negative functions ψ_a that depend on x through x_a only, and there exists a product measure $\mu = \otimes_{\alpha \in V} \mu_\alpha$ on $\mathcal{X}$, such that P has density f with respect to μ where f has the form

$$f(x) = \prod_{a \text{ complete}} \psi_a(x). \tag{3.11}$$

The functions ψ_a are not uniquely determined. There is arbitrariness in the choice of μ, but also groups of functions ψ_a can be multiplied together or split up in different ways. In fact one can without loss of generality assume — although this is not always practical — that only cliques appear as the sets a, i.e. that

$$f(x) = \prod_{c \in \mathcal{C}} \psi_c(x), \tag{3.12}$$

where $\mathcal{C}$ is the set of cliques of $\mathcal{G}$. If P factorizes, we say that P has property (F) and the set of such probability measures is denoted by $M_F(\mathcal{G})$. We have

Proposition 3.8 *For any undirected graph $\mathcal{G}$ and any probability distribution on $\mathcal{X}$ it holds that*

$$(F) \implies (G) \implies (L) \implies (P).$$

Proof: We only have to show that (F) implies (G) as the remaining implications are given in (3.9). Let (A, B, S) be any triple of disjoint subsets such that S separates A from B. Let $\tilde{A}$ denote the connectivity components in $\mathcal{G}_{V \setminus S}$ which contain A and let $\tilde{B} = V \setminus (\tilde{A} \cup S)$. Since A and B are separated by S, their elements are in different connectivity components of $\mathcal{G}_{V \setminus S}$ and any clique of $\mathcal{G}$ is either a subset of $\tilde{A} \cup S$ or of $\tilde{B} \cup S$. If $\mathcal{C}_A$ denotes the cliques contained in $\tilde{A} \cup S$, we thus obtain from (3.12) that

$$f(x) = \prod_{c \in \mathcal{C}} \psi_c(x) = \prod_{c \in \mathcal{C}_A} \psi_c(x) \prod_{c \in \mathcal{C} \setminus \mathcal{C}_A} \psi_c(x) = h(x_{\tilde{A} \cup S}) k(x_{\tilde{B} \cup S}).$$

Hence (3.6) gives that $\tilde{A} \perp\!\!\!\perp \tilde{B} \,|\, S$. Applying (C2) twice gives the desired independence. □

In the case where P has a positive and continuous density we can use the Möbius inversion lemma to show that (P) implies (F), and thus all Markov properties are equivalent. This result seems to have been discovered in various forms by a number of authors (Speed 1979) but is usually attributed to Hammersley and Clifford (1971) who proved the result in the discrete case. The condition that the density be continuous can probably be considerably relaxed (Koster 1994), whereas the positivity is essential. More precisely, we have

Theorem 3.9 (Hammersley and Clifford) *A probability distribution P with positive and continuous density f with respect to a product measure μ satisfies the pairwise Markov property with respect to an undirected graph $\mathcal{G}$ if and only if it factorizes according to $\mathcal{G}$.*

Proof: If P factorizes, it is pairwise Markov as shown in Proposition 3.8, so we just have to show that (P) implies (F).

Since the density is positive, we may take logarithms on both sides of (3.11). Hence this equation can be rewritten as

$$\log f(x) = \sum_{a: a \subseteq V} \phi_a(x), \qquad (3.13)$$

where $\phi_a(x) = \log \psi_a(x)$ and $\phi_a \equiv 0$ unless a is a complete subset of V.

Assume then that P is pairwise Markov and choose a fixed but arbitrary element $x^* \in \mathcal{X}$. Define for all $a \subseteq V$

$$H_a(x) = \log f(x_a, x^*_{a^c}),$$

where $(x_a, x^*_{a^c})$ is the element y with $y_\gamma = x_\gamma$ for $\gamma \in a$ and $y_\gamma = x^*_\gamma$ for $\gamma \notin a$. Since x^* is fixed, H_a depends on x through x_a only. Let further for all $a \subseteq V$

$$\phi_a(x) = \sum_{b: b \subseteq a} (-1)^{|a \setminus b|} H_b(x).$$

From this relation it is also clear that ϕ_a depends on x through x_a only. Next we can apply Lemma A.2 (Möbius inversion) to obtain that

$$\log f(x) = H_V(x) = \sum_{a: a \subseteq V} \phi_a(x)$$

such that we have proved the theorem if we can show that $\phi_a \equiv 0$ whenever a is not a complete subset of V. So let us assume that $\alpha, \beta \in a$ and $\alpha \not\sim \beta$. Let further $c = a \setminus \{\alpha, \beta\}$. If we write H_a as short for $H_a(x)$ we have

$$\phi_a(x) = \sum_{b: b \subseteq c} (-1)^{|c \setminus b|} \left\{ H_b - H_{b \cup \{\alpha\}} - H_{b \cup \{\beta\}} + H_{b \cup \{\alpha, \beta\}} \right\}. \qquad (3.14)$$

Let $d = V \setminus \{\alpha, \beta\}$. Then, by the pairwise Markov property and (3.7), we have

$$\begin{aligned}
H_{b \cup \{\alpha, \beta\}}(x) - H_{b \cup \{\alpha\}}(x) &= \log \frac{f(x_b, x_\alpha, x_\beta, x^*_{d \setminus b})}{f(x_b, x_\alpha, x^*_\beta, x^*_{d \setminus b})} \\
&= \log \frac{f(x_\alpha \mid x_b, x^*_{d \setminus b}) f(x_\beta, x_b, x^*_{d \setminus b})}{f(x_\alpha \mid x_b, x^*_{d \setminus b}) f(x^*_\beta, x_b, x^*_{d \setminus b})}
\end{aligned}$$

$$= \log \frac{f(x_\alpha^* \mid x_b, x_{d\setminus b}^*) f(x_\beta, x_b, x_{d\setminus b}^*)}{f(x_\alpha^* \mid x_b, x_{d\setminus b}^*) f(x_\beta^*, x_b, x_{d\setminus b}^*)}$$

$$= \log \frac{f(x_b, x_\alpha^*, x_\beta, x_{d\setminus b}^*)}{f(x_b, x_\alpha^*, x_\beta^*, x_{d\setminus b}^*)}$$

$$= H_{b\cup\{\beta\}}(x) - H_b(x).$$

Thus all terms in the curly brackets in (3.14) add to zero and henceforth the entire sum is zero. This completes the proof. □

The expression inside curly brackets in (3.14) is the logarithm of what is known as the *partial cross-product ratio* so that we can alternatively write

$$\phi_a(x) = \sum_{b:b\subseteq c} (-1)^{|c\setminus b|} \log \operatorname{cpr}(x_\alpha, x_\beta; x_\alpha^*, x_\beta^* \mid x_b, x_{d\setminus b}^*).$$

The pairwise Markov property ensures that all these partial cross-product ratios are equal to 1.

Example 3.10 The following example is due to Moussouris (1974) and shows that the global Markov property (G) may not imply the factorization property (F) without positivity assumptions on the density.

The example is concerned with the distribution of four binary random variables, denoted by (X_1, X_2, X_3, X_4). The following combinations are assumed to have positive probabilities, in fact each of them given a probability equal to 1/8:

$$(0,0,0,0) \quad (1,0,0,0) \quad (1,1,0,0) \quad (1,1,1,0)$$
$$(0,0,0,1) \quad (0,0,1,1) \quad (0,1,1,1) \quad (1,1,1,1).$$

The distribution so specified satisfies the global Markov property (G) with respect to the chordless four-cycle

with the vertices identified cyclically with the random variables. This is seen as follows.

For example, if we consider the conditional distribution of (X_1, X_3), given that $(X_2, X_4) = (0, 1)$, we find

$$P\{X_1 = 0 \mid (X_2, X_4) = (0,1)\} = 1.$$

Since the conditional distribution of X_1 is degenerate, it is trivially independent of X_3. All other combinations of conditions on (X_2, X_4) give in a similar way degenerate distributions for one of the remaining variables and this picture is repeated when conditioning on (X_1, X_3). Hence, we have

$$X_1 \perp\!\!\!\perp X_3 \mid (X_2, X_4) \text{ and } X_2 \perp\!\!\!\perp X_4 \mid (X_1, X_3),$$

which shows that the distribution is globally Markov with respect to the graph displayed.

But the density does not factorize. This is seen by an indirect argument. Assume the density factorizes. Then

$$0 \neq 1/8 = f(0,0,0,0) = \phi_{\{1,2\}}(0,0)\phi_{\{2,3\}}(0,0)\phi_{\{3,4\}}(0,0)\phi_{\{4,1\}}(0,0).$$

But also

$$0 = f(0,0,1,0) = \phi_{\{1,2\}}(0,0)\phi_{\{2,3\}}(0,1)\phi_{\{3,4\}}(1,0)\phi_{\{4,1\}}(0,0),$$

whereby

$$\phi_{\{2,3\}}(0,1)\phi_{\{3,4\}}(1,0) = 0.$$

Since now

$$0 \neq 1/8 = f(0,0,1,1) = \phi_{\{1,2\}}(0,0)\phi_{\{2,3\}}(0,1)\phi_{\{3,4\}}(1,1)\phi_{\{4,1\}}(1,0),$$

this implies $\phi_{\{2,3\}}(0,1) \neq 0$ and hence $\phi_{\{3,4\}}(1,0) = 0$, which contradicts that

$$0 \neq 1/8 = f(1,1,1,0) = \phi_{\{1,2\}}(1,1)\phi_{\{2,3\}}(1,1)\phi_{\{3,4\}}(1,0)\phi_{\{4,1\}}(0,1).$$

Hence the density cannot factorize. □

In general none of the Markov properties are preserved under weak limits. This is because weak convergence of joint distributions does not imply convergence of conditional distributions. This fact is illustrated in the following example.

Example 3.11 Let $Y = (Y_1, Y_2, Y_3)^\top$ be a trivariate normal random variable with mean zero and covariance matrix

$$\Sigma = \begin{pmatrix} 1 & \frac{1}{\sqrt{n}} & \frac{1}{2} \\ \frac{1}{\sqrt{n}} & \frac{2}{n} & \frac{1}{\sqrt{n}} \\ \frac{1}{2} & \frac{1}{\sqrt{n}} & 1 \end{pmatrix}.$$

Using Proposition C.5 the conditional distribution of $(Y_1, Y_3)^\top$ given Y_2 is bivariate normal with covariance matrix

$$\Sigma_{13|2} = \begin{pmatrix} 1 & \frac{1}{2} \\ \frac{1}{2} & 1 \end{pmatrix} - \begin{pmatrix} \frac{1}{\sqrt{n}} \\ \frac{1}{\sqrt{n}} \end{pmatrix} \left(\frac{n}{2}\right) \left(\frac{1}{\sqrt{n}}, \frac{1}{\sqrt{n}}\right) = \begin{pmatrix} \frac{1}{2} & 0 \\ 0 & \frac{1}{2} \end{pmatrix}$$

and hence $Y_1 \perp\!\!\!\perp Y_3 \mid Y_2$, which means that Y satisfies the global Markov property on the graph

```
o———o———o
1   2   3
```

As we shall see later in Proposition 5.2, this can alternatively be seen from inspection of the inverse covariance matrix

$$K = \Sigma^{-1} = \begin{pmatrix} 2 & -\sqrt{n} & 0 \\ -\sqrt{n} & \frac{3n}{2} & -\sqrt{n} \\ 0 & -\sqrt{n} & 2 \end{pmatrix}.$$

The distribution of Y converges weakly to the normal distribution with covariance matrix

$$\bar{\Sigma} = \begin{pmatrix} 1 & 0 & \frac{1}{2} \\ 0 & 0 & 0 \\ \frac{1}{2} & 0 & 1 \end{pmatrix}$$

which is not Markov on the graph considered, since Y_2 is degenerate and

$$\bar{\Sigma}_{13|2} = \bar{\Sigma}_{13} = \begin{pmatrix} 1 & \frac{1}{2} \\ \frac{1}{2} & 1 \end{pmatrix}.$$

Hence the Markov property is not in general stable under weak limits. □

There are, however, important exceptions where the Markov property is preserved under limits and the case of discrete sample space is one.

Proposition 3.12 *Assume that a sequence of probability distributions P_n on a discrete sample space converges to P. If P_n all satisfy any of the Markov properties* (P), (L), *or* (G), *then P satisfies the same Markov property.*

Proof: The density p of the limiting distribution is given as $p = \lim p_n$, where p_n is the density of P_n. Let A, B, C be disjoint and assume that $A \perp\!\!\!\perp B \mid C$ for all P_n. If $p(i_C) > 0$ we get

$$p(i_A, i_B, i_C) = \lim_{n \to \infty} p_n(i_A, i_B, i_C) = \lim_{n \to \infty} \frac{p_n(i_A, i_C) p_n(i_B, i_C)}{p_n(i_C)}$$

$$= \frac{p(i_A, i_C) p(i_B, i_C)}{p(i_C)},$$

whereby $A \perp\!\!\!\perp B \mid C$ also in the limit. Thus we have shown that all conditional independences are preserved, as desired. □

Observe that the property (F) is not preserved under weak limits, even in the discrete case, as the following example shows.

Example 3.13 Consider four binary random variables (X_1, X_2, X_3, X_4) and let
$$f_n(x_1, x_2, x_3, x_4) = \frac{n^{x_1 x_2 + x_2 x_3 + x_3 x_4 - x_1 x_4 - x_2 - x_3 + 1}}{8 + 8n}.$$
Direct calculations show that, as $n \to \infty$, f_n converges to the distribution in Example 3.10. Hence we deduce that (F) is not stable under limits. □

It holds however, in the discrete case, that any probability distribution which satisfies (F) can be obtained as a limit of positive probabilities that factorize. This is true because if p factorizes we get

$$p(i) = \prod_a \psi_a(i) = \lim_{\epsilon \to 0} \frac{\prod_a \{\psi_a(i) + \epsilon\}}{\sum_j \prod_a \{\psi_a(j) + \epsilon\}} = \lim_{\epsilon \to 0} p_\epsilon(i),$$

where p_ϵ clearly is positive and factorizes.

In the discrete case we refer to limits of positive Markov probabilities as *extended Markov* probabilities and denote these by $M_E(\mathcal{G})$. The argument above shows that $M_F(\mathcal{G}) \subseteq M_E(\mathcal{G})$ but the inclusion is strict for a general graph $\mathcal{G}$.

Distributions in $M_E(\mathcal{G})$ are identified through their clique marginals when their state space is finite. This fact is a special case of the following lemma (when $\mathcal{A}$ is taken to be the set of cliques of the graph $\mathcal{G}$).

Lemma 3.14 *Let $\mathcal{A}$ be a set of subsets of a finite set Δ and P and Q be probability mesures on the product space $\mathcal{I} = \times_{\delta \in \Delta} \mathcal{I}_\delta$ where, for each δ, $\mathcal{I}_\delta$ is a finite set. If P and Q have identical marginals to the sets $a \in \mathcal{A}$ and both are limits of measures whose densities with respect to a product measure μ factorize over $\mathcal{A}$, then $P = Q$.*

Proof: It is no restriction to assume that μ is the counting measure on $\mathcal{I}$ such that we have

$$p(i) = \lim_{n \to \infty} \prod_{a \in \mathcal{A}} \phi_a^n(i_a), \quad q(i) = \lim_{n \to \infty} \prod_{a \in \mathcal{A}} \psi_a^n(i_a).$$

Then we get

$$\begin{aligned}
\sum_i p(i) \log q(i) &= \sum_i p(i) \log \left\{ \lim_{n \to \infty} \prod_{a \in \mathcal{A}} \psi_a^n(i_a) \right\} \\
&= \lim_{n \to \infty} \sum_i p(i) \sum_{a \in \mathcal{A}} \log \psi_a^n(i_a) \\
&= \lim_{n \to \infty} \sum_{a \in \mathcal{A}} \sum_i p(i) \log \psi_a^n(i_a) \\
&= \lim_{n \to \infty} \sum_{a \in \mathcal{A}} \sum_i q(i) \log \psi_a^n(i_a)
\end{aligned}$$

$$= \lim_{n\to\infty} \sum_i q(i) \sum_{a\in\mathcal{A}} \log \psi_a^n(i_a)$$

$$= \sum_i q(i) \log q(i),$$

where we have exploited that $\sum_i p(i) \log \psi_a^n(i_a)$ depends on P through its a-marginal only and therefore is equal to $\sum_i q(i) \log \psi_a^n(i_a)$.

Interchanging the role of p and q in the above calculations yields

$$\sum_i q(i) \log p(i) = \sum_i p(i) \log p(i).$$

Combining these identities with the information inequality (A.4) gives

$$\sum_i p(i) \log p(i) = \sum_i q(i) \log p(i) \le \sum_i q(i) \log q(i)$$
$$= \sum_i p(i) \log q(i) \le \sum_i p(i) \log p(i).$$

Since (A.4) is strict unless $p(i) \equiv q(i)$, we must have $P = Q$ and the lemma is proved. □

Lemma 3.14 still holds in the finite case for a general μ and a general system of subsets $\mathcal{A}$. For countable state spaces and general μ it does not hold, i.e. we may have two different probability distributions P and Q that both have factorizing densities (H. G. Kellerer, personal communication). This fact seems to contradict a remark in Csiszár (1975).

If $\mathcal{A}$ is the set of cliques of a decomposable graph, the lemma holds under quite general circumstances (Kellerer 1964a, 1964b). It is not known whether the conclusion of the lemma holds for factorizing probability distributions in the case of a product measure μ and countable (or more general) state spaces. The critical point in the proof is the interchange of the two sums.

The following example is due to Matúš and Studený (1995) and shows that not all global Markov probabilities can be obtained as limits of factorizing distributions.

Example 3.15 Let $\mathcal{G}$ be the four-cycle as in Example 3.10, but let all four variables have three possible values a, b, and c. Let P be the distribution with each of the following nine states having probability equal to $1/9$:

(a,a,a,a) (b,a,b,c) (c,a,c,b)
(a,b,b,b) (b,b,c,a) (c,b,a,c)
(a,c,c,c) (b,c,a,b) (c,c,b,a)

This distribution is globally Markov with respect to $\mathcal{G}$ because the conditional distribution of (X_1, X_3) given (X_2, X_4) is always degenerate and this is also true for the conditional distribution of (X_2, X_4) given (X_1, X_3). It cannot be obtained as a limit of factorizing distributions, as the argument below shows.

All pairs (X_i, X_{i+1}) are marginally independent in P (where we have let $X_5 = X_1$). Hence P has the same clique marginals as the distribution Q defined by $X_1, \ldots, X_4$ being mutually independent and uniformly distributed on the space $\{a, b, c\}$. Clearly $Q \in M_E(\mathcal{G})$. By Lemma 3.14 there is only one element in $M_E(\mathcal{G})$ with these marginals so we must have $P \notin M_E(\mathcal{G})$. □

We introduce the notation $M^+(\mathcal{G})$ for the positive Markov probabilities, $M_P(\mathcal{G})$ for the pairwise, $M_L(\mathcal{G})$ for the local, $M_G(\mathcal{G})$ for the global, and recall that we use $M_F(\mathcal{G})$ for those that factorize. We can then summarize the previous considerations as follows. In the general case we have the chain of inclusions

$$M^+(\mathcal{G}) \subset M_F(\mathcal{G}) \subseteq M_G(\mathcal{G}) \subseteq M_L(\mathcal{G}) \subseteq M_P(\mathcal{G}). \quad (3.15)$$

In the discrete case we have further that $M_G(\mathcal{G})$, $M_L(\mathcal{G})$, and $M_P(\mathcal{G})$ are all closed under limits and

$$M_F(\mathcal{G}) \subseteq M_E(\mathcal{G}) \subseteq M_G(\mathcal{G}). \quad (3.16)$$

All inclusions in (3.15) and (3.16) are strict for a general graph $\mathcal{G}$ as shown through various examples.

When (A, B, S) form a weak decomposition of $\mathcal{G}$ the Markov properties decompose accordingly. This is expressed formally in three propositions below.

Proposition 3.16 *Assume that (A, B, S) decompose $\mathcal{G} = (V, E)$ weakly. Then a probability distribution P factorizes with respect to $\mathcal{G}$ if and only if both its marginal distributions $P_{A\cup S}$ and $P_{B\cup S}$ factorize with respect to $\mathcal{G}_{A\cup S}$ and $\mathcal{G}_{B\cup S}$ respectively and the densities f satisfy*

$$f(x)f_S(x_S) = f_{A\cup S}(x_{A\cup S})f_{B\cup S}(x_{B\cup S}). \quad (3.17)$$

Proof: Suppose that p factorizes with respect to $\mathcal{G}$ such that

$$f(x) = \prod_{c \in \mathcal{C}} \psi_c(x).$$

Since (A, B, S) decomposes $\mathcal{G}$, all cliques are either subsets of $A \cup S$ or of $B \cup S$. Let $\mathcal{A}$ denote the cliques that are subsets of $A \cup S$ and $\mathcal{B}$ those that are subsets of $B \cup S$. Then

$$f(x) = \prod_{c \in \mathcal{A}} \psi_c(x) \prod_{c \in \mathcal{B} \setminus \mathcal{A}} \psi_c(x) = h(x_{A\cup S})k(x_{B\cup S}).$$

By direct integration we find

$$f_{A\cup S}(x_{A\cup S}) = h(x_{A\cup S})\bar{k}(x_S)$$

where

$$\bar{k}(x_S) = \int k(x_{B\cup S})\mu_B(dx_B),$$

and similarly with the other marginals. This gives (3.17) as well as the factorizations of both marginal densities.

Conversely, assume that (3.17) holds and that $f_{A\cup S}$ and $f_{B\cup S}$ factorize. Then let

$$\psi_S(x_S) = \begin{cases} \frac{1}{f_S(x_S)} & \text{if } f_S(x_S) \neq 0 \\ 0 & \text{otherwise.} \end{cases}$$

Since f_S is obtained from integration of f, the latter must also be almost everywhere zero when f_S is. Hence

$$\tilde{f}(x) = f_{A\cup S}(x_{A\cup S}) f_{B\cup S}(x_{B\cup S}) \psi_S(x_S)$$

is a density for P and P factorizes. □

The analogous result is also true for the global Markov property (G), as we shall now show.

Proposition 3.17 *Assume that (A, B, S) decompose $\mathcal{G} = (V, E)$ weakly. Then a probability distribution P is globally Markov with respect to $\mathcal{G}$ if and only if both marginal distributions $P_{A\cup S}$ and $P_{B\cup S}$ are globally Markov with respect to $\mathcal{G}_{A\cup S}$ and $\mathcal{G}_{B\cup S}$ respectively and the densities satisfy (3.17).*

Proof: Suppose that a probability distribution P satisfies the global Markov property with respect to $\mathcal{G}$. Then (3.17) follows because S separates A from B. To show that $P_{A\cup S}$ is globally Markov with respect to $\mathcal{G}_{A\cup S}$ we show that for any triple $(\tilde{A}, \tilde{B}, \tilde{S})$ of disjoint subsets of $A \cup S$ with $\tilde{S}$ separating $\tilde{A}$ from $\tilde{B}$ in $\mathcal{G}_{A\cup S}$, $\tilde{S}$ separates $\tilde{A}$ from $\tilde{B}$ in $\mathcal{G}$. So let $(\tilde{A}, \tilde{B}, \tilde{S})$ be such a triple and let $\alpha \in \tilde{A}$ and $\beta \in \tilde{B}$. Because S is complete, at least one of them, say α, is in A. If there were a path from α to β avoiding $(A \cup S) \setminus \{\alpha, \beta\}$ it would go via B, contradicting that A is separated from B by S.

Next, assume that $P_{A\cup S}$ and $P_{B\cup S}$ both satisfy the global Markov property and the joint density f factorizes as in (3.17). Let U, W and C be disjoint subsets of V such that C separates U from W and assume first that $V = U \cup W \cup C$.

Since S is complete and C separates U from W, either $U \cap S$ or $W \cap S$ must be empty; we assume the former. The entire set of variables is then partitioned into eight disjoint subsets

$$U \cap A, \; C \cap A, \; W \cap A, \; C \cap S, \; W \cap S, \; U \cap B, \; C \cap B, \; W \cap B$$

and we denote the corresponding subsets of random variables by $X_1, \ldots, X_8$; see the diagram below.

	U	C	W
A	X_1	X_2	X_3
S		X_4	X_5
B	X_6	X_7	X_8

It must hold that $C \cap (A \cup S)$ separates $U \cap A$ from $W \cap (A \cup S)$ in $\mathcal{G}_{A \cup S}$, for otherwise C would not separate U from W as assumed. Hence, from the global Markov property of $P_{A \cup S}$ and (3.6) we must have the factorization

$$f_{A \cup S}(x_{A \cup S}) = h_1(x_1, x_2, x_4) k_1(x_2, x_3, x_4, x_5).$$

By symmetry we further find

$$f_{B \cup S}(x_{B \cup S}) = h_2(x_4, x_6, x_7) k_2(x_4, x_5, x_7, x_8).$$

Combined with (3.17) this gives

$$f(x) f_S(x_S) = h_1(x_1, x_2, x_4) k_1(x_2, x_3, x_4, x_5) h_2(x_4, x_6, x_7) k_2(x_4, x_5, x_7, x_8).$$

Using that $x_S = (x_4, x_5)$ and letting $h = h_1 h_2$ and $k = k_1 k_2 / f_S$ yields

$$f(x) = h(x_1, x_2, x_4, x_6, x_7) k(x_2, x_3, x_4, x_5, x_7, x_8) = h(x_{U \cup C}) k(x_{W \cup C}),$$

which shows that $U \perp\!\!\!\perp W \mid C$.

If $U \cup W \cup C$ is not the entire set of variables V we proceed as in the proof of Proposition 3.8. Details are omitted. □

In the case of discrete sample spaces we have the analogous result for extended Markov probabilities.

Proposition 3.18 *Assume that (A, B, S) decompose $\mathcal{G} = (V, E)$ weakly and the sample space is discrete. Then a probability distribution P is extended Markov with respect to $\mathcal{G}$ if and only if both marginal distributions $P_{A \cup S}$ and $P_{B \cup S}$ are extended Markov with respect to $\mathcal{G}_{A \cup S}$ and $\mathcal{G}_{B \cup S}$ respectively and the densities satisfy*

$$p(i) p(i_S) = p(i_{A \cup S}) p(i_{B \cup S}). \tag{3.18}$$

Proof: Assume that P is extended Markov. Then

$$p(i) = \lim_{n \to \infty} p_n(i),$$

where p_n are positive and factorize. By Proposition 3.16 we therefore have

$$p_n(i) p_n(i_S) = p_n(i_{A \cup S}) p_n(i_{B \cup S}),$$

where the factors on the right-hand side factorize on $\mathcal{G}_{A\cup S}$ and $\mathcal{G}_{B\cup S}$ respectively. Hence their limits are extended Markov. Letting n tend to infinity in the equation yields thus the desired factorization.

Suppose conversely that (3.18) holds with the factors being extended Markov. Then

$$p(i_{A\cup S}) = \lim_{n\to\infty} q_n(i_{A\cup S}), \quad p(i_{B\cup S}) = \lim_{n\to\infty} r_n(i_{B\cup S}), \qquad (3.19)$$

where q_n and r_n factorize appropriately. Let now

$$\pi_n(i) = \frac{q_n(i_{A\cup S})}{q_n(i_S)} r_n(i_{B\cup S}). \qquad (3.20)$$

Then π_n defines a probability distribution that factorizes. We need to show that

$$p(i) = \lim_{n\to\infty} \pi_n(i).$$

This is trivial if $p(i_S) \neq 0$. If $p(i_S) = 0$ then both $p(i) = 0$ and $p(i_{B\cup S}) = 0$. Hence, by (3.19), the second factor in (3.20) converges to zero. The first factor in (3.20) is bounded above by one and therefore the product converges to zero as desired. □

As we have seen, the various Markov properties are different for general, undirected graphs, unless the densities are all positive. However, in the special case of decomposable graphs some Markov properties coincide.

Proposition 3.19 *Let $\mathcal{G}$ be weakly decomposable. Then*

$$M_F(\mathcal{G}) = M_G(\mathcal{G}).$$

Proof: The proof is by induction on $|\mathcal{C}|$, the number of cliques in the graph $\mathcal{G}$.

The result is trivial for $|\mathcal{C}| = 1$, so let us assume the equation to be true for all graphs with at most n cliques and let $\mathcal{G}$ have $n+1$ cliques.

We only have to show that $M_G(\mathcal{G}) \subseteq M_F(\mathcal{G})$. Assume that P is globally Markov with respect to $\mathcal{G}$ and let (A, B, S) be a proper weak decomposition of $\mathcal{G}$. By Proposition 3.17, both marginals $P_{A\cup S}$ and $P_{B\cup S}$ are globally Markov and (3.17) holds. By the induction assumption the marginal distributions both factorize. The full factorization of P is now obtained from (3.17) by dividing with f_S. □

And a direct consequence of this and (3.16) is the corollary below.

Corollary 3.20 *Assume that $\mathcal{G}$ is weakly decomposable and the state space is discrete. Then*

$$M_F(\mathcal{G}) = M_E(\mathcal{G}) = M_G(\mathcal{G}).$$

It is an easy consequence of Example 3.10 that the converse to Proposition 3.19 holds in the sense that if the graph is not decomposable, it has a chordless cycle and one can by analogy construct a distribution which does not factorize but is globally Markov.

The global Markov property is the strongest of the Markov properties in the sense that the associated list of conditional independence statements strictly contains the statements associated with the other properties. Moreover, it cannot be further strengthened. For example it holds (Frydenberg 1990b) that if all state spaces are binary, i.e. $\mathcal{X}_\alpha = \{1, -1\}$, then

$$A \perp\!\!\!\perp B \mid S \text{ for all } P \in M_F(G) \iff S \text{ separates } A \text{ from } B. \quad (3.21)$$

In other words, if A and B are not separated by S then there is a factorizing distribution that makes them conditionally dependent. Geiger and Pearl (1993) conjecture that to any undirected graph $\mathcal{G}$ and fixed state space $\mathcal{X}$ one can find a single $P \in M_F(G)$ such that for this P it holds that

$$A \perp\!\!\!\perp B \mid S \iff S \text{ separates } A \text{ from } B.$$

This is clearly a stronger statement but no proof is known.

3.2.2 Markov properties on directed acyclic graphs

Before we proceed to the case of a general chain graph we consider the same setup as in the previous subsection, only now the graph $\mathcal{G}$ is assumed to be directed and acyclic. The Markov property on a directed acyclic graph was first studied systematically in Kiiveri *et al.* (1984) but see also for example Pearl and Verma (1987), Verma and Pearl (1990a), J.Q. Smith (1989), Geiger and Pearl (1990), Lauritzen *et al.* (1990) and other references given below.

We say that a probability distribution P admits a *recursive factorization* according to $\mathcal{G}$, if there exist non-negative functions, henceforth referred to as *kernels*, $k^\alpha(\cdot, \cdot), \alpha \in V$ defined on $\mathcal{X}_\alpha \times \mathcal{X}_{\text{pa}(\alpha)}$, such that

$$\int k^\alpha(y_\alpha, x_{\text{pa}(\alpha)}) \mu_\alpha(dy_\alpha) = 1$$

and P has density f with respect to μ, where

$$f(x) = \prod_{\alpha \in V} k^\alpha(x_\alpha, x_{\text{pa}(\alpha)}).$$

We then also say that P has property (DF). It is an easy induction argument to show that if P admits a recursive factorization as above, then the kernels $k^\alpha(\cdot, x_{\text{pa}(\alpha)})$ are densities for the conditional distribution of X_α, given $X_{\text{pa}(\alpha)} = x_{\text{pa}(\alpha)}$. Also it is immediate that if we form the undirected moral graph $\mathcal{G}^m$ (marrying parents and deleting directions) such as described towards the end of Section 2.1.1, we have

MARKOV PROPERTIES

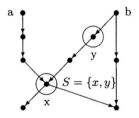

Fig. 3.1. The directed global Markov property. Is $a \perp\!\!\!\perp b \mid S$?

Lemma 3.21 *If P admits a recursive factorization according to the directed, acyclic graph $\mathcal{G}$, it factorizes according to the moral graph $\mathcal{G}^m$ and obeys therefore the global Markov property relative to $\mathcal{G}^m$.*

Proof: The factorization follows from the fact that, by construction, the sets $\{\alpha\} \cup \text{pa}(\alpha)$ are complete in $\mathcal{G}^m$ and we can therefore let $\psi_{\{\alpha\}\cup\text{pa}(\alpha)} = k^\alpha$. The remaining part of the statement follows from the fact that (F) implies (G) in the undirected case; see Proposition 3.8. □

It clearly also holds that

Proposition 3.22 *If P admits a recursive factorization according to the directed, acyclic graph $\mathcal{G}$ and A is an ancestral set, then the marginal distribution P_A admits a recursive factorization according to $\mathcal{G}_A$.*

From this it directly follows that

Corollary 3.23 *Let P factorize recursively according to $\mathcal{G}$. Then*

$$A \perp\!\!\!\perp B \mid S$$

whenever A and B are separated by S in $(\mathcal{G}_{\text{An}(A\cup B\cup S)})^m$, the moral graph of the smallest ancestral set containing $A \cup B \cup S$.

The property in Corollary 3.23 will be referred to as the *directed global Markov property* (DG). The directed global Markov property has the same role as the global Markov property has in the case of an undirected graph, in the sense that it gives the sharpest possible rule for reading conditional independence relations off the directed graph. The procedure is illustrated in the following example.

Example 3.24 Consider a directed Markov field on the graph in Fig. 3.1 and the problem of deciding whether $a \perp\!\!\!\perp b \mid S$. The moral graph of the smallest ancestral set containing all the variables involved is shown in Fig. 3.2. It is immediate that S separates a from b in this graph, implying $a \perp\!\!\!\perp b \mid S$. □

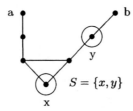

Fig. 3.2. The moral graph of the smallest ancestral set in the graph of Fig. 3.1 containing $\{a\} \cup \{b\} \cup S$. Clearly S separates a from b in this graph, implying $a \perp\!\!\!\perp b \mid S$.

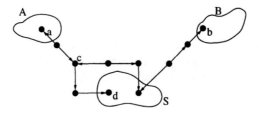

Fig. 3.3. Example of an active chain from A to B. The path from c to d is not part of the chain, but indicates that c must have descendants in S.

An alternative formulation of the directed global Markov property was given by Pearl (1986a, 1986b) with a full formal treatment in Verma and Pearl (1990a, 1990b). A chain π from a to b in a directed, acyclic graph $\mathcal{G}$ is said to be *blocked* by S, if it contains a vertex $\gamma \in \pi$ such that either

- $\gamma \in S$ and arrows of π do not meet head-to-head at γ, or

- $\gamma \notin S$ nor has γ any descendants in S, and arrows of π do meet head-to-head at γ.

A chain that is not blocked by S is said to be *active*. Two subsets A and B are now said to be *d-separated* by S if all chains from A to B are blocked by S. We then have

Proposition 3.25 *Let A, B and S be disjoint subsets of a directed, acyclic graph $\mathcal{G}$. Then S d-separates A from B if and only if S separates A from B in $(\mathcal{G}_{\mathrm{An}(A \cup B \cup S)})^m$.*

Proof: Suppose S does not d-separate A from B. Then there is an active chain from A to B such as, for example, indicated in Fig. 3.3. All vertices in this chain must lie within $\mathrm{An}(A \cup B \cup S)$. This follows because if the arrows

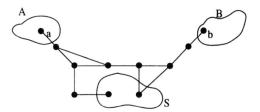

Fig. 3.4. The moral graph corresponding to the active chain in $\mathcal{G}$.

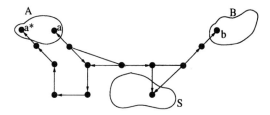

Fig. 3.5. The chain in the graph $(\mathcal{G}_{\mathrm{An}(A \cup B \cup S)})^m$ makes it possible to construct an active chain in $\mathcal{G}$ from A to B.

meet head-to-head at some vertex γ, either $\gamma \in S$ or γ has descendants in S. And if not, either of the subpaths away from γ either meets another arrow, in which case γ has descendants in S, or leads all the way to A or B. Each of these head-to-head meetings will give rise to a marriage in the moral graph such as illustrated in Fig. 3.4, thereby creating a chain from A to B in $(\mathcal{G}_{\mathrm{An}(A \cup B \cup S)})^m$, circumventing S.

Suppose conversely that A is not separated from B in $(\mathcal{G}_{\mathrm{An}(A \cup B \cup S)})^m$. Then there is a chain in this graph that circumvents S. The chain has pieces that correspond to edges in the original graph and pieces that correspond to marriages. Each marriage is a consequence of a meeting of arrows head-to-head at some vertex γ. If γ is in S or it has descendants in S, the meeting does not block the chain. If not, γ must have descendants in A or B, since the ancestral set was smallest. In the latter case, a new chain can be created with one head-to-head meeting fewer, using the line of descent, such as illustrated in Fig. 3.5. Continuing this substitution process eventually leads to an active chain from A to B and the proof is complete. □

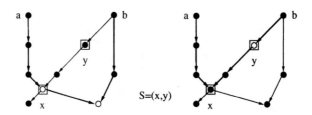

Fig. 3.6. Illustration of Pearl's separation criterion. There are two chains from a to b, drawn with bold lines. Both are blocked, but different vertices γ, indicated with open circles, play the role of blocking vertices.

We illustrate the concept of d-separation by applying it to the query of Example 3.24. As Fig. 3.6 indicates, all chains between a and b are blocked by S, whereby the global Markov property gives that $a \perp\!\!\!\perp b \mid S$.

Geiger and Pearl (1990) show in their Theorem 5 that the criterion of d-separation cannot be improved, in the sense that for any given directed acyclic graph $\mathcal{G}$, one can find state spaces $\mathcal{X}_\alpha, \alpha \in V$ and a probability P such that

$$A \perp\!\!\!\perp B \mid S \iff S \text{ d-separates } A \text{ from } B. \qquad (3.22)$$

An argument analogous to that given in Frydenberg (1990b) shows that the analogue of (3.21) holds for d-separation in directed, acyclic graphs. Geiger and Pearl (1990) show that in the case of real sample spaces, a Gaussian distribution satisfying (3.22) exists and conjecture the similar result to be true also for the case where the state spaces all have two points, i.e. $\mathcal{X}_\alpha = \{1, -1\}$.

Also the directed case has analogues of the pairwise and local Markov properties. We say that P obeys the *directed pairwise Markov property* (DP) if for any pair (α, β) of non-adjacent vertices with $\beta \in \operatorname{nd}(\alpha)$,

$$\alpha \perp\!\!\!\perp \beta \mid \operatorname{nd}(\alpha) \setminus \{\beta\}.$$

Similarly P obeys the *directed local Markov property* (DL) if any variable is conditionally independent of its non-descendants, given its parents:

$$\alpha \perp\!\!\!\perp \operatorname{nd}(\alpha) \mid \operatorname{pa}(\alpha).$$

As $\operatorname{pa}(\alpha) \subseteq \operatorname{nd}(\alpha)$, it follows from properties (C2) and (C3) of conditional independence that the directed local Markov property (DL) implies the pairwise (DP). The converse is not true in general, as the following example shows.

Fig. 3.7. The local and pairwise directed Markov properties are not equivalent in general. This is illustrate by the directed acyclic graph above.

Example 3.26 Let $X = Y = Z$ and W be independent of X with

$$P\{X = 1\} = P\{X = 0\} = P\{W = 1\} = P\{W = 0\} = 1/2.$$

This distribution is pairwise Markov with respect to the graph in Fig. 3.7. However, it is not locally Markov, as this would imply $X \perp\!\!\!\perp (Y, Z) \mid W$, which clearly is not the case. A consequence of Theorem 3.27 is that the density in this example does not admit a recursive factorization. □

In contrast to the undirected case we have that the remaining three properties (DF), (DL) and (DG) are equivalent, just assuming the existence of the density f.

Theorem 3.27 *Let $\mathcal{G}$ be a directed, acyclic graph. For a probability distribution P on $\mathcal{X}$ which has density with respect to a product measure μ, the following conditions are equivalent:*

(DF) *P admits a recursive factorization according to $\mathcal{G}$;*

(DG) *P obeys the directed global Markov property, relative to $\mathcal{G}$;*

(DL) *P obeys the directed local Markov property, relative to $\mathcal{G}$.*

Proof: That (DF) implies (DG) is Corollary 3.23. That (DG) implies (DL) follows by observing that $\{\alpha\} \cup \mathrm{nd}(\alpha)$ is an ancestral set and that $\mathrm{pa}(\alpha)$ obviously separates $\{\alpha\}$ from $\mathrm{nd}(\alpha) \setminus \mathrm{pa}(\alpha)$ in $(\mathcal{G}_{\{\alpha\} \cup \mathrm{nd}(\alpha)})^m$. The final implication is shown by induction on the number of vertices $|V|$ of $\mathcal{G}$. Let α_0 be a terminal vertex of $\mathcal{G}$. Then we can let k^{α_0} be the conditional density of X_{α_0}, given $X_{V \setminus \{\alpha_0\}}$, which by (DL) can be chosen to depend on $x_{\mathrm{pa}(\alpha_0)}$ only. The marginal distribution of $X_{V \setminus \{\alpha_0\}}$ trivially obeys the directed local Markov property and admits a factorization by the inductive assumption. Combining this factorization with k^{α_0} yields the factorization for P. This completes the proof. □

In fact (DL) and (DG) are equivalent, even without assuming the existence of a density with respect to a product measure (Lauritzen *et al.* 1990).

If the probability distribution P satisfies (3.10), for example if the density is positive, then it is not difficult to see that (DP) implies (DL) and all directed Markov properties are equivalent.

Since the three conditions in Theorem 3.27 are all equivalent, we choose to speak of a *directed Markov distribution* as one where any of the conditions (DF), (DL) or (DG) is satisfied. The set of such distributions is denoted by $M(\mathcal{G})$ where $\mathcal{G}$ is a directed and acyclic graph.

It follows as in Proposition 3.12 that in the case of a discrete sample space, the directed Markov properties are preserved under weak limits. So there is no need for a special term or symbol for extended Markov probabilities on a directed acyclic graph.

In the particular case when the directed acyclic graph $\mathcal{G}$ is perfect (see Section 2.1.3) the directed Markov property on $\mathcal{G}$ and the factorization Markov property on its undirected version $\mathcal{G}^\sim$ coincide. This is contained in the following

Proposition 3.28 *Let $\mathcal{G}$ be a perfect directed acyclic graph and $\mathcal{G}^\sim$ its undirected version. Then P admits a recursive factorization with respect to $\mathcal{G}$ if and only if it factorizes according to $\mathcal{G}^\sim$.*

Proof: That the graph is perfect means that $\mathrm{pa}(\alpha)$ is complete for all $\alpha \in V$. Hence $\mathcal{G}^m = \mathcal{G}^\sim$. From Lemma 3.21 it then follows that any $P \in M(\mathcal{G})$ also factorizes with respect to $\mathcal{G}^\sim$.

The reverse inclusion is established by induction on the number of vertices $|V|$ of $\mathcal{G}$. For $|V| = 1$ there is nothing to show. For $|V| = n + 1$ let $P \in M_F(\mathcal{G}^\sim)$ and find a terminal vertex $\alpha \in V$. This vertex has $\mathrm{pa}_\mathcal{G}(\alpha) = \mathrm{bd}_{\mathcal{G}^\sim}(\alpha)$ and, since $\mathcal{G}$ is perfect, this set is complete in both graphs as well. Hence $(V \setminus \{\alpha\}, \{\alpha\}, \mathrm{bd}(\alpha))$ is a weak decomposition of $\mathcal{G}^\sim$ and Proposition 3.16 gives the factorization

$$f(x) = f(x_{V \setminus \{\alpha\}}) f(x_{\mathrm{cl}(\alpha)}) / f(x_{\mathrm{bd}(\alpha)}),$$

where the first factor factorizes according to $\mathcal{G}^\sim_{V \setminus \{\alpha\}}$. Using the induction assumption on this factor gives the full recursive factorization of P. $\square$

In the discrete case Proposition 3.18 implies that the above result can be strengthened to comprise the extended Markov property.

Proposition 3.29 *Let $\mathcal{G}$ be a perfect directed acyclic graph and $\mathcal{G}^\sim$ its undirected version. If all vertices are discrete and hence the total sample space, then*

$$M(\mathcal{G}) = M_F(\mathcal{G}^\sim) = M_E(\mathcal{G}^\sim).$$

In particular, any extended Markov probability factorizes.

Proof: The proof that $M(\mathcal{G}) = M_E(\mathcal{G}^\sim)$ is completely analogous to the proof of the preceding proposition. Since $M(\mathcal{G}) \subseteq M_F(\mathcal{G}^\sim) \subseteq M_E(\mathcal{G}^\sim)$ the result follows. □

3.2.3 Markov properties on chain graphs

In the present section we deal with the Markov properties for a general chain graph $\mathcal{G} = (V, E)$ thereby unifying the directed and undirected cases. A detailed study of the Markov property on chain graphs can be found in Frydenberg (1990a). Here we give the most basic results and definitions.

The factorization in the case of a chain graph is more complex and involves two parts. Denote the set of chain components of the graph by $\mathcal{T}$.

Then we first assume a factorization of the density as in the directed acyclic case:
$$f(x) = \prod_{\tau \in \mathcal{T}} f(x_\tau \mid x_{\text{pa}(\tau)}). \qquad (3.23)$$

But there is one more assumption. To explain this assumption, let τ_* denote the undirected graph that has $\tau \cup \text{pa}(\tau)$ as nodes and undirected edges between a pair (α, β) if either both of these are in $\text{pa}(\tau)$ or there is an edge, directed or undirected, between them in the chain graph $\mathcal{G}$. Thus for chain components that are singletons, τ_* is complete. This is the case for all chain components if $\mathcal{G}$ is a directed acyclic graph. In an undirected graph, τ_* is just the subgraph induced by the connected component τ.

The second requirement in the chain graph factorization is then that the factors in (3.23) can be further factorized as
$$f(x_\tau \mid x_{\text{pa}(\tau)}) = \prod_a \phi_a(x), \qquad (3.24)$$

where a varies over all subsets of $\tau \cup \text{pa}(\tau)$ that are complete in τ_*, and $\phi_a(x)$ as usual represent functions that depend on x through x_a only.

This factorization clearly unifies the directed and undirected cases. A probability distribution P on $\mathcal{X}$ that has a density f with respect to a product measure which satisfies (3.23) and (3.24) is said to *factorize* according to $\mathcal{G}$, and we also say that P has property (FC).

The chain graph factorization has several equivalent formulations. To describe these, let the vertex set be partitioned in a dependence chain as $V = V(1) \cup \cdots \cup V(T)$ such that each of the sets $V(t)$ only has lines between vertices, and arrows point from vertices in sets with lower number to those with higher number. The set of *concurrent* variables relative to this partitioning is next defined to be the set $C(t) = V(1) \cup \cdots \cup V(t)$. Let $\mathcal{G}^*(t)$ be the undirected graph with vertex set $C(t)$ and α adjacent to β in $\mathcal{G}^*(t)$ if either $(\alpha, \beta) \in E$ or $(\beta, \alpha) \in E$ or if $\{\alpha, \beta\} \subseteq C(t-1)$, i.e. $C(t-1)$

is made complete in $\mathcal{G}^*(t)$ by adding all missing edges between these and directions on existing edges are ignored. Let further $B(t) = \text{pa}\{V(t)\}$ and $\mathcal{G}_*(t) = \mathcal{G}^*(t)_{V(t) \cup B(t)}$. Then we have the following equivalent formulations of the chain graph factorization. There are obviously others, for example described through conditioning. We abstain from listing them all.

Proposition 3.30 *For a probability distribution P on $\mathcal{X}$ that admits a positive density with respect to a product measure, the following conditions are equivalent:*

(i) *P factorizes according to $\mathcal{G}$;*

(ii) *P admits a density that factorizes as*

$$f(x) = \prod_{\tau \in \mathcal{T}} \frac{f\left(x_{\tau \cup \text{pa}(\tau)}\right)}{f\left(x_{\text{pa}(\tau)}\right)} \qquad (3.25)$$

and each of the numerators factorizes on the graph τ_*;

(iii) *for any dependence chain, P admits a density that factorizes as*

$$f(x) = \prod_{t=1}^{T} \frac{f\left(x_{C(t)}\right)}{f\left(x_{C(t-1)}\right)} \qquad (3.26)$$

and each of the numerators factorizes on the graph $\mathcal{G}^*(t)$;

(iv) *for any dependence chain, P admits a density that factorizes as*

$$f(x) = \prod_{t=1}^{T} \frac{f\left(x_{V(t) \cup B(t)}\right)}{f\left(x_{B(t)}\right)} \qquad (3.27)$$

and each of the numerators factorizes on the graph $\mathcal{G}_*(t)$;

(v) *for some dependence chain, P admits a density that factorizes as in either of* (iii) *or* (iv).

Proof: We leave the details of this to the reader. □

Also in the case of chain graphs, there are pairwise, local and global Markov properties, in an even greater variety. We say that a probability P satisfies

(PB) the *pairwise block-recursive Markov property* relative to the dependence chain $V(1), \ldots, V(T)$, if for any pair (α, β) of non-adjacent vertices we have

$$\alpha \perp\!\!\!\perp \beta \mid C(t^*) \setminus \{\alpha, \beta\},$$

where t^* is the smallest t that has $\alpha, \beta \in C(t)$;

(PC) *the pairwise chain Markov property*, relative to $\mathcal{G}$, if for any pair (α, β) of non-adjacent vertices with $\beta \in \mathrm{nd}(\alpha)$

$$\alpha \perp\!\!\!\perp \beta \mid \mathrm{nd}(\alpha) \setminus \{\beta\};$$

(LC) *the local chain Markov property*, relative to $\mathcal{G}$, if for any vertex $\alpha \in V$

$$\alpha \perp\!\!\!\perp \mathrm{nd}(\alpha) \mid \mathrm{bd}(\alpha);$$

(GC) *the global chain Markov property*, relative to $\mathcal{G}$, if for any triple (A, B, S) of disjoint subsets of V such that S separates A from B in $(\mathcal{G}_{\mathrm{An}(A \cup B \cup S)})^m$, the moral graph of the smallest ancestral set containing $A \cup B \cup S$, we have

$$A \perp\!\!\!\perp B \mid S.$$

In general, the property (PB) depends on the particular dependence chain.

Recently Bouckaert and Studený (1995) have given a separation criterion which is analogous to d-separation and equivalent to the global chain Markov property (GC). They also show that any conditional independence statement that is derivable from the local chain Markov property and the properties (C1)–(C5) is represented in the global chain Markov property (GC). Further, they show that probability distributions exist that satisfy the local chain Markov property and violate all conditional independence statements that are not of the form given by (GC). In this sense, the global chain Markov property is therefore the strongest possible.

Note that the Markov properties unify the corresponding properties for the directed and undirected cases. For in the undirected case we have that $\mathrm{nd}(\alpha) = V \setminus \{\alpha\}$ and $\mathcal{G} = (\mathcal{G}_{\mathrm{An}(A \cup B \cup S)})^m$. And in the directed case $\mathrm{bd}(\alpha) = \mathrm{pa}(\alpha)$.

Generally all these Markov properties are therefore different without additional assumptions, just as in the undirected case. In the remaining part of the section we assume that all probability measures have positive densities, implying that all five of the basic properties of conditional independence (C1)–(C5) hold. As we shall see, this implies that all Markov properties are equivalent. We first need a few lemmas.

Lemma 3.31 *If $V(1), \ldots, V(T)$ is a dependence chain for $\mathcal{G}$ and and P satisfies the pairwise block-recursive Markov property relative to $\mathcal{G}$, then $P_{C(T-1)}$ satisfies the pairwise block-recursive Markov property relative to $\mathcal{G}^*(T-1)$.*

Proof: This is trivial, as the pairwise block-recursive Markov property is defined to be identical to the combination of the pairwise undirected

Markov properties for the marginal distributions $P_{C(t)}$ of the concurrent variables $C(t)$ on the graphs $\mathcal{G}^*(t)$, for $t = 1, \ldots, T$. □

In the case of the pairwise chain graph Markov property, the corresponding result needs an extra condition.

Lemma 3.32 *If C is a terminal chain component in $\mathcal{G}$ and P satisfies (3.10) and the pairwise chain Markov property relative to $\mathcal{G}$, then $P_{V \setminus C}$ satisfies the pairwise chain Markov property relative to $\mathcal{G}_{V \setminus C}$.*

Proof: Let $\alpha, \beta \in V \setminus C$ be non-adjacent in $\mathcal{G}$ and assume $\beta \in \mathrm{nd}(\alpha)$. We have to show that $\alpha \perp\!\!\!\perp \beta \mid (\mathrm{nd}(\alpha) \setminus (C \cup \{\beta\}))$, since $\mathrm{nd}(\alpha) \setminus C$ is the set of non-descendants of α in $\mathcal{G}_{V \setminus C}$. If $C \cap \mathrm{nd}(\alpha) = \emptyset$ this is a direct consequence of the fact that (PC) holds for $\mathcal{G}$. Else we must have $C \subseteq \mathrm{nd}(\alpha)$. Since no pair δ, γ with $\delta \in \mathrm{de}(\alpha) \cup \{\alpha\}$ and $\gamma \in C$ can be adjacent and $\mathrm{nd}(\gamma) = V \setminus \{\gamma\}$ for all such γ, the pairwise chain Markov property implies that $\delta \perp\!\!\!\perp \gamma \mid V \setminus \{\delta, \gamma\}$ for any such pair. Repeated use of (3.10) yields
$$\mathrm{de}(\alpha) \cup \{\alpha\} \perp\!\!\!\perp C \mid \mathrm{nd}(\alpha) \setminus C$$
and thus by (C2),
$$\alpha \perp\!\!\!\perp C \mid \mathrm{nd}(\alpha) \setminus C. \tag{3.28}$$
The pairwise property gives directly that
$$\alpha \perp\!\!\!\perp \beta \mid \mathrm{nd}(\alpha) \setminus \{\beta\}. \tag{3.29}$$
Thus (3.10) used on (3.28) and (3.29) yields
$$\alpha \perp\!\!\!\perp (C \cup \{\beta\}) \mid \mathrm{nd}(\alpha) \setminus (C \cup \{\beta\})$$
and the result follows from (C2). □

Lemma 3.33 *If P satisfies the pairwise chain Markov property (PC) with respect to $\mathcal{G}$ as well as (3.10), it satisfies the pairwise undirected Markov property (P) with respect to $\mathcal{G}^m$.*

Proof: We use induction on the number of chain components. If there is only one chain component in $\mathcal{G}$, it is undirected and connected, $\mathcal{G}^m = \mathcal{G}$ and there is nothing to show.

Assume then the lemma to hold for all graphs with n or fewer chain components and let $\mathcal{G}$ have $n+1$ components. Let α, β be non-adjacent in $\mathcal{G}^m$. Without loss of generality we assume that $\beta \in \mathrm{nd}(\alpha)$. If $V \setminus \{\alpha\} = \mathrm{nd}(\alpha)$, the conditional independence $\alpha \perp\!\!\!\perp \beta \mid V \setminus \{\alpha, \beta\}$ follows directly. Else there must be a terminal chain component $C \subseteq (V \setminus (\{\alpha\} \cup \mathrm{nd}(\alpha))$.

Because α and β are non-adjacent in $\mathcal{G}^m$ they will also be in $(\mathcal{G}_{V\setminus C})^m$. The induction assumption together with Lemma 3.32 gives that

$$\alpha \perp\!\!\!\perp \beta \mid V \setminus (C \cup \{\alpha, \beta\}). \tag{3.30}$$

Also at least one of α and β, say the former, cannot be in $\mathrm{bd}(C)$, since otherwise they would be adjacent in the moral graph. Thus

$$\alpha \perp\!\!\!\perp \gamma \mid V \setminus \{\alpha, \gamma\} \text{ for all } \gamma \in C. \tag{3.31}$$

Repeated use of property (3.10) of conditional independence on (3.31) yields

$$\alpha \perp\!\!\!\perp C \mid V \setminus (C \cup \{\alpha\}). \tag{3.32}$$

We can now use (C4) on (3.30) and (3.32) to obtain that

$$\alpha \perp\!\!\!\perp \{\beta\} \cup C \mid V \setminus (C \cup \{\alpha, \beta\}).$$

From (C3) and (C2) it follows that $\alpha \perp\!\!\!\perp \beta \mid V \setminus \{\alpha, \beta\}$, which was to be shown. □

Clearly, from arguments analogous to the directed and undirected cases, we have in general that

(FC) $\implies$ (GC) $\implies$ (PC) $\implies$ (PB) for any dependence chain,

but if we assume (3.10), all Markov properties are equivalent.

Theorem 3.34 *Assume that P is such that (3.10) holds for disjoint subsets of V; then*

(GC) $\iff$ (LC) $\iff$ (PC) $\iff$ (PB) *for some dependence chain.*

Proof: We must show that the pairwise block-recursive property for any dependence chain implies the global chain Markov property (GC). We argue first that (PB) implies (PC) and then that (PC) implies (GC).

The first part is shown by using induction on the number of chain components of $\mathcal{G}$. If there is only one chain component the statement is trivially true. Assume that (PB) implies (PC) for all chain graphs with at most n chain components and let $\mathcal{G}$ have $n+1$ chain components. Assume then that P satisfies (PB) with respect to a given dependence chain and let α and β be non-adjacent vertices.

If $t^* < T$, marginalization to $C(t^*)$ and the inductive assumption will give that

$$\alpha \perp\!\!\!\perp \beta \mid \mathrm{nd}(\alpha) \setminus \{\beta\}.$$

If $t^* = T$ and $V(T)$ has only one chain component, the same conclusion can be drawn because then $\mathrm{nd}(\alpha) = C(T) \setminus \{\alpha\}$.

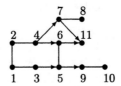

Fig. 3.8. A chain graph. The chain components are $\{1,2,3,4\}$, $\{5,6\}$, $\{7,8\}$, $\{9,10\}$, $\{11\}$. Is $3 \perp\!\!\!\perp 8 \mid \{2,5\}$? Is $3 \perp\!\!\!\perp 8 \mid 2$?

Fig. 3.9. The moral graph of the smallest ancestral set in the graph of Fig. 3.8 containing $\{2,3,5,8\}$. A connection between 3 and 4 has been introduced since these both have children in the same chain component $\{5,6\}$. We cannot conclude $3 \perp\!\!\!\perp 8 \mid \{2,5\}$.

If there is more than one chain component in $V(T)$, Theorem 3.7 yields that P is globally Markov with respect to $\mathcal{G}^*(T)$ and hence the same conclusion holds. Thus P satisfies (PC).

Assume next that P satisfies (PC). If R is any ancestral set in $\mathcal{G}$, it can be obtained from V by stepwise removal of terminal chain components. By Lemma 3.32 each removal preserves the pairwise property. Combining with Lemma 3.33 we find that P_R is pairwise Markov with respect to $(\mathcal{G}_R)^m$. Theorem 3.7 yields that P_R is globally Markov relative to the same graph. Thus, the result follows by letting $R = \mathrm{An}(A \cup B \cup S)$. □

Example 3.35 As an illustration of the global chain Markov property, consider the graph in Fig. 3.8 and the question of deciding whether it holds that $3 \perp\!\!\!\perp 8 \mid \{2,5\}$. The smallest ancestral set containing these variables is the set $\{1,2,3,4,5,6,7,8\}$. The moral graph of this adds an edge between 3 and 4, because these both have children in the chain component $\{5,6\}$. Thus the graph in Fig. 3.9 appears. Since there is a path between 3 and 8 circumventing 2 and 5 in this graph, we cannot conclude that $3 \perp\!\!\!\perp 8 \mid \{2,5\}$.

If we instead consider the question whether $3 \perp\!\!\!\perp 8 \mid 2$, the smallest ancestral set becomes $\{1,2,3,4,7,8\}$, no edge has to be added between 3 and 4 and Fig. 3.10 reveals that $3 \perp\!\!\!\perp 8 \mid 2$. □

Fig. 3.10. The moral graph of the smallest ancestral set in the graph of Fig. 3.8 containing $\{2,3,8\}$. We conclude that $3 \perp\!\!\!\perp 8 \mid 2$.

We cannot expect factorization results to be more general for chain graphs than for undirected graphs, since the chain graphs contain these as special cases. But there is a result analogous to Theorem 3.9.

Theorem 3.36 *A probability distribution with strictly positive and continuous density f satisfies the pairwise block-recursive Markov property with respect to $\mathcal{G}$ if and only if it factorizes according to $\mathcal{G}$.*

Proof: That any chain graph Markov density factorizes as in (3.26) is immediate. That the numerators also factorize appropriately is seen as follows. Since the density is assumed positive and continuous, (3.10) holds and the different Markov properties are equivalent. From the pairwise block-recursive Markov property it follows that any two variables α and β that are not adjacent in $\mathcal{G}^*(t)$ are conditionally independent given the remaining concurrent variables $C(t)$. But since $V(t) \perp\!\!\!\perp C(t-1) \mid B(t)$ we have that

$$\mathcal{L}\left(X_\alpha, X_\beta \mid X_{C(t)\setminus\{\alpha,\beta\}}\right) = \mathcal{L}\left(X_\alpha, X_\beta \mid X_{(V(t)\cup B(t))\setminus\{\alpha,\beta\}}\right)$$

and hence the marginal distribution of the variables in $V(t) \cup B(t)$ is pairwise Markov with respect to the undirected graph $\mathcal{G}^*(t)$. Hence the densities factorize by Theorem 3.9.

Conversely, assume a factorization given with the stated properties and let α, β be non-adjacent in $\mathcal{G}$. Let t^* be the smallest t with $\alpha, \beta \in C(t)$. Clearly, if the density factorizes, so does the marginal density of $C(t)$ for all t and hence we can assume that $t^* = T$. If α and β are not both in $V(T) \cup B(T)$, the expression (3.27) directly gives that α and β are independent given the remaining variables in $C(t^*) = C(T) = V$. Else if $\alpha, \beta \in V(T) \cup B(T)$ they are not both in $B(T)$ since $t^* = T$. Therefore they are also non-adjacent in $\mathcal{G}^*(T)$. Hence $f(x_{V(T)\cup B(T)})$ is a product of functions, one of which does not depend on β and one of which does not depend on α. Since this is the only factor in (3.27) containing both variables, the same is true for the full joint density. Hence α and β are conditionally independent given the remaining variables, and the pairwise block-recursive Markov property has been established. □

If P is not strictly positive but extended Markov, i.e. P is the limit of positive chain graph Markov distributions, there is a modified version of Theorem 3.36:

Corollary 3.37 *A probability distribution on a discrete sample space satisfies the extended Markov property with respect to a chain graph $\mathcal{G}$ if and only if it factorizes as*

$$p(i) = \prod_{t=1}^{T} \frac{p\left(i_{V(t)\cup B(t)}\right)}{p\left(i_{B(t)}\right)}, \qquad (3.33)$$

where $0/0 = 0$ and each of the terms in the numerator are extended Markov with respect to the graph $\mathcal{G}^*(t)$.

Proof: This follows as in the undirected case from Theorem 3.36 by taking limits. □

Also this factorization can be written in a number of equivalent ways.

3.3 Notes

There are several aspects of graphs and conditional independence that have not been covered here. Cox and Wermuth (1993) discuss alternative ways of encoding conditional independence assumptions through graphs. In Andersson and Perlman (1993), conditional independence models for the multivariate normal distribution are determined through certain distributive lattices of subspaces. The precise relation between such conditional independence restrictions and those given by graphs is discussed in Andersson et al. (1995b). Basically, conditional independence restrictions given by lattices correspond to those given by directed acyclic graphs that are transitive, i.e. they satisfy that $\alpha \mapsto \beta$ implies $\alpha \to \beta$.

An important question is related to the notion of *Markov equivalence*. Two graphs are called Markov equivalent if they induce the same conditional independence restrictions. Conditions for two chain graphs to be Markov equivalent were given by Frydenberg (1990a) and further results related to this notion are given in Andersson et al. (1995a, 1996).

Recently, Koster (1996) has studied Markov properties of *reciprocal graphs*. These generalize chain graphs but permit directed cycles and appear therefore to be natural objects for studying systems with feedback, such as encountered in connection with structural equation models (Goldberger and Duncan 1973; Jöreskog 1977). Paz and Geva (1996) study *annotated graphs* that are even more general.

Another recent development is the systematic use of directed acyclic graphs for the interpretation, conjecture and discovery of causal relations

(Spirtes *et al.* 1993; Pearl 1995; Shafer 1996). This development is a natural extension of earlier work by Wright (1921), Wold (1954, 1960), Blalock (1971), and others.

4
Contingency tables

4.1 Examples

Before we begin developing the formal theory, it seems appropriate to consider a few examples that illustrate some basic points.

Example 4.1 Consider the data in Table 4.1, taken from an investigation of 237 Danish women performed by the Gallup Institute. To shorten notation we introduce three variables:

1. childhood experience of physical punishment, denoted by E and having states {yes, no};

2. use of physical punishment, denoted by U and having states {yes, no};

3. political affiliation, denoted by A and having states $\{l, s, r\}$; the three states correspond to political affiliations to either the Danish Social Democratic Party (s) or to parties which are considered politically to the left (l) or right (r) of this party.

There are many alternative ways of summarizing fundamental aspects of the data in Table 4.1. One way is to look at all pairwise marginals, displayed in Table 4.2. From direct inspection of the numbers in this table, some

Table 4.1. Relation between use of physical punishment, political affiliation and childhood experience of physical punishment in 237 Danish women.

Childhood experience and		Political affiliation		
use of physical punishment (PP)		Left	Soc.Dem.	Right
Has experienced PP	Uses PP	12	27	58
	Does not use PP	7	28	30
Has not experienced PP	Uses PP	9	5	9
	Does not use PP	19	15	18

Table 4.2. Tables describing marginal associations between childhood experience (E) of physical punishment, use (U) of physical punishment and political affiliation (A).

	U			A				A		
E	yes	no	E	l	s	r	U	l	s	r
yes	97	65	yes	19	55	88	yes	21	32	67
no	23	52	no	28	20	27	no	26	43	48

association between use of physical punishment and childhood experience seems apparent, and both of these seem also positively associated to having a political affiliation to the right of the Social Democratic Party. But this is all vague and relationships are clearly somewhat complex.

Alternatively, let us formulate the following model. The basic data are considered as a sample of 237 independent and identically distributed triplets (E, U, A) where p_{eua} denotes the probability that a randomly selected woman belongs to category (e, u, a). The cell entries in the given table are obtained from the basic data by counting the number N_{eua} belonging to category (e, u, a). Under this assumption, the counts follow a multinomial distribution.

Table 4.2 indicates that no pair of variables under investigation are independent. But let us formulate the hypothesis that, conditional on childhood experience, there is no association between use of physical punishment and political affiliation. This corresponds to independence in each of the two subtables of Table 4.1 obtained when dividing the material into groups that share childhood experience. In symbols, it would correspond to

$$\frac{p_{eua}}{p_{e++}} = \frac{p_{eu+}}{p_{e++}} \frac{p_{e+a}}{p_{e++}}, \qquad (4.1)$$

where $+$ denotes summation over an index. The relation (4.1) is clearly equivalent to

$$p_{eua} = \frac{p_{eu+} p_{e+a}}{p_{e++}},$$

and, as we shall later show, the maximum likelihood estimate of the probabilities under this assumption is given as

$$\hat{p}_{eua} = \frac{n_{eu+} n_{e+a}}{n_{e++} n}.$$

Table 4.3 shows the expected numbers $n\hat{p}_{eua}$ when conditional independence is assumed. The numbers do not deviate strongly from the observations although some discrepancy seems observable.

Table 4.3. Expected numbers of women grouped after use of physical punishment, political affiliation, and childhood experience of physical punishment, under the assumption of independence between use and affiliation, conditionally on childhood experience.

Childhood experience and use of physical punishment (PP)		Political affiliation		
		Left	Soc.Dem.	Right
Has experienced PP	Uses PP	11.4	32.9	52.7
	Does not use PP	7.6	22.1	35.3
Has not experienced PP	Uses PP	8.6	6.1	8.3
	Does not use PP	19.4	13.9	18.7

The relation (4.1) is equivalent to the existence of numbers $\{\alpha_{eu}\}$ and $\{\beta_{ea}\}$ such that
$$p_{eua} = \alpha_{eu}\beta_{ea}.$$
Taking logarithms we can write this as
$$\log p \in F_{EU} + F_{EA}, \qquad (4.2)$$
where F_{EU} is the linear subspace of tables of numbers, with entries depending only on the variables E and U, i.e. tables as

$\alpha_{yes,yes}$	$\alpha_{yes,yes}$	$\alpha_{yes,yes}$
$\alpha_{yes,no}$	$\alpha_{yes,no}$	$\alpha_{yes,no}$
$\alpha_{no,yes}$	$\alpha_{no,yes}$	$\alpha_{no,yes}$
$\alpha_{no,no}$	$\alpha_{no,no}$	$\alpha_{no,no}$

and F_{EA} is defined similarly.

The relation (4.2) identifies the model of conditional independence as a *log–linear model*. It can also be convenient to represent the model of conditional independence by the picture

$$\bullet\!\!-\!\!\!-\!\!\bullet\!\!-\!\!\!-\!\!\bullet$$
$$U \quad E \quad A$$

which is a reason for calling this model a *graphical model*. Alternatively, the association structure can be graphically represented as

$$\bullet\!\!\leftarrow\!\!\!-\!\!\bullet\!\!-\!\!\!\rightarrow\!\!\bullet$$
$$U \quad E \quad A$$

indicating that the variable E could be conceived to be a basic explanation for the association between U and A. Such a model is a *recursive graphical model*. □

Table 4.4. Death sentences in 4863 murder cases in Florida, partitioned after colour of alleged murderer and colour of victim.

	Sentence	
Murderer	Death	Other
Black	59	2547
White	72	2185

Victim	Murderer	Sentence	
		Death	Other
Black	Black	11	2309
	White	0	111
White	Black	48	238
	White	72	2074

In Example 4.1 we have seen that indication of a strong association in a marginal table (use of physical punishment and political affiliation) can be present, even when there is no or only weak association between the variables, conditionally on childhood experience. This is an instance of a phenomenon called the Yule–Simpson paradox (Yule 1903; Simpson 1951). As a consequence it is quite important whether independence or dependence is measured in marginal or conditional tables.

The situation can be much worse than in the previous example, as the next example illustrates.

Example 4.2 Table 4.4 describes the sentences in 4863 murder cases in Florida over the six years 1973-78 (Range 1979). It is registered whether there was a death sentence, and the colour of the skin, both of the victim and the murderer. Note that the marginal table, where the colour of the victim is ignored, shows a greater proportion of white murderers receiving death sentence than black (3.2% vs. 2.3%), whereas the table for given colour of victim shows a very different picture: both for black and white victims, there is a much higher proportion of black murderers receiving death sentences. In particular, note that 111 white murderers killed black victims and none were sentenced to death. Both tables are true representations of the same numerical facts. But a careless researcher may jump to unwarranted conclusions. □

The next example is more complex as it is concerned with rates of occurrence rather than the total number of occurrences of an event of a certain type.

Table 4.5. Observed number of lung cancer cases in the male population between 1968 and 1971 in four Danish cities distributed over age groups.

Age group	City			
	Fredericia	Horsens	Kolding	Vejle
40–54	11	13	4	5
55–59	11	6	8	7
60–64	11	15	7	10
65–69	10	10	11	14
70–74	11	12	9	8
> 75	10	2	12	7

Table 4.6. Approximate number of male inhabitants in four Danish cities in the period 1968–1971, distributed over age groups.

Age group	City			
	Fredericia	Horsens	Kolding	Vejle
40–54	3059	2875	3142	2520
55–59	800	1083	1050	878
60–64	710	923	895	839
65–69	581	834	702	631
70–74	509	634	535	539
> 75	605	782	659	619

Example 4.3 The data in Table 4.5 were at a certain point the focus of Danish public interest. They were reported in Clemmensen et al. (1974) and raised suspicion of an increased risk of lung cancer in Fredericia. In this case, the issue is not independence between age and city. It is rather a question of whether the rates n_{ij}/q_{ij} of cases per inhabitant in age group i and city j are higher for some combinations of ij than others, apart from random fluctuations. Here n_{ij} denotes the number of lung cancer cases and q_{ij} the number of inhabitants. Table 4.6 gives the number of inhabitants in the various age groups.

E.B. Andersen (1977) suggested the following model for these data. The number of lung cancer cases N_{ij} in age group i and city j are independent and Poisson-distributed with expectation $m_{ij} = \lambda_{ij} q_{ij}$. Further, the rates

Table 4.7. Expected number of lung cancer cases between 1968 and 1971 in the male population of four Danish cities distributed over age groups, assuming the mutiplicative Poisson model.

Age group	City			
	Fredericia	Horsens	Kolding	Vejle
40–54	10.96	7.40	7.76	6.88
55–59	8.62	8.39	7.80	7.20
60–64	11.61	10.85	10.09	10.45
65–69	12.19	12.58	10.16	10.08
70–74	11.67	10.45	8.46	9.41
> 75	8.96	8.33	6.73	6.98

λ_{ij} have the following multiplicative structure:

$$\lambda_{ij} = \alpha_i \beta_j.$$

The model has a reasonable fit, and the question whether there is a geographical dependence of lung cancer risk can now be phrased as a question of β_j being constant. Taking logarithms, we find for the expectation

$$\log m_{ij} = \log \alpha_i + \log \beta_j + \log q_{ij}.$$

Because of the last term, the logarithm of the expectation is in an affine rather than linear subspace, and the model is a *log-affine model*. The theory developed in this chapter will also cover this type of model. The expected number of cases under the model is displayed in Table 4.7. The most notable feature of the data is the very few observed cases in Horsens among citizens over 75. □

4.2 Basic facts and concepts

4.2.1 Notation and terminology

A contingency table is formed by classifying a number of objects according to a set of criteria and counting the number of objects in each classification.

We express this formally by introducing a finite set Δ of *classification criteria* and for each $\delta \in \Delta$ a finite set $\mathcal{I}_\delta$ of possible *levels* of these. We often refer to the criteria as *variables*. The *cells* of the table are the elements $i = (i_\delta)_{\delta \in \Delta}$ of the product $\mathcal{I}$ of the level sets

$$i \in \mathcal{I} = \times_{\delta \in \Delta} \mathcal{I}_\delta.$$

Data typically appear in two different forms: as a *list* of $|n|$ objects $(i^1, \ldots, i^{|n|})$, where each entry identifies which cell a given object belongs to, or as a *contingency table* of *counts* $n = \{n(i)\}_{i \in \mathcal{I}}$. Here $|n| = \sum_{i \in \mathcal{I}} n(i)$. If we introduce the indicator functions

$$\chi^i(j) = \begin{cases} 1 & \text{if } j = i \\ 0 & \text{otherwise,} \end{cases}$$

the counts are given as

$$n(i) = \sum_{\nu=1}^{|n|} \chi^i(j^\nu).$$

The table has *dimension* equal to the number $|\Delta|$ of variables.

Basic manipulations with contingency tables are those of forming marginal tables and of forming slices. An *a-marginal table* is for $a \subset \Delta$ obtained by only classifying the objects according to the criteria in a, i.e. by only considering the variables in a. It has *marginal cells*

$$i_a \in \mathcal{I}_a = \times_{\delta \in a} \mathcal{I}_\delta.$$

For an arbitrary vector $x = \{x(i)\}_{i \in \mathcal{I}}$ in $\mathcal{R}^\mathcal{I}$ we define its *a-marginal* as

$$x(i_a) = \sum_{j: j_a = i_a} x(j) = \sum_{j_{\Delta \setminus a} \in \mathcal{I}_{\Delta \setminus a}} x(i_a, j_{\Delta \setminus a})$$

and $|x| = \sum_{i \in \mathcal{I}} |x(i)|$. For $a = \emptyset$ we write $x(i_\emptyset) = \sum_{i \in \mathcal{I}} x(i)$. Note that when $x(i) \geq 0$ for all i, then $|x| = x(i_\emptyset)$. The *marginal counts* are the quantities $n(i_a)$. Again, if we let

$$\chi^{i_a}(j) = \begin{cases} 1 & \text{if } j_a = i_a \\ 0 & \text{otherwise.} \end{cases} \tag{4.3}$$

we have

$$n(i_a) = \sum_{\nu=1}^{|n|} \chi^{i_a}(j^\nu) = \sum_{j \in \mathcal{I}} n(j) \chi^{i_a}(j).$$

For the marginal corresponding to the empty set we get $n(i_\emptyset) = |n|$, the total number of observations.

For $b \subset \Delta$ and $i_b \in \mathcal{I}_b$ the i_b-*slice* of the table is obtained by only classifying those objects that belong to the marginal cell i_b or, in other words, by studying the table for fixed level of the variables in b. Formally the i_b-slice has cells $\mathcal{I}_a$ where $a = \Delta \setminus b$, and counts $n^{i_b}(i_a) = n(i_a, i_b) = n(i)$.

The notation described above is convenient when dealing with the general, unspecified case and is therefore practical for theoretical developments. When working on particular tables it performs less elegantly. In such cases the subscript notation described below performs better.

Example 4.4 In the case where we have three criteria $\Delta = \{\alpha, \beta, \gamma\}$ having levels

$$\mathcal{I}_\alpha = \{1, \ldots, R\}, \quad \mathcal{I}_\beta = \{1, \ldots, S\}, \quad \mathcal{I}_\gamma = \{1, \ldots, T\},$$

we say that we have a three-dimensional $R \times S \times T$ *table*. Instead of $\{(n(r, s, t)\}$ we write (n_{rst}) for the table of counts. For the marginal counts we write for example

$$n_{r+t} = \sum_{s=1}^{S} n_{rst}, \quad n_{+s+} = \sum_{r=1}^{R} \sum_{t=1}^{T} n_{rst}$$

instead of

$$n\{(r,t)_{\{\alpha,\gamma\}}\}, \quad n\{(s)_{\{\beta\}}\}$$

and so on. The $(r, s, t)_{\{\alpha,\gamma\}}$-slice becomes the $r+t$-slice and consists of the cells (r, s, t) with $s = 1, \ldots, S$. □

It seems most convenient to use this double notation system since the subscript notation gets completely out of hand when the general case is treated. Thus we shall in the following switch between notation systems as we switch between general theory and concrete examples. We hope that this will not confuse the reader, but rather bridge the gap between the general notation used in the book and the notation most common in the applied literature.

Data for contingency tables are collected under various different sampling schemes, of which we here consider three:

1. all cell counts and the total number of observations are random;

2. the total number of observations is fixed but cell counts are otherwise random;

3. the number of observations $n(i_b)$ in each i_b-slice is fixed for some $b \subset \Delta$.

Data of the first type appear typically when counting the number of events in fixed time periods such as, for example, cancer incidents as in Example 4.3 or traffic accidents, and classifying them according to type of road, geographical region, season etc. The basic model for this sampling scheme assumes that the cell counts are realizations of independent and Poisson-distributed random variables $\{N(i)\}_{i \in \mathcal{I}}$. Thus letting $\mathbf{E}N(i) = \lambda(i)$ the joint distribution of the counts becomes

$$P\{N(i) = n(i), i \in \mathcal{I}\} = \prod_{i \in \mathcal{I}} \frac{\lambda(i)^{n(i)}}{n(i)!} e^{-\lambda(i)}. \tag{4.4}$$

In case 2 where a fixed number n of objects is classified, each of these is supposed independently to belong to a given cell i with probability $p(i) \geq 0$ where $|p| = 1$, such that the counts follow a multinomial distribution

$$P\{N(i) = n(i), i \in \mathcal{I}\} = \frac{|n|!}{\prod_{i \in \mathcal{I}} n(i)!} \prod_{i \in \mathcal{I}} p(i)^{n(i)}. \qquad (4.5)$$

Here it deserves notice that the distribution (4.5) can be obtained from (4.4) by conditioning upon the total number of observations $|N|$ being equal to $|n|$ and letting $p(i) = \lambda(i)/|\lambda|$. Conversely (4.4) appears when randomizing the total number of observations in (4.5) to become Poisson-distributed with mean equal to $|\lambda|$.

Finally, in the case where the numbers in each slice are fixed, we assume the counts in the slices to be independent and multinomially distributed as above with cell probabilities in slice i_b equal to $p(i_a \,|\, i_b)$. Thus the joint distribution of the table becomes

$$\begin{aligned} P\{N(i) = n(i), i \in \mathcal{I}\} &= \prod_{i_b \in \mathcal{I}_b} \left\{ \frac{n(i_b)!}{\prod_{i_a \in \mathcal{I}_a} n^{i_b}(i_a)!} \prod_{i_a \in \mathcal{I}_a} p(i_a \,|\, i_b)^{n^{i_b}(i_a)} \right\} \\ &= \prod_{i_b \in \mathcal{I}_b} \left\{ \frac{n(i_b)!}{\prod_{i_a \in \mathcal{I}_a} n(i)!} \prod_{i_a \in \mathcal{I}_a} p(i_a \,|\, i_b)^{n(i)} \right\}. \end{aligned} \qquad (4.6)$$

In this case we can obtain the distribution (4.6) from (4.4) by conditioning upon $N(i_b) = n(i_b), i_b \in \mathcal{I}_b$, but also by the same conditioning in (4.5).

In the following we refer to the three cases as the *Poisson*, *multinomial* and *restricted multinomial* sampling schemes.

4.2.2 Saturated models

In general we say that a model is a *saturated model* if its parameters are only constrained by restrictions that are due to the sampling procedure. If we consider the vector of *expected cell counts* and denote this by $\{m(i)\}_{i \in \mathcal{I}}$ we have in the three sampling situations that

$$m(i) = \lambda(i), \quad m(i) = |n|p(i), \quad m(i) = n(i_b)p(i_a \,|\, i_b), \qquad (4.7)$$

respectively. The mean vector must satisfy different constraints in the three cases.

1. In the Poisson case the restriction is

$$m(i) \geq 0 \qquad (4.8)$$

and there are no further restrictions.

2. In the multinomial case, the restrictions $p(i) \geq 0$ and $|p| = 1$ translate through (4.7) to
$$m(i) \geq 0, \quad |m| = |n|. \tag{4.9}$$

3. In the restricted sampling we similarly translate $p(i_a \,|\, i_b) \geq 0$ and $\sum_{i_a} p(i_a \,|\, i_b) = 1$ to
$$m(i) \geq 0, \quad m(i_b) = n(i_b) \quad \text{for all } i_b \in \mathcal{I}_b. \tag{4.10}$$

Exploiting these restrictions we find that in the multinomial and restricted sampling cases we have
$$\prod_{i \in \mathcal{I}} e^{m(i)} = e^{|m|} = e^{|n|},$$
which is constant since n is fixed. It follows that in all three cases the likelihood function L becomes proportional to
$$L(m) \propto \prod_{i \in \mathcal{I}} m(i)^{n(i)} e^{-m(i)}. \tag{4.11}$$

From (A.2) we get
$$L(m) \leq L(n)$$
with equality if and only if $m(i) = n(i)$ for all $i \in \mathcal{I}$. On the other hand, if $m \equiv n$, then m satisfies the constraints in all three cases, such that we have

Proposition 4.5 *In the saturated models, the maximum likelihood estimate of the mean vector is given by*
$$\hat{m}(i) = n(i), \quad i \in \mathcal{I},$$
implying for the three cases that
$$\hat{\lambda}(i) = n(i), \quad \hat{p}(i) = n(i)/|n|, \quad \hat{p}(i_a \,|\, i_b) = n(i)/n(i_b).$$

This result just confirms the natural fact that the counts are estimates for their means if no artificial restrictions are imposed.

4.2.3 Log–affine and log–linear models

Maximum likelihood estimates

We next investigate the problem of maximizing the likelihood function in the case where we further restrict the mean vector to be of the form $q\exp(v)$, where q is fixed and not identically zero and $v \in H$, where H is a linear subspace of $\mathcal{R}^{\mathcal{I}}$ that contains the constant function $e(i) \equiv 1$. The symbol $\mathcal{R}^{\mathcal{I}}$ denotes as usual the vector space of real-valued functions on $\mathcal{I}$.

We denote by $M(q, H)$ the set of mean vectors of the form $q \exp(v)$ with $v \in H$ that also satisfy the appropriate restrictions due to the sampling scheme (4.8), (4.9) or (4.10). Strictly speaking we would need different symbols for the three sampling cases, but it will usually be clear from the context which restrictions apply.

A model that restricts the mean vector in this way is called a *log–affine model*. This reflects the fact that if $q(i) > 0$ for all i, then

$$m \in M(q, H) \iff \log m \in \log q + H$$

and m satisfies the appropriate sampling restrictions. The cells where $q(i) = 0$ are called *structural zeros*.

Further we let the *extended log–affine model* consist of all pointwise limits of vectors in the log–affine model and denote this by $\bar{M}(q, H)$. The extended log–affine model will typically contain elements where $m(i) = 0$ for some cells i with $q(i) > 0$. The extended model is convenient because, as we shall see later, it ensures the existence of the maximum likelihood estimate, whether or not the counts of the observed table satisfy constraints such as being strictly positive.

In many interesting cases we consider log–affine models with $q(i) = 1$ for all i or, equivalently, log–affine models with $\log q \in H$. We call such models *log–linear*. In log–linear models we usually write $M(H)$ instead of $M(1, H)$ and so on. Note in particular that the saturated models are special cases of extended log–linear models. More precisely, the saturated models correspond to log–linear models $\bar{M}(H)$ with $H = \mathcal{R}^{\mathcal{I}}$.

Example 4.6 Consider the three-dimensional $R \times S \times T$ table under multinomial sampling with the model of conditional independence of R and T for given S

$$p_{rst} = \frac{p_{rs+} p_{+st}}{p_{+s+}}. \tag{4.12}$$

If we assume that the probabilities are all positive it is readily seen that the factorization in (4.12) is equivalent to the logarithm of the expectations having an expansion as

$$\log m_{rst} = \alpha_{rs} + \beta_{st}$$

for suitably chosen α and β. In other words, the model of conditional independence is a log–linear model $M(H)$ with

$$H = F_{RS} + F_{ST},$$

where F_{RS} and F_{ST} are the vector spaces of functions of (r, s, t) that depend only on (r, s) and (s, t) respectively. The extended log–linear model $\bar{M}(H)$

BASIC FACTS AND CONCEPTS

can be seen to coincide exactly with the set of probabilities that factorize as in (4.6), without necessarily satisfying the positivity restriction. □

In the following we show, first for the Poisson case, that the maximum likelihood estimate in the extended log–affine model exists uniquely and can be found by solving the likelihood equation (4.13).

We begin by establishing the existence and uniqueness. We assume without loss of generality that $n(i) = 0$ for all i with $q(i) = 0$, since the latter means that data in cell i are impossible and the model therefore would be absurd otherwise. Recall also that the likelihood function is given by (4.11) in all three cases and the difference between the three problems are concerned with the restrictions imposed upon the mean vector m.

Proposition 4.7 *In the extended log–affine model $\bar{M}(q, H)$ under Poisson sampling, the maximum likelihood estimate $\hat{m}$ of the mean vector m exists and is unique.*

Proof: We first show existence and consider initially the compact subset $\bar{M}_{|n|}$ of $\bar{M}(q, H)$ having $|m| = |n|$. Since the likelihood function is continuous it must attain its maximum over that set at $\tilde{m}$, say. We then have for an arbitrary $m \in \bar{M}(q, H)$ that

$$\begin{aligned} L(m) &= L\left(\frac{|n|m}{|m|}\right) |n|^{-|n|} e^{|n|} |m|^{|n|} e^{-|m|} \\ &\leq L(\tilde{m}) |n|^{-|n|} e^{|n|} |m|^{|n|} e^{-|m|} \leq L(\tilde{m}), \end{aligned}$$

where we have used that H contains the constant to ensure that $|n|m/|m| \in \bar{M}_{|n|}$, the particular expression (4.11) for the likelihood function, as well as the inequality (A.2). Thus L attains its global maximum at $\tilde{m}$.

To show uniqueness we exploit the log-convexity of $\bar{M}(q, H)$. Assume that the maximum is attained both at m_1 and m_2 and let $\bar{m} = \sqrt{m_1 m_2}$. Then $\bar{m} \in \bar{M}(q, H)$ and we have

$$\begin{aligned} L(\bar{m}) &\leq L(m_1) = \sqrt{L(m_1)L(m_2)} = L(\bar{m}) e^{|\bar{m}| - (|m_1| + |m_2|)/2} \\ &= L(\bar{m}) \exp\left[-\sum_i \left\{(m_1(i))^{1/2} - m_2(i)^{1/2}\right\}^2 / 2\right], \end{aligned}$$

whereby we must have $m_1 \equiv m_2$ and the result follows. □

Next, let us recall that $\mathcal{R}^{\mathcal{I}}$ is a Euclidean space when equipped with the standard inner product

$$\langle x, y \rangle = \sum_{i \in \mathcal{I}} x(i) y(i).$$

If H is a linear subspace of $\mathcal{R}^{\mathcal{I}}$, Π_H denotes the orthogonal projection onto H with respect to this inner product, and for a being a subset of Δ, the *factor subspace* F_a is the subspace of functions that only depend on i through i_a, i.e.

$$x \in F_a \iff x(i) = x(j) \quad \text{for all } i, j \text{ with } i_a = j_a.$$

These subspaces are studied in some detail in Section B.2. We have

Theorem 4.8 *The maximum likelihood estimate $\hat{m}$ of the mean vector in the extended log–affine model determined by q and a linear subspace H containing the constant function is the unique element of $\bar{M}(q, H)$ that satisfies the likelihood equation*

$$\Pi_H \hat{m} = \Pi_H n. \tag{4.13}$$

This holds under Poisson as well as under multinomial sampling. If H contains the factor subspace F_b, $\hat{m}$ is also the estimate under restricted multinomial sampling with fixed b-marginals.

Proof: We have already established the existence and uniqueness of the estimate under Poisson sampling and need only to show that the estimate can be found by solving the likelihood equation (4.13). Observe first that this equation is equivalent to the system of equations

$$\sum_{i \in \mathcal{I}} n(i) v(i) = \sum_{i \in \mathcal{I}} \hat{m}(i) v(i) \quad \text{for all } v \in H. \tag{4.14}$$

First assume that $\hat{m}$ is the maximum likelihood estimate in the Poisson case. Let $v \in H$ be arbitrary but fixed. For any $t \in \mathcal{R}$ let $m_t = \hat{m} \exp(tv)$ and

$$h(t) = \log L(m_t) = \log L(\hat{m}) + t \sum_{i \in \mathcal{I}} n(i) v(i) - \sum_{i \in \mathcal{I}} \hat{m}(i) \left\{ e^{tv(i)} - 1 \right\}.$$

Obviously, $m_t \in \bar{M}(q, H)$ for all t and the function h is differentiable at 0 and attends its maximum there. Thus its derivative vanishes at 0 and we get

$$0 = h'(0) = 0 + \sum_{i \in \mathcal{I}} n(i) v(i) - \sum_{i \in \mathcal{I}} \hat{m}(i) v(i),$$

which implies (4.14).

Assume then that $\hat{m}$ satisfies the equation (4.13) and let further $m = \lim_{k \to \infty} m_k \in \bar{M}(q, H)$ be arbitrary with $m_k = q \exp v_k$ and $v_k \in H$, and let similarly $\hat{m} = \lim_{l \to \infty} r_l$ with $r_l = q \exp w_l$ and $w_l \in H$. Using the identity (4.14) twice and the inequality (A.2) once, we obtain

$$\begin{aligned}
L(m) &= \lim_{k\to\infty} L(m_k) = \lim_{k\to\infty} \prod_{i\in\mathcal{I}} \{q(i)\exp v_k(i)\}^{n(i)} e^{-m_k(i)} \\
&= \lim_{k\to\infty} \prod_{i\in\mathcal{I}} m_k(i)^{\hat{m}(i)} e^{-m_k(i)} \prod_{i\in\mathcal{I}} q(i)^{n(i)-\hat{m}(i)} \\
&\leq \prod_{i\in\mathcal{I}} \hat{m}(i)^{\hat{m}(i)} e^{-\hat{m}(i)} \prod_{i\in\mathcal{I}} q(i)^{n(i)-\hat{m}(i)} \\
&= \lim_{l\to\infty} \prod_{i\in\mathcal{I}} r_l(i)^{\hat{m}(i)} e^{-r_l(i)} \prod_{i\in\mathcal{I}} q(i)^{n(i)-\hat{m}(i)} \\
&= \lim_{l\to\infty} \prod_{i\in\mathcal{I}} r_l(i)^{n(i)} e^{-r_l(i)} = \prod_{i\in\mathcal{I}} \hat{m}(i)^{n(i)} e^{-\hat{m}(i)} = L(\hat{m}).
\end{aligned}$$

Hence, if (4.13) is satisfied, L is maximal at $\hat{m}$.

This takes care of the Poisson case. Because H contains the constant function, it follows from (4.13) that $\hat{m}$ automatically satisfies the constraint $|\hat{m}| = |n|$ and, similarly, $\hat{m}(i_b) = n(i_b)$ when $F_b \subseteq H$. For (4.14) gives

$$|\hat{m}| = \sum_{i\in\mathcal{I}} \hat{m}(i) = \langle \hat{m}, e\rangle = \langle n, e\rangle = \sum_{i\in\mathcal{I}} n(i) = |n|,$$

where $e(i) \equiv 1$, and since $\chi^{i_b} \in F_b$, where χ^{i_b} was defined in (4.3),

$$\hat{m}(i_b) = \sum_{j\in\mathcal{I}} \hat{m}(j)\chi^{i_b}(j) = \langle \hat{m}, \chi^{i_b}\rangle = \langle n, \chi^{i_b}\rangle = \sum_{j\in\mathcal{I}} n(j)\chi^{i_b}(j) = n(i_b).$$

Since the three sampling cases have identical likelihood functions, the estimate given by (4.13) maximizes the likelihood in the other sampling cases as well. Thus the theorem has been proved. □

The conclusion of Theorem 4.8 holds for the Poisson sampling case even when H does not contain the constant, but the proof of Proposition 4.7 becomes more involved. The theorem naturally contains Proposition 4.5 concerning estimation in saturated models as a special case if we let $q(i) \equiv 1$ and $H = \mathcal{R}^\mathcal{I}$.

If q is strictly positive, existence of the maximum likelihood estimate within the log–affine model itself can be established in a simple fashion:

Corollary 4.9 *If $q(i) > 0$ for all $i \in \mathcal{I}$, it holds in all three sampling situations that the maximum likelihood estimate exists in the log-affine model $M(q, H)$ if and only if $\hat{m}(i) > 0$ for all i, where $\hat{m}$ is the estimate in the extended model $\bar{M}(q, H)$.*

Proof: If $\hat{m}(i) > 0$, we can take logarithms in the expression

$$\hat{m}(i) = \lim_{k\to\infty} q(i)\exp v_k(i)$$

to obtain
$$\lim_{k\to\infty} v_k(i) = \log \hat{m}(i) - \log q(i).$$

As subspaces are closed sets, we must have $\lim_{k\to\infty} v_k(i) \in H$ and hence also $\hat{m} \in M(q, H)$ whereby the maximum likelihood estimate exists in $M(q, H)$.

Assume conversely that $\hat{m}(i) = 0$ for some $i \in \mathcal{I}$, implying that $\hat{m} \notin M(q, H)$, and further that L attains its maximum over $M(q, H)$ at $\tilde{m}$, say, which is then different from $\hat{m}$.

There is a sequence (m_k) of elements in $M(q, H)$ converging to $\hat{m}$. Since L is continuous we find
$$L(\hat{m}) = \lim_{k\to\infty} L(m_k) \leq L(\tilde{m}).$$

Because $\hat{m}$ maximizes L we must have
$$L(\hat{m}) = L(\tilde{m}),$$

contradicting the uniqueness of $\hat{m}$. Hence L does not attain its maximum in $M(q, H)$. □

A log–affine model is, in the cases we consider, a regular exponential model. Therefore maximum likelihood estimation can be treated using the results of Appendix D. We have however chosen to derive the estimation results directly, in particular to be able to handle cases with observations $n(i) = 0$ for some cells.

Asymptotic distribution of estimates

The exact distribution of the maximum likelihood estimate in a log–affine model is in general intractable and one has to rely on the asymptotic results from Section D.2. So let us first consider a log–affine model $M(q, H)$ in the Poisson case with $H = \mathcal{R}^\mathcal{I}$, i.e. we consider a log–affine saturated model as the base model. We might as well assume that $q(i) > 0$ for all $i \in \mathcal{I}$ since we otherwise only have to consider the cells with this satisfied. To make the connection to exponential families clear, we write

$$\begin{aligned}
f\{n(i)_{i\in\mathcal{I}}, \theta)\} &= \prod_{i\in\mathcal{I}} \frac{\{q(i)e^{\theta(i)}\}^{n(i)}}{n(i)!} \exp\left\{-q(i)e^{\theta(i)}\right\} \\
&= \exp\left\{\sum_i \theta(i)n(i) - \sum_i q(i)e^{\theta(i)}\right\} \prod_{i\in\mathcal{I}} \frac{q(i)^{n(i)}}{n(i)!} \\
&= \exp\left(\langle \theta, n\rangle - \langle q, e^\theta\rangle\right) \mu(i),
\end{aligned}$$

where $\theta = \log(m/q)$ is the canonical parameter which obviously varies in $\Theta = H = \mathcal{R}^{\mathcal{I}}$. Thus the canonical statistic is the set of counts, the cumulant function is $\psi(\theta) = \langle q, e^\theta \rangle$, and the base measure is

$$\mu(i) = \prod_{i \in \mathcal{I}} \frac{q(i)^{n(i)}}{n(i)!}.$$

From elementary properties of the Poisson-distribution we immediately find the mean value, covariance and inverse covariance to be given by

$$\tau(\theta)_i = m(i) = q(i)e^{\theta(i)}, \quad v(\theta)_{ij} = m(i)\delta_{ij}, \quad v(\theta)_{ij}^{-1} = m(i)^{-1}\delta_{ij},$$

where δ is the Kronecker delta. In the case of multinomial sampling we have the same canonical statistic, but now we are in the singular case, since $\sum_i n(i) = |n|$, where $|n|$ is fixed. The joint density becomes

$$\begin{aligned}
f\{n(i)_{i \in \mathcal{I}}, \theta\} &= \frac{|n|!}{\prod_{i \in \mathcal{I}} n(i)!} \prod_{i \in \mathcal{I}} \left\{ \frac{q(i)e^{\theta(i)}}{\sum_{j \in \mathcal{I}} q(j)e^{\theta(j)}} \right\}^{n(i)} \\
&= \exp\{\langle \theta, n \rangle - \psi_0(\theta)\} |n|! \prod_{i \in \mathcal{I}} \frac{q(i)^{n(i)}}{n(i)!}.
\end{aligned}$$

Thus the canonical parameter and statistic are the same whereas the base measure and log-normalizing constant are different. Using standard properties of the multinomial distribution we find the mean and covariance to be

$$\tau_0(\theta)_i = m(i) = |n| \frac{q(i)e^{\theta(i)}}{\sum_{j \in \mathcal{I}} q(j)e^{\theta(j)}}, \quad v_0(\theta)_{ij} = m(i)\delta_{ij} - m(i)m(j)/|n|.$$

Alternatively we might write $v_0 = v - m \otimes m/|n|$. A direct calculation using that $|m| = |n|$ shows that v^{-1} is a generalized inverse to v_0.

Finally in the case of fixed b-marginals we find by a similar calculation that the set of counts is the canonical statistic and we get for the mean

$$\tau_b(\theta)_i = m(i) = n(i_b) \frac{q(i)e^{\theta(i)}}{\sum_{j:j_b=i_b} q(j)e^{\theta(j)}}$$

and covariance

$$v_b(\theta)_{ij} = \begin{cases} m(i)\delta_{ij} - m(i)m(j)/n(i_b) & \text{if } i_b = j_b \\ 0 & \text{otherwise.} \end{cases}$$

Even in this case it is true that v^{-1} is a generalized inverse to v_b. This also follows directly using that $m(i_b) = n(i_b)$. For later reference we formulate this as a lemma.

Lemma 4.10 *Let v, v_0 and v_b denote the covariance of the canonical statistic in the saturated log–linear model in the three basic sampling cases. Then, in all cases, v^{-1} is a generalized inverse such that*

$$v_0 v^{-1} v_0 = v_0, \quad v_b v^{-1} v_b = v_b.$$

Proof: Direct verification, using $|m| = |n|$ in the case of v_0 and $m(i_b) = n(i_b)$ in the case of v_b. □

The asymptotic distribution of $\hat{m}$ under a reduced log–affine model $M(q, H_0)$, where H_0 is a proper subspace of $H = \mathcal{R}^{\mathcal{I}}$, is therefore from (D.11) in the three cases

$$\hat{m}_0 \overset{a}{\sim} \mathcal{N}(m, \Pi_0^{v\top} v \Pi_0^v), \tag{4.15}$$

$$\hat{m}_0 \overset{a}{\sim} \mathcal{N}(m, \Pi_0^{v\top} v_0 \Pi_0^v), \tag{4.16}$$

$$\hat{m}_0 \overset{a}{\sim} \mathcal{N}(m, \Pi_0^{v\top} v_b \Pi_0^v). \tag{4.17}$$

Here Π_0^v is the orthogonal projection onto H_0 with respect to the inner product determined by v. We have used that this is also the orthogonal projection onto H_0 with respect to v_0 and v_b, the latter assuming $F_b \subseteq H_0$ as usual.

To see that this is indeed the case, recall that A is an orthogonal projection with respect to the inner product determined by v if and only if it is idempotent and self-adjoint with respect to v, i.e.

$$vA = A^\top v. \tag{4.18}$$

Thus, if we let $A = \Pi_0^v$ we must have

$$m(i) a_{ij} = a_{ji} m(j). \tag{4.19}$$

Since H contains the constant function we must also have $Ae = e$ and thus

$$\sum_k a_{ik} = 1. \tag{4.20}$$

For the multinomial sampling case we want to show that

$$(v - m \otimes m/|n|)A = A^\top (v - m \otimes m/|n|)$$

and because of (4.18) this is equivalent to showing

$$(m \otimes m)A = A^\top (m \otimes m).$$

Using (4.19) and (4.20) we get

$$\{A^\top (m \otimes m)\}_{ij} = \sum_k a_{ki} m(k) m(j) = \sum_k a_{ik} m(i) m(j) = m(i) m(j) \tag{4.21}$$

and similarly also that $(m \otimes m)A = m \otimes m$. The case of fixed b-marginals is analogous.

Deviances of log-affine models

Consider an extended log–affine model $\bar{M}_0 = \bar{M}(q, H_0)$ and a submodel thereof $\bar{M}_1$, determined in a similar way by H_1, where H_1 is a linear subspace of H_0. In the case of restricted sampling with fixed b-marginals we will further assume that the factor subspace F_b is contained in H_k for $k = 0, 1$. If we denote the maximum likelihood estimates of the mean vectors in the two cases by $\hat{m}_0$ and $\hat{m}_1$, these will then both exist and be strictly positive on the set $\mathcal{I}^+ = \{i \mid n(i) > 0\}$. The deviances to the saturated model with estimate $\hat{m}$ are for $k = 0, 1$

$$\begin{aligned} d_k &= -2\log \frac{L(\hat{m}_k)}{L(\hat{m})} \\ &= 2\sum_{i \in \mathcal{I}^+} n(i) \log \frac{n(i)}{\hat{m}_k(i)} + \sum_{i \in \mathcal{I}} \{\hat{m}_k(i) - n(i)\} \\ &= 2\sum_{i \in \mathcal{I}^+} n(i) \log \frac{n(i)}{\hat{m}_k(i)}, \end{aligned}$$

where we have exploited that $\sum \hat{m}_k(i) = |n|$. Thus the deviance statistic between the models is the difference

$$d_{01} = d_1 - d_0 = -2\log \frac{L(\hat{m}_1)}{L(\hat{m}_0)} = 2\sum_{i \in \mathcal{I}^+} n(i) \log \frac{\hat{m}_0(i)}{\hat{m}_1(i)}.$$

A Taylor expansion of $f(x) = x\log(x/a)$ around $x = a$ yields

$$x\log(x/a) = x - a + \frac{(x-a)^2}{2a} + o\{(x-a)^2\},$$

such that when $n(i)$ is close to $\hat{m}_k(i)$ we have

$$d_k \approx \sum_{i \in \mathcal{I}^+} \frac{(n(i) - \hat{m}_k(i))^2}{\hat{m}_k(i)}.$$

Alternatively it is common to use Pearson's X^2-statistic $\tilde{d}_k$:

$$\tilde{d}_k = \sum_{i \in \mathcal{I}_k^*} \frac{(n(i) - \hat{m}_k(i))^2}{\hat{m}_k(i)},$$

where $\mathcal{I}_k^* = \{i \mid \hat{m}_k(i) > 0\}$. Note that if we let

$$\tilde{d}_{01} = \sum_{i \in \mathcal{I}_1^*} \frac{(\hat{m}_0(i) - \hat{m}_1(i))^2}{\hat{m}_1(i)},$$

we only have approximate additivity $\tilde{d}_{01} \approx \tilde{d}_0 - \tilde{d}_1$. Also, since the diagonal matrix with entries m^{-1} is generalized inverse to the covariance matrix in all three sampling cases as shown in Lemma 4.10, the statistic $\tilde{d}_{01}$ is one of the quadratic approximations to the deviance described in Section D.2. More precisely,

$$\tilde{d}_{01} = (\hat{m}_0 - \hat{m}_1)^\top v(\hat{\theta}_1)^- (\hat{m}_0 - \hat{m}_1).$$

Generally and ideally it seems appropriate to judge the deviance d_{01}, or any of its approximate equivalents, in its exact conditional distribution given the estimate under the smaller of the two models $\bar{M}_1$; see Section D.1. Note in particular that *this distribution will be the same in the three sampling cases.*

The exact distribution is in general intractable and one has to rely on asymptotic and approximate results, exploiting that we deal with exponential models. When calculating the degrees of freedom for the approximating χ^2-distribution, special care has to be taken because we allow means at particular cells to be zero.

Since the distribution to be considered does not depend on which sampling case we are working with, we might as well assume the simpler Poisson case. Let now $M_k^* = M(q^*, H_k), k = 0, 1$, where

$$q^*(i) = \begin{cases} q(i) & \text{if } \hat{m}_1(i) > 0 \\ 0 & \text{otherwise.} \end{cases}$$

Thus q^* keeps track of $\mathcal{I}_1^*$. Note in particular that $\hat{m}_1$ is also the estimate of m under the model M_1^* and $\hat{m}_0$ is the estimate under the extended model $\bar{M}_0^*$. The results of Section D.2 therefore yield that, under the hypothesis $m \in M_1^*$, the deviance is asymptotically χ^2-distributed with degrees of freedom equal to

$$f = \dim M_0^* - \dim M_1^*,$$

where $\dim M_k^*$ is the dimension of the vector space H_k^* of equivalence classes under the relation

$$v \overset{*}{\sim} w \iff v(i) = w(i) \quad \text{for all } i \in \mathcal{I}_1^*.$$

Equivalently $\dim M_k^* = \dim H_k - \dim N_k$, where N_k is the null space of the linear projection $v \to vq^*$. The dimensions can in general be quite difficult to compute in a given case. If all estimates of cell means are positive, i.e. $\mathcal{I}_1^* = \mathcal{I}$, we have $H_k = H_k^*$ and if $H_0 = \mathcal{R}^\mathcal{I}$, i.e. if the model space is unrestricted, then $\dim M_0^* = |\mathcal{I}_1^*|$, but otherwise special care has to be taken. An algorithm for calculating this dimension for hierarchical models based on Borosh and Fraenkel (1966) is described in Haberman (1974).

Concerning the quality of the χ^2-approximation, very little is known in general apart from the fact that the approximation demands $m(i)$ to be rather large for all $i \in \mathcal{I}_1^*$.

The asymptotic χ^2-distribution of the deviance can sometimes be improved by using so-called Bartlett corrections. However, some investigations indicate that no improvement is gained in the cases considered (Frydenberg and Jensen 1989).

An alternative approach is based on the family of *power divergence* statistics given as

$$d_k^\lambda = \frac{2}{\lambda(\lambda+1)} \sum_{i \in \mathcal{I}^+} n(i) \left\{ \left(\frac{n(i)}{\hat{m}_k(i)} \right)^\lambda - 1 \right\}$$

for $0 \leq \lambda \leq 1$. Then $d_k = d_k^0$ and, if $\mathcal{I}_k^* = \mathcal{I}^+$, $\tilde{d}_k = d_k^1$. These statistics are all asymptotically χ^2-distributed and by choosing appropriate values of λ, improved approximations may be achieved. This is discussed in detail by Read and Cressie (1988).

4.3 Hierarchical models

We shall in this section investigate log–affine models determined by subspaces of the form

$$H_\mathcal{A} = \sum_{a \in \mathcal{A}} F_a,$$

where $\mathcal{A}$ is a class of subsets of Δ and F_a are the factor subspaces as described in Section B.2. Model subspaces of this form are called *hierarchical model subspaces* and log–affine (or log–linear) models determined by such spaces are *hierarchical log–affine models*. We mostly use the term *hierarchical model* for short. Note that saturated models are hierarchical models with $\mathcal{A} = \{\Delta\}$. The model of conditional independence in a three-way table, see Example 4.6, is another hierarchical model.

If $F_\delta \subseteq H_\mathcal{A}$ for all $\delta \in \Delta$, all main effects are present in the hierarchical model with generating class $\mathcal{A}$. Hierarchical models not having this property are less interesting since the variables without main effects are uniformly distributed and completely independent of the remaining variables. The statistical analysis can then be performed without taking these variables into account. Therefore we will in the following assume all main effects to be present unless the opposite is explicitly stated.

If $\mathcal{A}$ is a collection of subsets of Δ we let red $\mathcal{A}$ denote the elements of $\mathcal{A}$ that are maximal with respect to inclusion, i.e. red $\mathcal{A}$ is obtained from $\mathcal{A}$ by throwing away all subsets that are contained in other subsets in $\mathcal{A}$. Since

$$H_\mathcal{A} = H_{\text{red}\,\mathcal{A}},$$

this reduced collection is the most economical way of specifying the model space. If no sets in $\mathcal{A}$ are subsets of other sets, $\mathcal{A}$ is reduced already, and the collection is called a *generating class*.

The hierarchical model subspaces are typically not direct sums of the factor subspaces in the generating class and therefore the logarithm of the expectation vector can be decomposed into terms from the factor spaces in many different ways. In this way various decompositions into interactions appear.

It follows that the permissible interactions for mean vectors in a hierarchical log–linear model with generating class $\mathcal{A}$ are exactly those among factors in b, where b is a subset of a for some $a \in \mathcal{A}$. Thus the generating class is a list of maximal permissible interactions. The term 'hierarchical' refers to the fact that if interactions are permitted between factors in b and if $c \subset b$, interactions will also be permitted between factors in c.

4.3.1 Estimation in hierarchical log–affine models

The likelihood equations

Consider a hierarchical log–affine model, i.e. a log–affine model of the form $M(q, H)$, where the space H is a hierarchical model subspace with generating class $\mathcal{A}$. We shall also use the notation $M(q, \mathcal{A})$ or, in the log–linear case where $q \equiv 1$, just $M(\mathcal{A})$. The general result about estimation in log–affine models can be used to obtain

Theorem 4.11 *The maximum likelihood estimate $\hat{m}$ of the mean vector in the extended hierarchical log–affine model is the unique element of $\bar{M}(q, \mathcal{A})$ that satisfies the likelihood equation*

$$\hat{m}(i_a) = n(i_a), \quad i_a \in \mathcal{I}_a, a \in \mathcal{A}. \tag{4.22}$$

This holds under Poisson as well as under multinomial sampling. If $b \subseteq a$ for some $a \in \mathcal{A}$, $\hat{m}$ is also the estimate under restricted multinomial sampling with fixed b-marginals.

Proof: From Lemma B.3 it follows that (4.22) is a translation of (4.13). Thus the result is a special case of Theorem 4.8. □

Theorem 4.11 identifies which equations are to be solved but gives no advice in doing so. In general the equations have to be solved by iterative methods. Below we shall describe one of these.

Iterative proportional scaling

This method is a special case of the method described in Section A.4 and also known as the Deming–Stephan algorithm (Deming and Stephan 1940) or the IPS-algorithm. The algorithm was described in detail in Darroch and Ratcliff (1972).

It consists of iteratively and successively adjusting the marginal counts appearing in (4.22).

Let a table of counts $\{n(i)\}_{i \in \mathcal{I}}$ be given and let $M = M(q, \mathcal{A})$ be a hierarchical log–affine model. Let M^* be the set of mean vectors such that $n(i) > 0 \implies m(i) > 0$. Denote finally the extended model by $\bar{M} = \bar{M}(q, \mathcal{A})$. Note that M^* depends on the observed counts and that

$$M^* \cap \bar{M} = \{m \in \bar{M} \mid L(m) > 0\}, \qquad (4.23)$$

where L is the likelihood function. Thus $\hat{m} \in M^*$.

Define for $m \in M^*$ and $a \in \mathcal{A}$ the operation of 'adjusting the a-marginal' by

$$(T_a m)(i) = m(i) \frac{n(i_a)}{m(i_a)}, \quad \text{where } 0(0/0) = 0.$$

Note that

$$(T_a m)(i_a) = n(i_a) \text{ and } |T_a m| = |n|. \qquad (4.24)$$

Some important properties of T_a are collected in the following lemma:

Lemma 4.12 *The transformation T_a satisfies for all $a \in \mathcal{A}$*

(i) T_a is continuous on M^*;

(ii) $T_a m$ is the uniquely determined maximum likelihood estimate of the mean in the extended hierarchical model $\bar{M}(m, \{a\})$;

(iii) $L(T_a m) \geq L(m)$ with equality if and only if $m(i_a) = n(i_a)$, for all $i_a \in \mathcal{I}_a$, which happens if and only if $T_a m = m$;

(iv) $T_a(M^* \cap \bar{M}) \subseteq M^* \cap \bar{M}$.

Proof: To show (i), let $(m_k) \in M^*$ with $m_k \to m \in M^*$. Then, if $m(i_a) \neq 0$ we trivially have $T_a m_k \to T_a m$. If $m(i_a) = 0$, $m(j) = 0$ for all j with $j_a = i_a$. But then, since $m \in M^*$, we must have $n(j) = 0$ for all such j, whereby $n(i_a) = 0$. Thus

$$(T_a m_k)(i) = 0 = (T_a m)(i) \quad \text{for all } k$$

and (i) follows.

The assertion (ii) of the lemma is a consequence of Theorem 4.11, since $T_a m \in \bar{M}(m, a)$ and $T_a m$ satisfies (4.24).

The statement (iii) follows directly from (ii). To show (iv), note that for $\epsilon > 0$ and for all $m \in M^*$ we have

$$(T_a m)(i) = \lim_{\epsilon \to 0} m(i) \frac{n(i_a) + \epsilon}{m(i_a) + \epsilon},$$

since $m(i_a) = 0 \implies m(i) = 0$. Thus, it follows that

$$T_a(M^* \cap M) \subseteq M^* \cap \bar{M}.$$

The continuity of T_a combined with (iii) and (4.23) now gives (iv). □

Note that the map T_a cannot in general be continuously extended to all of $\bar{M}$.

We are then ready to construct the iterative procedure for maximizing the likelihood function. Choose any ordering $(a_1, \ldots, a_k)$ of the generators in $\mathcal{A}$ and let $T_\nu = T_{a_\nu}$ for $\nu = 1, \ldots, k$. Choose further an arbitrary starting value $m_0 \in M$ and define recursively for $r = 0, 1, \ldots$

$$m_{r+1} = (T_1 \cdots T_k) m_r. \qquad (4.25)$$

Then we have

Theorem 4.13 *For any starting value $m_0 \in M$ it holds that*

$$\hat{m} = \lim_{r \to \infty} m_r.$$

Proof: We just have to realize that this is a special instance of iterative partial maximization, covered by Proposition A.3 of Section A.4. To do this, we let

$$\Theta = \{m \in \bar{M} \mid L(m) \geq L(m_0) \text{ and } |m| = |n|\}.$$

Then Θ is clearly compact and, by Lemma 4.12 and (4.24), T_i are all continuous transformations of Θ into itself. Further they maximize L over the sections

$$\Theta_i(m) = \bar{M}(m, \{a_i\}) \cap \Theta,$$

also by Lemma 4.12. Since we know already that the global maximum of L is uniquely determined, the result follows. □

Direct join of hierarchical models

When the generating class of a hierarchical log–linear model is the direct join (see Section 2.2) of generating classes, then the calculation of maximum likelihood estimates can usually be simplified. This has been exploited in a fundamental fashion in the program CoCo (Badsberg 1991, 1995).

Proposition 4.14 *Consider the extended hierarchical log–linear model with generating class $\mathcal{C}$, where $\mathcal{C}$ is the direct join of generating classes $\mathcal{A}$ and $\mathcal{B}$. Let $\hat{m}$ denote the maximum likelihood estimate of m in this model and* let $\hat{m}_1, \hat{m}_2$ be the corresponding estimates in $\bar{M}(\mathcal{A})$ and $\bar{M}(\mathcal{B})$ respectively. Then, with the usual convention $(0/0 = 0)$, we have

$$\hat{m}(i) = \frac{\hat{m}_1(i_A)\hat{m}_2(i_B)}{n(i_{A\cap B})}. \tag{4.26}$$

Proof: Let
$$\tilde{m}(i) = \frac{\hat{m}_1(i_A)\hat{m}_2(i_B)}{n(i_{A\cap B})}.$$

We clearly have $\tilde{m} \in \bar{M}$. According to Theorem 4.11 we just need to show that the marginals fit.

Since the join is direct, $\mathcal{A} \wedge \mathcal{B} = \{h\}$, where $h = A \cap B$ is contained in generators $a \in \mathcal{A}$ and $b \in \mathcal{B}$. Thus we must have

$$\hat{m}_1(i_{A\cap B}) = \hat{m}_2(i_{A\cap B}) = n(i_{A\cap B}).$$

Now let $c \in \mathcal{C}$. Then either $c \in \mathcal{A}$ or $c \in \mathcal{B}$. Suppose the former and calculate the marginal in steps by first adding over $B \setminus A$:

$$\tilde{m}(i_A) = \frac{\hat{m}_1(i_A)\hat{m}_2(i_{A\cap B})}{n(i_{A\cap B})} = \hat{m}_1(i_A).$$

Adding over $A \setminus c$ we then find

$$\tilde{m}(i_c) = \hat{m}_1(i_c) = n(i_c)$$

and the marginal to c fits. □

Note in particular from the proof that

$$\hat{m}(i_A) = \hat{m}_1(i_A) \quad \text{and} \quad \hat{m}(i_B) = \hat{m}_2(i_B). \tag{4.27}$$

A consequence of the proposition is the conditional independence

$$\{\hat{m}(i_A)\}_{i_A \in \mathcal{I}_A} \perp\!\!\!\perp \{\hat{m}(i_B)\}_{i_B \in \mathcal{I}_B} \mid \{\hat{m}(i_{A\cap B})\}_{i_{A\cap B} \in \mathcal{I}_{A\cap B}}, \tag{4.28}$$

which is a version of the hyper Markov property (Dawid and Lauritzen 1993).

4.3.2 Test in hierarchical models

Essentially the general discussion of the asymptotic theory for log–affine models in Section 4.2.3 applies here. If just the estimate under the smaller of two models is everywhere positive, the deviance will be approximately χ^2-distributed with degrees of freedom equal to

$$f = \dim H_{\mathcal{A}_0} - \dim H_{\mathcal{A}_1},$$

where $\mathcal{A}_0$ and $\mathcal{A}_1$ are the generating classes for the two models. Thus the technical problem is to calculate these dimensions. Below we shall develop recursive formulae for this.

Dimensions of hierarchical model subspaces

First we note that the factor subspaces have dimension equal to
$$\dim F_a = |\mathcal{I}_a| = \prod_{\delta \in a} |\mathcal{I}_\delta|.$$

Next let us examine how the spaces are recursively constructed. Let $b \subseteq \Delta$ and let
$$\mathcal{A}^* = \text{red}\{b \cap a \mid a \in \mathcal{A}\}.$$
Then, since $\mathcal{A}^* = \{b\} \wedge \mathcal{A}$ we get from Proposition B.5 that
$$F_b \cap H_\mathcal{A} = H_{\mathcal{A}^*}.$$

This gives the following recursive formula for calculating the dimension of hierarchical model subspaces:
$$\begin{aligned} \dim H_{\{b\} \cup \mathcal{A}} &= \dim F_b + \dim H_\mathcal{A} - \dim H_{\mathcal{A}^*} \\ &= \prod_{\delta \in b} |\mathcal{I}_\delta| + \dim H_\mathcal{A} - \dim H_{\mathcal{A}^*}. \end{aligned} \quad (4.29)$$

From Proposition B.5 we deduce that more generally we have that
$$\dim H_{\mathcal{A} \vee \mathcal{B}} = \dim H_\mathcal{A} + \dim H_\mathcal{B} - \dim H_{\mathcal{A} \wedge \mathcal{B}}.$$

Example 4.15 To illustrate the use of (4.29), consider four variables $\Delta = \{1, 2, 3, 4\}$ and
$$\mathcal{A} = \{\{1, 2\}, \{2, 3\}, \{1, 3, 4\}\}.$$
Letting $\mathcal{B} = \{\{2, 3\}, \{1, 3, 4\}\}$ and $b = \{1, 2\}$ we have $\mathcal{B}^* = \{\{1\}, \{2\}\}$ and
$$\dim H_\mathcal{A} = |\mathcal{I}_1||\mathcal{I}_2| + \dim H_\mathcal{B} - \dim H_{\mathcal{B}^*}$$

and further,
$$\begin{aligned} \dim H_\mathcal{A} &= |\mathcal{I}_1||\mathcal{I}_2| + (|\mathcal{I}_2||\mathcal{I}_3| + |\mathcal{I}_1||\mathcal{I}_3||\mathcal{I}_4| - |\mathcal{I}_3|) - (|\mathcal{I}_1| + |\mathcal{I}_2| - 1) \\ &= |\mathcal{I}_1||\mathcal{I}_2| + |\mathcal{I}_2||\mathcal{I}_3| + |\mathcal{I}_1||\mathcal{I}_3||\mathcal{I}_4| - |\mathcal{I}_1| - |\mathcal{I}_2| - |\mathcal{I}_3| + 1, \end{aligned}$$

where we have used that $\dim F_\emptyset = 1$. □

Partitioning the deviance

In the case when the models to be compared have generating classes which are direct joins of others, in such a way that the restrictions in the smaller of the two models are confined to each of the components in the join, the deviance statistic can be partitioned.

More precisely, consider the extended hierarchical log–linear models $\bar{M}_k = \bar{M}(\mathcal{C}_k), k = 0, 1$ where the generating classes $\mathcal{C}_k$ are the direct join of the generating classes $\mathcal{A}_k$ and $\mathcal{B}_k$ in such a way that their meet

$$\mathcal{A}_k \wedge \mathcal{B}_k = \{h\}$$

is independent of k. Then the decomposition of the maximum likelihood estimate in (4.26) implies that the deviance decomposes as

$$d = d_\mathcal{A} + d_\mathcal{B}, \qquad (4.30)$$

where d is the deviance for testing $\bar{M}_1$ under the assumption of $\bar{M}_0$, $d_\mathcal{A}$ is the deviance for testing $\bar{M}(\mathcal{A}_1)$ under $\bar{M}(\mathcal{A}_0)$ and $d_\mathcal{B}$ is the deviance for testing $\bar{M}(\mathcal{B}_1)$ under $\bar{M}(\mathcal{B}_0)$. This is seen by direct verification. It follows from (4.28) that $d_\mathcal{A}$ and $d_\mathcal{B}$ are independent in the conditional distribution used to judge them. A similar decomposition is only approximately true for the χ^2-statistic $\tilde{d}$.

The decomposition has several important features. Firstly it simplifies the calculation of the deviances. Secondly, each of the deviances in the decomposition can be assigned to a particular type of deviation. Finally, and perhaps most importantly, each of the deviance components is based on marginal tables with a smaller number of variables than the total deviance. These marginal tables will have larger expected cell counts, which generally has a positive effect on the quality of the χ^2-approximation.

Exact tests in hierarchical models

In general, exact methods for testing hypotheses concerning hierarchical models are not available unless the models are decomposable (see below). The problem is that it is difficult to compute or simulate from the conditional distribution of the contingency table given the marginals.

However, there are exceptional cases. For example, Morgan and Blumenstein (1991) give an algorithm that allows exact p-values to be computed for tests in hierarchical models. However, the algorithm is based upon complete enumeration of all possible cases. In large tables this will not in general be feasible.

Besag and Clifford (1989) describe a Markov chain Monte Carlo method for calculating p-values for testing the hypothesis of vanishing second-order interaction in a $2 \times S \times T$ table. This is based on making random changes to the table of counts that preserve the marginals. More precisely, consider a $2 \times 2 \times 2$ subtable obtained by varying each of r, s, and t at two levels. If 1 is added to any of these cells, the pairwise marginals can be preserved by adding or subtracting 1 in each of the other seven cells in a unique fashion. Choosing such 'invariant moves' with appropriate probabilities gives a simulated Markov chain that has the conditional distribution of interest as its unique equilibrium distribution.

In the case of a general $R \times S \times T$ table the invariant moves above are not sufficient to make the Markov chain irreducible (Glonek 1987) and therefore the p-values obtained from this simulation are not those desired.

Diaconis and Sturmfels (1994) demonstrate how to use computational algebraic geometry to find a complete system of invariant moves. For example, in a $3 \times 3 \times 3$ table, the moves of the type described need to be supplemented with moves that modify the margins as

0	0	0	0	+	-	0	-	+
0	0	0	-	0	+	+	0	-
0	0	0	+	-	0	-	+	0

There is a total of 54 such moves, obtained by renaming the various levels and variables.

The computation involved in finding these moves is quite heavy and even though the method so described is promising, it is not clear whether it will be feasible as a general practical method for computing p-values in hierarchical models.

4.3.3 Interaction graphs and graphical models

Consider a hierarchical log–linear model $M(\mathcal{A})$ and assume as usual that all main effects are included, i.e.

$$\Delta = \cup_{a \in \mathcal{A}} a.$$

With this model we associate as in Section 2.2.2 the undirected graph $\mathcal{G} = \mathcal{G}(\mathcal{A})$ with edges given as

$$\delta_1 \sim \delta_2 \iff \{\delta_1, \delta_2\} \subseteq a \quad \text{for some } a \in \mathcal{A}.$$

In words, two variables are neighbours in the graph if and only if interaction is permitted between the variables; cf. Section B.2. This graph is called the *interaction graph* of the hierarchical model.

In the multinomial sampling case this gives rise to a particular interpretation of the hierarchical model. More precisely, Theorem 3.9 combined with Theorem 3.7 ensures that the interaction graph of a hierarchical model can be used to read off conditional independences implied by the model: whenever two groups of variables A and B are separated by the variables S in this interaction graph, they will be conditionally independent given the variables in S. From Proposition 3.12 it follows that this will continue to hold also in the extended hierarchical model $\bar{M}$.

In particular, different connected components of the interaction graph $\mathcal{G}(\mathcal{A})$ correspond to variables that are mutually independent.

On the other hand, different hierarchical models may well have the same interaction graph. Consider for example the cases $\Delta = \{1, 2, 3\}$ with $\mathcal{A}_1 = \{\{1, 2, 3\}\}$ and $\mathcal{A}_2 = \{\{1, 2\}, \{1, 3\}, \{2, 3\}\}$. Both have the complete 3-graph as their interaction graph and no conditional independences apply.

Thus the interaction graph does not in general give a complete picture of the structure of a hierarchical model. Models where it does are termed *graphical interaction models*. More precisely, a hierarchical model is *graphical* if its generating class is a graphical hypergraph, i.e. if its generating class $\mathcal{A}$ exactly consists of the cliques of its interaction graph. Thus $\mathcal{A}_1$ above is graphical whereas $\mathcal{A}_2$ is not.

A graphical model is therefore particularly simple to interpret. Its interaction graph is an example of such an interpretation, displaying all conditional independences through graph separation.

If a model is not graphical, its interaction graph $\mathcal{G}(\mathcal{A})$ is still helpful for the interpretation and the graphical generating class $\mathcal{A}^* = \mathcal{C}(\mathcal{G}(\mathcal{A}))$ is generating class for the smallest graphical model containing $M(\mathcal{A})$.

Note that saturated models are graphical models with the complete graph as their interaction graph.

In the special case of direct joins we obtain the useful

Lemma 4.16 *If the generating class $\mathcal{C}$ of a hierarchical model is the direct join of generating classes $\mathcal{A}$ and $\mathcal{B}$ with $\mathcal{A} \wedge \mathcal{B} = \{A \cap B\}$, we have that*

$$m(i) = \frac{m(i_A)m(i_B)}{m(i_{A \cap B})} \qquad (4.31)$$

for all $m \in \bar{M}(\mathcal{C})$.

Proof: From Lemma 2.23 we have that the direct join corresponds to a decomposition of the interaction graphs. Proposition 3.16 gives then the factorization of the corresponding extended Markov probabilities. The result now follows from the relation $m(i) = |n|p(i)$. □

Example 4.17 The model for conditional independence in a three-way table described in Example 4.6 is a hierarchical model with generating class $\mathcal{C} = \{\{R, S\}, \{S, T\}\}$. Since this generating class is the direct join of two simple generating classes

$$\mathcal{C} = \{\{R, S\}\} \vee \{\{S, T\}\},$$

(4.31) gives the factorization

$$m_{rst} = \frac{m_{rs+}m_{+st}}{m_{+s+}},$$

which is essentially identical to (4.12). □

4.4 Decomposable models

In the present section we study the special features of graphical interaction models whose interaction graphs are decomposable. We shall always work in the extended models since we are able to obtain exact results throughout, limiting the need for asymptotic theory. Such models are called *decomposable* graphical models.

Theorem 2.25 implies that these models are built up from saturated models by successive direct joins. In other words, a decomposable model is an extended hierarchical log–linear model whose generating class is a decomposable hypergraph.

4.4.1 Basic factorizations

In Proposition 3.18 a formula (3.18) was given that describes how the extended Markov property behaves across decompositions of the graph. Using this factorization an especially simple expression for the vector of means can be obtained, as we shall now show.

Consider a graphical model with a decomposable graph. As shown in Proposition 2.17 we can number the cliques to form a perfect sequence, i.e. a sequence $C_1, \ldots, C_k$ where each combination of subgraphs induced by $H_{j-1} = C_1 \cup \cdots \cup C_{j-1}$ and C_j is a decomposition. Repeated use of (3.18) gives the formula

$$p(i) = \frac{\prod_{j=1}^k p(i_{C_j})}{\prod_{j=2}^k p(i_{S_j})}, \qquad (4.32)$$

where $S_j = H_{j-1} \cap C_j$ is the sequence of separators. Alternatively we can collect the terms in the denominator in groups and obtain the formula

$$p(i) = \frac{\prod_{C \in \mathcal{C}} p(i_C)}{\prod_{S \in \mathcal{S}} p(i_S)^{\nu(S)}}, \qquad (4.33)$$

where $\nu(S)$ is an index that counts the number of times a given separator S occurs in a perfect sequence. Here it is important to remember to count the empty set as a separator, whenever this appears. Rewriting this in terms of the expectation vector yields

$$m(i) = \frac{\prod_{j=1}^k m(i_{C_j})}{\prod_{j=2}^k m(i_{S_j})} \qquad (4.34)$$

or, alternatively,

$$m(i) = \frac{\prod_{C \in \mathcal{C}} m(i_C)}{\prod_{S \in \mathcal{S}} m(i_S)^{\nu(S)}}. \qquad (4.35)$$

The two last formulae then hold for decomposable models under all three basic sampling situations.

The factorization of means induces a structural decomposition of the joint density of the entire contingency table of counts. By direct calculation we find that the sampling distribution of the table of counts in the Poisson case has probability function

$$e^{-|m|} \left\{ \prod_{i \in \mathcal{I}} \frac{1}{n(i)!} \right\} \frac{\prod_{C \in \mathcal{C}} \prod_{i_C \in \mathcal{I}_C} m(i_C)^{n(i_C)}}{\prod_{S \in \mathcal{S}} \prod_{i_S \in \mathcal{I}_S} m(i_S)^{n(i_S)\nu(S)}}. \qquad (4.36)$$

Similarly, in the case of multinomial sampling the counts have density equal to

$$\frac{|n|!}{|n|^{|n|}} \left\{ \prod_{i \in \mathcal{I}} \frac{1}{n(i)!} \right\} \frac{\prod_{C \in \mathcal{C}} \prod_{i_C \in \mathcal{I}_C} m(i_C)^{n(i_C)}}{\prod_{S \in \mathcal{S}} \prod_{i_S \in \mathcal{I}_S} m(i_S)^{n(i_S)\nu(S)}} \qquad (4.37)$$

and in the case of restricted multinomial sampling with fixed b-marginals we obtain

$$\prod_{i_b \in \mathcal{I}_b} \frac{n(i_b)!}{n(i_b)^{n(i_b)}} \left\{ \prod_{i \in \mathcal{I}} \frac{1}{n(i)!} \right\} \frac{\prod_{C \in \mathcal{C}} \prod_{i_C \in \mathcal{I}_C} m(i_C)^{n(i_C)}}{\prod_{S \in \mathcal{S}} \prod_{i_S \in \mathcal{I}_S} m(i_S)^{n(i_S)\nu(S)}}. \qquad (4.38)$$

4.4.2 Maximum likelihood estimation

Exact results

In Proposition 4.14 we derived a formula for combining maximum likelihood estimates of mean vectors in two hierarchical log–linear models to find the estimate in the model formed by their direct join.

Combining this with the fact that in the saturated models the estimates are simply given as $\hat{m}(i) = n(i)$, it follows as above that similar explicit formulae for the maximum likelihood estimate in a decomposable graphical model exist. More precisely we find by repeated use of (4.26) that

$$\hat{m}(i) = \frac{\prod_{j=1}^{k} n(i_{C_j})}{\prod_{j=2}^{k} n(i_{S_j})}. \qquad (4.39)$$

Using the alternative formula where the separators and cliques are not numbered, we get

Proposition 4.18 *In a decomposable graphical model with graph $\mathcal{G}$ the maximum likelihood estimate of the mean vector is given as*

$$\hat{m}(i) = \frac{\prod_{C \in \mathcal{C}} n(i_C)}{\prod_{S \in \mathcal{S}} n(i_S)^{\nu(S)}}, \qquad (4.40)$$

where $\mathcal{C}$ is the set of cliques of $\mathcal{G}$ and $\mathcal{S}$ the set of separators with multiplicities ν in any perfect sequence.

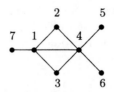

Fig. 4.1. Graph of a decomposable model for a 7-dimensional table.

Example 4.19 Consider the case where $\Delta = \{\alpha, \beta\}$, $E = \emptyset$ and $\mathcal{I}_\alpha = \mathcal{I}_\beta = \{1, 2\}$. Assume that the counts N_{rs} are independent and Poisson-distributed and that we have observations $N_{rs} = n_{rs}$. Thus we are considering the multiplicative Poisson model for a 2×2 table.

$$\mathbf{E} N_{rs} = m_{rs} = \frac{m_{r+} m_{+s}}{m_{++}}, \quad r, s = 1, 2.$$

The set of cliques is $\mathcal{C} = \{\{\alpha\}, \{\beta\}\}$ and $\mathcal{S} = \{\emptyset\}$ with $\nu(\emptyset) = 1$. We get from (4.40) that

$$\hat{m}_{rs} = \frac{n_{r+} n_{+s}}{|n|}.$$

This formula is of course well known. □

Example 4.20 As a more complicated example, consider the graphical model for a 7-dimensional contingency table with the graph in Fig. 4.1. We have $\Delta = \{1, 2, 3, 4, 5, 6, 7\}$ and the five cliques are

$$C_1 = \{1, 2, 4\},\ C_2 = \{1, 3, 4\}, C_3 = \{4, 5\},\ C_4 = \{4, 6\},\ C_5 = \{1, 7\},$$

where the numbering is perfect with separators

$$S_2 = \{1, 4\},\ S_3 = S_4 = \{4\},\ S_5 = \{1\}.$$

Thus the multiplicities are

$$\nu(\{1, 4\}) = \nu(\{1\}) = 1,\ \nu(\{4\}) = 2.$$

We therefore find that the mean factorizes as

$$m_{rstuvwz} = \frac{m_{rs+u+++}\, m_{r+tu+++}\, m_{+++uv++}\, m_{+++u+w+}\, m_{r+++++z}}{m_{r++u+++}\, (m_{+++u+++})^2\, m_{r++++++}}.$$

The estimate $\hat{m}$ factorizes in the same way, just that m should be replaced with n everywhere on the right-hand side. □

In the decomposable case we can obtain a closed form expression for the distribution of the maximum likelihood estimate. Observe first that there is a one-to-one correspondence between $\hat{m}$ and the set of clique marginals, so the problem is identical to finding the joint distribution of these. From (4.36), (4.37) and (4.38) it follows that the essential problem is to find the sum

$$f\Big(\big(n(i_C), i_C \in \mathcal{I}_C\big)_{C\in\mathcal{C}}\Big) = \sum_{x: x(i_C) = n(i_C), i_C \in \mathcal{I}_C, C \in \mathcal{C}} \prod_{i\in\mathcal{I}} \frac{1}{x(i)!}.$$

This combinatorial problem was first solved by Sundberg (1975), who showed the following

Lemma 4.21 *For any table of counts $\big(n(i)\big)_{i\in\mathcal{I}}$ and any decomposable graph with cliques $\mathcal{C}$ it holds that*

$$f\Big\{\big(n(i_C), i_C \in \mathcal{I}_C\big)_{C\in\mathcal{C}}\Big\} = \frac{\prod_{S\in\mathcal{S}}\big\{\prod_{i_S \in \mathcal{I}_S} n(i_S)!\big\}^{\nu(S)}}{\prod_{C\in\mathcal{C}} \prod_{i_C \in \mathcal{I}_C} n(i_C)!}. \tag{4.41}$$

The easiest way of proving the lemma is to find the distribution of the maximum likelihood estimate in the Poisson case. The lemma is in fact equivalent to

Theorem 4.22 *In a decomposable graphical model the distribution of the maximum likelihood estimate under Poisson sampling has density*

$$e^{-|m|}\frac{\prod_{C\in\mathcal{C}}\prod_{i_C\in\mathcal{I}_C} m(i_C)^{n(i_C)}/n(i_C)!}{\prod_{S\in\mathcal{S}}\prod_{i_S\in\mathcal{I}_S}\big(m(i_S)^{n(i_S)}/n(i_S)!\big)^{\nu(S)}}. \tag{4.42}$$

The corresponding expressions for the density in the other sampling cases are obtained by replacing the factor $e^{-|m|}$ by $\frac{|n|!}{|n|^{|n|}}$ and $\prod_{i_b} n(i_b)!/n(i_b)^{n(i_b)}$ respectively.

Proof: It is clearly enough to consider the Poisson case. The proof is by induction on the number $|\mathcal{C}|$ of cliques in $\mathcal{G}$. Consider first a saturated model. Here the estimate is the set of counts itself, there are no separators and (4.42) reduces to

$$e^{-|m|}\prod_{i\in\mathcal{I}} m(i)^{n(i)}/n(i)!,$$

which is exactly the density of the Poisson distribution.

In the non-saturated case we order the cliques of $\mathcal{G}$ to to form a perfect sequence (see Section 2.1.3) and let $\mathcal{A} = \{C_1, \ldots, C_{k-1}\}$ and $\mathcal{B} = \{C_k\}$. Further let $S = S_k = A \cap B$. From (4.28) we learn that the estimates of the marginal probabilities $\hat{m}_A$ and $\hat{m}_B$ are conditionally independent given the marginal counts $\big(n(i_S)\big)$. Since the model determined by $\mathcal{A}$ is a decomposable model with one clique fewer, we obtain by the induction assumption that

$$P(\hat{m}_A) = e^{-m(i_A)} \frac{\prod_{j=1}^{k-1} \prod_{i_{C_j} \in \mathcal{I}_{C_j}} m(i_{C_j})^{n(i_{C_j})}/n(i_{C_j})!}{\prod_{j=2}^{k-1} \prod_{i_{S_j} \in \mathcal{I}_{S_j}} \left(m(i_{S_j})^{n(i_{S_j})}/n(i_{S_j})!\right)},$$

where $P(\hat{m}_A)$ has been used as generic symbol for the density of $\hat{m}_A$. Since the model determined by $\mathcal{B}$ is saturated, we also have

$$P(\hat{m}_B) = e^{-m(i_B)} \prod_{i_B \in \mathcal{I}_B} m(i_B)^{n(i_B)}/n(i_B)!$$

as well as

$$P(n_S) = e^{-m(i_S)} \prod_{i_S \in \mathcal{I}_S} m(i_S)^{n(i_S)}/n(i_S)!.$$

The conditional independence then implies that

$$P(\hat{m}_A, \hat{m}_B, n_S) = P(\hat{m}_A)P(\hat{m}_B)/P(n_S).$$

The result finally follows by collecting terms and noting that $\hat{m}$ is in a one-to-one correspondence with the triple $(\hat{m}_A, \hat{m}_B, n_S)$. □

The explicit expression for the distribution of the maximum likelihood estimate is useful, but perhaps it is even more interesting to note that the distribution obeys fundamental conditional independences.

As used in the proof, it follows from (4.28) that whenever (A, B, S) form a decomposition of $\mathcal{G}$ we have

$$\{\hat{m}(i_{A \cup S})\} \perp\!\!\!\perp \{\hat{m}(i_{B \cup S})\} \mid \{\hat{m}(i_S)\}, \tag{4.43}$$

where each expression inside curly brackets stands for the entire collection of corresponding quantities. It can be shown (Dawid and Lauritzen 1993) that (4.43) also holds when just S separates A from B. This property of the distribution of the maximum likelihood estimate is the so-called *hyper Markov property*.

Consequently the distribution with density (4.42) is called the *hyper Poisson distribution*. In the multinomial sampling case we correspondingly have that the maximum likelihood estimate follows a *hyper multinomial distribution*. Under restricted multinomial sampling the distribution is the product of hyper multinomial distributions.

Example 4.23 Continuation of Example 4.19. In the 2×2 case considered earlier, (4.42) becomes

$$e^{-m_{++}} \frac{\left\{(m_{1+}^{n_{1+}}/n_{1+}!)(m_{2+}^{n_{2+}}/n_{2+}!)\right\}\left\{(m_{+1}^{n_{+1}}/n_{+1}!)(m_{+2}^{n_{+2}}/n_{+2}!)\right\}}{m_{++}^{|n|}/|n|!}. \tag{4.44}$$

The hyper Markov property (4.43) expresses that the two sets of marginals are conditionally independent given the total

$$\{N_{1+}, N_{2+}\} \perp\!\!\!\perp \{N_{+1}, N_{+2}\} \mid N_{++},$$

which, of course, can also be seen directly from (4.44). □

Asymptotic results

Even though exact results are available in the decomposable case, it is sometimes convenient to have the asymptotic results as well. These also are simple and explicit in the decomposable case. We shall first derive the asymptotic distribution of $\hat{m}$ using (4.15). Thus the essential problem is to find a convenient expression for the projection onto the hierarchical model subspace with respect to the inner product determined by v, which in the Poisson case is a diagonal matrix with entries $m(i)$.

First, let Π_a^v denote the orthogonal projection onto the factor subspace F_a with respect to v and, similarly $\Pi_{\mathcal{A}}^v$ onto the hierarchical model subspace $H_{\mathcal{A}}$. We want to find $\Pi_{\mathcal{C}}^v$. Clearly, the projection Π_a^v is given as the weighted average

$$(\Pi_a^v x)(i) = \frac{(\Pi_a m x)(i)}{(\Pi_a m)(i)} = \frac{\sum_{j:j_a = i_a} m(j) x(j)}{m(i_a)}. \tag{4.45}$$

In general it is not true that these weighted projections commute as in (B.18). However, in the case of direct joins of generating classes we have

Lemma 4.24 *Assume that the generating class $\mathcal{C}$ is the direct join of generating classes $\mathcal{A}$ and $\mathcal{B}$ with $\mathcal{A} \wedge \mathcal{B} = \{A \cap B\}$. Then, if $m \in \bar{M}(\mathcal{C})$,*

$$\Pi_{\mathcal{A}}^v \Pi_{\mathcal{B}}^v = \Pi_{\mathcal{B}}^v \Pi_{\mathcal{A}}^v = \Pi_{\mathcal{A} \cap \mathcal{B}}^v. \tag{4.46}$$

Proof: We first show that Π_A^v and Π_B^v commute using the explicit expression (4.45) for these as well as the factorization (4.31) of m. We find

$$\sum_{j:j_B = i_B} m(j) (\Pi_A^v u)(j) = \sum_{j:j_B = i_B} \frac{m(j_A) m(j_B)}{m(j_{A \cap B})} \frac{1}{m(j_A)} \sum_{k:k_A = j_A} m(k) u(k)$$

$$= \frac{m(i_B)}{m(i_{A \cap B})} \sum_{k:k_{A \cap B} = i_{A \cap B}} m(k) u(k)$$

$$= m(i_B) (\Pi_{A \cap B}^m u)(i).$$

In particular we have that

$$\Pi_A^v \Pi_B^v = \Pi_B^v \Pi_A^v = \Pi_{A \cap B}^v.$$

In the general case we first realize that $H_{\mathcal{A}} \subseteq F_A$ and therefore

$$\Pi_{\mathcal{A}}^v = \Pi_A^v \Pi_{\mathcal{A}}^v = \Pi_{\mathcal{A}}^v \Pi_A^v.$$

In fact this would be true for orthogonal projections with respect to any inner product. We then finally use the calculation

$$\Pi_A^v \Pi_B^v = \Pi_A^v \Pi_A^v \Pi_B^v \Pi_B^v = \Pi_A^v \Pi_{A\cap B}^v \Pi_B^v = \Pi_A^v \Pi_{A\cap B}^v = \Pi_{A\cap B}^v$$

and the lemma is established. □

An important consequence of this result is the formula

$$\Pi_C^v = \Pi_A^v + \Pi_B^v - \Pi_{A\cap B}^v. \qquad (4.47)$$

Note that it is not essential that $m \in \bar{M}(\mathcal{C})$. In fact in the proof we have only used a basic factorization of m, i.e. that $m \in \bar{M}(\{A, B\})$.

As a corollary to the lemma we obtain an explicit formula for the weighted projection in the decomposable case:

Corollary 4.25 *Let $\mathcal{C}$ be the cliques of a decomposable graph $\mathcal{G}$ and assume that $m \in \bar{M}(\mathcal{G})$. Then*

$$\Pi_\mathcal{C}^v = \sum_{C \in \mathcal{C}} \Pi_C^v - \sum_{S \in \mathcal{S}} \nu(S) \Pi_S^v. \qquad (4.48)$$

Proof: The result follows by ordering the cliques to form a perfect sequence and using (4.47) recursively. □

This expression enables us to find the asymptotic covariance matrix for $\hat{m}$. First we find by direct calculation from (4.45) that

$$\{(\Pi_a^v)^\top v(\Pi_a^v)\}_{ij} = \begin{cases} \dfrac{m(i)m(j)}{m(i_a)} & \text{if } i_a = j_a \\ 0 & \text{otherwise.} \end{cases}$$

Defining the symbols

$$\mathcal{C}[i,j] = \{C \in \mathcal{C} \mid i_C = j_C\}, \quad \mathcal{S}[i,j] = \{S \in \mathcal{S} \mid i_S = j_S\},$$

we get from (4.48) that under Poisson sampling

$$\overset{a}{\mathbf{V}}(\hat{m}(i), \hat{m}(j)) = m(i)m(j) \left\{ \sum_{C \in \mathcal{C}[i,j]} \frac{1}{m(i_C)} - \sum_{S \in \mathcal{S}[i,j]} \frac{\nu(S)}{m(i_S)} \right\}. \qquad (4.49)$$

As we have seen in (4.21) we also have

$$(\Pi_\mathcal{C}^v)^\top (m \otimes m)(\Pi_\mathcal{C}^v) = m \otimes m$$

such that, in the case of multinomial sampling,

$$\overset{a}{\mathbf{V}}_0 (\hat{m}(i), \hat{m}(j)) = \overset{a}{\mathbf{V}} (\hat{m}(i), \hat{m}(j)) - \frac{m(i)m(j)}{|n|} \qquad (4.50)$$

and when sampling with fixed b-marginals,

$$\overset{a}{V}_b\left(\hat{m}(i), \hat{m}(j)\right) = \begin{cases} \overset{a}{V}\left(\hat{m}(i), \hat{m}(j)\right) - \frac{m(i)m(j)}{n(i_b)} & \text{if } i_b = j_b \\ 0 & \text{otherwise.} \end{cases} \quad (4.51)$$

In the above formulae we notice in particular that if $i = j$ then $\mathcal{C}[i,j] = \mathcal{C}$ and $\mathcal{S}[i,j] = \mathcal{S}$. Thus, for example, the asymptotic variance in the Poisson case becomes simply

$$\overset{a}{V}\left(\hat{m}(i)\right) = m(i)^2 \left\{ \sum_{C \in \mathcal{C}} \frac{1}{m(i_C)} - \sum_{S \in \mathcal{S}} \frac{\nu(S)}{m(i_S)} \right\}.$$

Example 4.26 Continuation of Example 4.23. If we consider the case of a 2×2 table under Poisson sampling, we get for the diagonal elements of the asymptotic covariance matrix of $\hat{m}$ that

$$\overset{a}{V}(\hat{m}_{rs}) = (m_{rs})^2 \left(\frac{1}{m_{r+}} + \frac{1}{m_{+s}} - \frac{1}{m_{++}} \right) = \frac{m_{rs}(m_{r+} + m_{+s} - m_{rs})}{m_{++}}.$$

Similarly, for some of the covariances we get

$$\overset{a}{V}(\hat{m}_{11}, \hat{m}_{21}) = m_{11} m_{21} \left(\frac{1}{m_{+1}} - \frac{1}{m_{++}} \right), \quad \overset{a}{V}(\hat{m}_{21}, \hat{m}_{12}) = -\frac{m_{21} m_{12}}{m_{++}}.$$

Note that these are the asymptotic covariances and different from the exact ones. □

Example 4.27 Continuation of Example 4.20. In the more complex example of the 7-dimensional table, we get for the asymptotic variance in the multinomial sampling case

$$\frac{\overset{a}{V}(\hat{m}_{rstuvwz})}{(m_{rstuvwz})^2} = (m_{rs+u+++})^{-1} + (m_{r+tu+++})^{-1} + (m_{+++uv++})^{-1}$$
$$+ (m_{+++u+w+})^{-1} + (m_{r+++++z})^{-1}$$
$$- \left\{ (m_{r++u+++})^{-1} + 2(m_{+++u+++})^{-1} + (m_{r++++++})^{-1} \right\} - |n|^{-1}$$

and for the covariances for example

$$\frac{\overset{a}{V}(\hat{m}_{1111111}, \hat{m}_{2111121})}{m_{1111111} m_{2111121}} = (m_{+++11++})^{-1} - 2(m_{+++1+++})^{-1} - |n|^{-1}$$

since $\mathcal{C}[1111111, 2111121] = \{4, 5\}$ and $\mathcal{S}[1111111, 2111121] = \{4\}$. □

To obtain the asymptotic covariances of the canonical parameter $\log \hat{m}$ one can either apply the delta-method using $d\log x/dx = x^{-1}$ or repeat the earlier calculations. We find in the Poisson case

$$\overset{a}{V}\left(\log \hat{m}(i), \log \hat{m}(j)\right) = \sum_{C \in \mathcal{C}[i,j]} \frac{1}{m(i_C)} - \sum_{S \in \mathcal{S}[i,j]} \frac{\nu(S)}{m(i_S)} \qquad (4.52)$$

and in the other cases

$$\overset{a}{V}_0 \left(\log \hat{m}(i), \log \hat{m}(j)\right) = \overset{a}{V}\left(\log \hat{m}(i), \log \hat{m}(j)\right) - |n|^{-1} \qquad (4.53)$$

$$\overset{a}{V}_b \left(\log \hat{m}(i), \log \hat{m}(j)\right) = \overset{a}{V}\left(\log \hat{m}(i), \log \hat{m}(j)\right) - n(i_b)^{-1}, (4.54)$$

the latter only when $i_b = j_b$, otherwise they are equal to zero.

4.4.3 Exact tests in decomposable models

The $R \times S$ table

Before we deal with the general case we study in some detail the problem of testing independence in a two-way table under multinomial sampling. As we show later, the general testing problem can be reduced to a combination of testing problems of this kind.

In an $R \times S$ table we have counts $N_{rs} = n_{rs}$ and under the assumption of independence of the criteria we have as in Example 4.19 that

$$\hat{m}_{rs} = \frac{n_{r+}n_{+s}}{|n|}, \quad r = 1, \ldots, R, \; s = 1, \ldots, S.$$

An exact test appears by judging the test statistic in the conditional distribution of the counts given the set of sufficient marginals. From (4.37) and (4.41) we find that the conditional distribution has density

$$P\left\{(n_{rs}) \mid (n_{r+}), (n_{+s})\right\} = \frac{\prod_{r=1}^{R} n_{r+}! \prod_{s=1}^{S} n_{+s}!}{|n|! \prod_{r=1}^{R} \prod_{s=1}^{S} n_{rs}!}. \qquad (4.55)$$

In the 2×2 table this reduces to the hypergeometric distribution

$$P\left\{(n_{11}, n_{12}, n_{21}, n_{22}) \mid (n_{1+}, n_{2+}), (n_{+1}, n_{+2})\right\} = \frac{\binom{n_{1+}}{n_{11}}\binom{n_{2+}}{n_{21}}}{\binom{|n|}{n_{+1}}}.$$

Even though we have this explicit expression for the conditional density, the actual evaluation of test probabilities is not so straightforward. To get the rest of the way we must further factorize (4.55). For the partial $k \times l$ tables for $2 \leq k \leq R$ and $2 \leq l \leq S$ consisting of the 'upper left corners' of the original table we let

$$n_{r+}^l = \sum_{s=1}^{l} n_{rs}, \quad n_{+s}^k = \sum_{r=1}^{k} n_{rs}, \quad n_{++}^{kl} = \sum_{r=1}^{k} \sum_{s=1}^{l} n_{rs}$$

be the corresponding row, column and total sums. We first quote a useful result obtained by S. Johansen, see Martin-Löf (1970).

Lemma 4.28 *The conditional distribution of the entries in the partial $k \times l$ table given the marginals $(n_{r+}), (n_{+s})$ has density*

$$\frac{1}{\prod_{r=1}^{k} \prod_{s=1}^{l} n_{rs}!} \prod_{r=1}^{k} \frac{n_{r+}!}{(n_{r+} - n_{r+}^{l})!} \prod_{s=1}^{l} \frac{n_{+s}!}{(n_{+s} - n_{+s}^{k})!}$$

$$\times \frac{(n_{++} - n_{++}^{kS})! \, (n_{++} - n_{++}^{Rl})!}{|n|! \, (|n| - n_{++}^{kS} - n_{++}^{Rl} + n_{++}^{kl})!}. \quad (4.56)$$

Proof: It is clearly enough to consider the Poisson case and we can also assume that $m_{rs} \equiv 1$.

First we partition the table into the four partial tables that appear as corners of the full table of which the upper left is the partial $k \times l$ table. Conditioning on the totals for the three remaining tables

$$(n_{++}^{kS} - n_{++}^{kl}, n_{++}^{Rl} - n_{++}^{kl}, |n| - n_{++}^{kS} - n_{++}^{Rl} + n_{++}^{kl})$$

and exploiting the conditional independence of the marginals given the totals, we find the joint conditional density of

$$\left\{ (N_{rs})_{r,s=1}^{k,l}, (N_{r+} - N_{r+}^{l})_{r=1}^{k}, (N_{+s} - N_{+s}^{k})_{s=1}^{l}, (N_{r+})_{r=k+1}^{R}, (N_{+s})_{s=l+1}^{S} \right\}$$

to be equal to

$$\frac{e^{-kl}}{\prod_{r=1}^{k} \prod_{s=1}^{l} n_{rs}!} \frac{(n_{++}^{kS} - n_{++}^{kl})! \, k^{(n_{++}^{kl} - n_{++}^{kS})}}{\prod_{r=1}^{k}(n_{r+} - n_{r+}^{l})!} \frac{(n_{++}^{Rl} - n_{++}^{kl})! \, l^{(n_{++}^{kl} - n_{++}^{Rl})}}{\prod_{s=1}^{l}(n_{+s} - n_{+s}^{k})!}$$

$$\times \frac{(|n| - n_{++}^{kS})! \, (R-k)^{(n_{++}^{kS} - |n|)}}{\prod_{r=k+1}^{R} n_{r+}!} \frac{(|n| - n_{++}^{Rl})! \, (S-l)^{(n_{++}^{Rl} - |n|)}}{\prod_{s=l+1}^{S} n_{+s}!}.$$

We now multiply with the joint density of the totals, which is equal to

$$\frac{(k(S-l))^{(n_{++}^{kS} - n_{++}^{kl})} e^{-k(S-l)}}{(n_{++}^{kS} - n_{++}^{kl})!} \frac{((R-k)l)^{(n_{++}^{Rl} - n_{++}^{kl})} e^{-(R-k)l}}{(n_{++}^{Rl} - n_{++}^{kl})!}$$

$$\times \frac{((R-k)(S-l))^{(|n| - n_{++}^{kS} - n_{++}^{Rl} + n_{++}^{kl})} e^{-(R-k)(S-l)}}{(|n| - n_{++}^{kS} - n_{++}^{Rl} + n_{++}^{kl})!},$$

to obtain the joint distribution of counts in the $k \times l$ table and marginals in the full table. Dividing by the joint density of the marginals obtained from (4.42) as

$$e^{-RS} \frac{|n|!}{\prod_{r=1}^{R} n_{r+}! \prod_{s=1}^{S} n_{+s}!}$$

and subsequently reducing, we obtain (4.56). This completes the proof of the lemma. □

Consider then for each k and l the 2×2 table with entries and marginals

$$
\begin{array}{cc|c}
n_{++}^{k-1,l-1} & n_{+l}^{k-1} & n_{++}^{k-1,l} \\
n_{k+}^{l-1} & n_{kl} & n_{k+}^{l} \\
\hline
n_{++}^{k,l-1} & n_{+l}^{k} & n_{++}^{kl}
\end{array}
\qquad (4.57)
$$

and let P_{kl} denote the corresponding conditional probabilities obtained from (4.55) as

$$ P_{kl} = \frac{n_{++}^{k-1,l}!\, n_{k+}^{l}!\, n_{++}^{k,l-1}!\, n_{+l}^{k}!}{n_{++}^{kl}!\, n_{++}^{k-1,l-1}!\, n_{+l}^{k-1}!\, n_{k+}^{l-1}!\, n_{kl}!}. $$

Then we have

Lemma 4.29 *The factor P_{kl} is the conditional probability that N_{kl} is equal to n_{kl} given the marginals $(n_{r+}), (n_{+s})$, the counts n_{kj} with $j = l+1, \ldots, S$ as well as n_{ij} with $i = k+1, \ldots, R$ and $j = 1, \ldots, S$.*

Proof: First we realize that conditioning with marginal totals of the full table as well as the counts in the lower $(R - k) \times S$ table is equivalent to conditioning with the marginals in the upper $k \times S$ table. The conditional density of $N_{kj}, j = s, \ldots, S$ given these marginals can therefore be found from (4.56) to be

$$ f(s) = \frac{1}{\prod_{j=s}^{S} n_{kj}!} \frac{n_{k+}!}{n_{k+}^{s-1}!} \prod_{j=s}^{S} \frac{n_{+j}^{k}!}{n_{+j}^{k-1}!} \cdot \frac{n_{++}^{k-1,S}!\, n_{++}^{k,s-1}!}{n_{++}^{kS}!\, n_{++}^{k-1,s-1}!}. $$

We find $P_{kS} = f(S)$ and for $l < S$ we have $P_{kl} = f(l)/f(l+1)$. This proves the result. □

As a consequence we get the factorization

$$ P = \prod_{k=2}^{R} \prod_{l=2}^{S} P_{kl}. $$

which can also be obtained by direct calculation. Similarly, if we let d_{kl} be the deviance for the hypothesis of independence, calculated in the 2×2 tables displayed in (4.57), we have

$$ d = \sum_{k=2}^{R} \sum_{l=2}^{S} d_{kl}, $$

also by elementary calculations. We shall refer to these partitionings of the deviance and probability density as the *Lancaster partitioning* since they were used first by Lancaster (1949). The components d_{kl} are the *Lancaster components*.

One way of carrying out the exact deviance test is then to partition the test into $(R-1)(S-1)$ exact tests in the 2×2 tables above and consider each of these separately.

An alternative way of exploiting the factorization is to use Lemma 4.29 for computer generation of random tables with the correct distribution. Random entries of the table are generated row by row from the lower right corner and up according to P_{kl}. This Monte Carlo algorithm is due to Patefield (1981).

Based on the computer-generated random tables, the test probabilities for any test statistic, including the deviance, can be calculated to a desired degree of accuracy.

This technique has been exploited in the programs MIM (Edwards 1995), DIGRAM (Kreiner 1989), and CoCo (Badsberg 1991, 1995), where simulated exact probabilities are calculated for several test statistics, in particular for the deviance, the χ^2-statistic, the power divergence statistics and Goodman–Kruskal's gamma.

The general case

Consider next the case where $\mathcal{G}$ is a decomposable graph and we wish to investigate the hypothesis that $m \in \bar{M}(\mathcal{G}_0)$, assuming that $m \in \bar{M}(\mathcal{G})$, where $\mathcal{G}_0 = (V_0, E_0)$ is a decomposable subgraph of $\mathcal{G}$ with the same vertices, i.e. $V = V_0 = \Delta$. First we assume that $\mathcal{G}_0$ has exactly one edge $e = \{\alpha, \beta\}$ less than $\mathcal{G}$. This means that we test the hypothesis of the additional conditional independence

$$H_0 : \alpha \perp\!\!\!\perp \beta \mid \Delta \setminus \{\alpha, \beta\}.$$

Consider first the case where the model $\bar{M}(\mathcal{G})$ is saturated, i.e. the graph $\mathcal{G}$ is complete and has only one clique which is Δ itself. Then the deviance statistic is equal to

$$d = \sum_{i \in \mathcal{I}^+} 2\, n(i) \log \frac{n(i_{\Delta \setminus \{\alpha,\beta\}}) n(i)}{n(i_{\Delta \setminus \{\alpha\}}) n(i_{\Delta \setminus \{\beta\}})}. \tag{4.58}$$

Partitioning this sum after $i_{\Delta \setminus \{\alpha,\beta\}}$ gives

$$d = \sum_{i_{\Delta \setminus \{\alpha,\beta\}} \in \mathcal{I}_{\Delta \setminus \{\alpha,\beta\}}} d^{i_{\Delta \setminus \{\alpha,\beta\}}},$$

where $d^{i_{\Delta \setminus \{\alpha,\beta\}}}$ are deviances for independence in the $i_{\Delta \setminus \{\alpha,\beta\}}$-slices of the table. In the conditional distribution given the $\Delta \setminus \{\alpha\}$ and $\Delta \setminus \{\beta\}$

marginals, the deviances for the slices are independent and distributed as in the case of a two-dimensional table.

In the case of a general decomposable graph $\mathcal{G}$, Lemma 2.19 ensures that e is a member of one clique C^* only. Moreover we have

Proposition 4.30 *The exact deviance test for the hypothesis H_0 of conditional independence is identical to the exact deviance test of*

$$H_0' : \alpha \perp\!\!\!\perp \beta \mid C^* \setminus \{\alpha, \beta\}$$

in the saturated model for the C^*-marginal table. Here C^* is the unique clique of $\mathcal{G}$ containing e. Moreover the deviance d can be partitioned as

$$d = \sum_{i_b \in \mathcal{I}_b} d^{i_b},$$

where d^{i_b} is the deviance for independence of α and β in the i_b-slice where $b = C^* \setminus \{\alpha, \beta\}$. In the conditional distribution considered, these components are independent and distributed as deviances for independence in two-dimensional tables.

Proof: We have to show that the deviance statistics in the two problems are identical, and also that their distributions agree.

First number the cliques of $\mathcal{G}$ to form a perfect sequence in $\mathcal{G}$ with $C_1 = C^*$. This can be done by Lemma 2.18. The hypothesis H_0 has generating class equal to

$$\mathcal{C}_0 = \text{red}\{C_0^*, C_1^*, C_2, \ldots, C_k\}$$

and by Lemma 2.20 we find that

$$C_0^*, C_1^*, C_2, \ldots, C_k$$

form a perfect sequence of complete sets. Hence we get from (4.30) that the deviance can be calculated as the deviance between the marginal graphical models with generating classes

$$\{C_0^*, C_1^*, C_2, \ldots, C_{k-1}\} \quad \text{and} \quad \{C_1, C_2, \ldots, C_{k-1}\}.$$

Continuing in this way we verify the first part of the statement.

By direct calculation using (4.42) we find that

$$\mathcal{L}(\hat{m}_{C^*} \mid \hat{m}_0) = \mathcal{L}(\hat{m}_{C^*} \mid n_{C_0^*}, n_{C_1^*})$$

as desired. The final statement in the proposition follows as in the case where $\mathcal{G}$ is complete. □

Note that a similar reduction does not necessarily hold for other statistics. It is, however, tempting to use a similar procedure by defining the test statistic for the general hypothesis to be the sum of those for the i_b-slices.

Example 4.31 Continuation of Example 4.20. Let $\mathcal{G}$ be the graph of Fig. 4.1. Removing the edge $\{1,4\}$ destroys the decomposability of the graph. All other single edges are members of only one clique. Thus the deviance test for removing the edge $\{1,2\}$ can be performed in the $\{1,2,4\}$ marginal table and the decomposition in Proposition 4.30 becomes

$$d = \sum_u d^u = \sum_u 2 \sum_{r,s} n_{rs+u+++} \log \frac{n_{rs+u+++} n_{+++u+++}}{n_{r++u+++} n_{+s+u+++}},$$

where the sum only extends over cells with positive marginal counts. □

Next we consider the situation where the decomposable submodel $\mathcal{G}_0$ has k edges less than $\mathcal{G}$. By Lemma 2.21 there is a sequence $\mathcal{G}_0 \subset \cdots \subset \mathcal{G}_k = \mathcal{G}$ that are decomposable and differ by one edge only. As a consequence we obtain the following

Proposition 4.32 *The deviance for testing that p belongs to a decomposable submodel $M(\mathcal{G}_0)$, assuming it belongs to the decomposable model $M(\mathcal{G})$, can be partitioned as*

$$d = -2 \log \frac{L(\mathcal{G}_0)}{L(\mathcal{G})} = \sum_{j=1}^k d_j$$

into deviances for conditional independence in suitable marginal saturated models.

Proof: With the notation above, we get for the likelihood ratio

$$Q = \frac{L(\mathcal{G}_0)}{L(\mathcal{G})} = \frac{L(\mathcal{G}_0)}{L(\mathcal{G}_1)} \frac{L(\mathcal{G}_1)}{L(\mathcal{G}_2)} \cdots \frac{L(\mathcal{G}_{k-1})}{L(\mathcal{G}_k)} = \prod_{j=1}^k Q_j.$$

Taking logarithms, we have the desired partitioning of the deviance. From Proposition 4.30 it follows that each of the terms is the likelihood ratio for conditional independence in the marginal saturated model given by the single clique of which the missing edge is a member. □

The components of the deviance will typically not be independent. Observe also that in general there might be several ways of partitioning the deviance. Nevertheless the combination of results in this section provides the possibility of detailed partitioning of the deviance statistic. First, d is partitioned into components that test single edges. Each of these can be partitioned into independence deviances for two-way tables. Finally the Lancaster partitioning gives a partitioning of these into deviances for 2×2 tables.

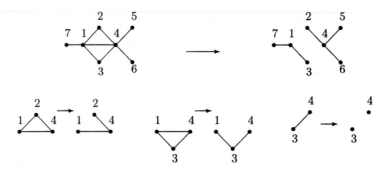

Fig. 4.2. Illustration of the decomposition of deviances in the 7-dimensional table. The likelihood ratio for the total reduction factorizes into products of likelihood ratios of stepwise reductions. Each of these is equal to the corresponding likelihood ratio for removing one edge in the indicated marginal saturated models.

Example 4.33 Continuation of Example 4.31. Suppose the additional conditional independences to be tested are specified by the absence of edges $\{1,2\}$, $\{1,4\}$, and $\{3,4\}$, i.e. the group of variables $\{1,3,7\}$ is independent of the variables $\{2,4,5,6\}$. The deviance can then be partitioned as $d = d_1 + d_2 + d_3$, where d_1 is the deviance for removing $\{1,2\}$ in the $\{1,2,4\}$ marginal as calculated before, d_2 is the deviance for removing $\{1,4\}$, supposing that $\{1,2\}$ has been removed already:

$$d_2 = 2 \sum_t \sum_{r,u} n_{r+tu+++} \log \frac{n_{r+tu+++}n_{++t++++}}{n_{r+t++++}n_{++tu+++}},$$

and d_3 is similarly the deviance for independence in the $\{3,4\}$ marginal table. Figure 4.2 illustrates this. Other partitions could be obtained by removing edges in different orders, only not beginning with removal of $\{1,4\}$, since this would create a 4-cycle. □

The Monte Carlo algorithm of Patefield (1981) can also be extended to the general case, even though the components of the deviance are not independent. For one can begin with the marginal tables that are sufficient under the model given by $\mathcal{G}_0$ and generate the necessary random tables in the relevant slices and compute d_1. This set of tables can then be used as input marginals for generating the tables needed for testing $\mathcal{G}_1$ assuming $\mathcal{G}_2$ and calculating d_2 and so on. The procedure will give the correct distribution for the deviance, since this is a function of $(\hat{m}_0, \ldots, \hat{m}_k)$ which form a Markov chain in the sense that $\hat{m}_j \perp\!\!\!\perp (\hat{m}_0, \ldots, \hat{m}_{j-2}) \,|\, \hat{m}_{j-1}$, and at each stage we simulate $\mathcal{L}(\hat{m}_j \,|\, \hat{m}_{j-1})$.

Example 4.34 Continuation of Example 4.33. To simulate the distribution for the exact test in this example, one proceeds as follows.

1. The marginal table $n_{\{3,4\}}$ is simulated using the observed marginal tables $n_{\{3\}}$ and $n_{\{4\}}$.

2. The marginal table $n_{\{1,3,4\}}$ is simulated using the observed marginal table $n_{\{1,3\}}$ and the simulated marginal table $n^*_{\{3,4\}}$.

3. The marginal table $n_{\{1,2,4\}}$ is simulated using the observed marginal table $n_{\{2,4\}}$ and the marginal table $n^*_{\{1,4\}}$, calculated from the simulated marginal table $n^*_{\{1,3,4\}}$.

As this illustrates, the typical feature is that whenever the next marginal table has to be simulated, a combination of observed and simulated marginal tables is used.

This simulation procedure has been implemented in the program CoCo (Badsberg 1991, 1995). □

It seems appropriate here to warn against misuse of the procedure of calculating overall significance levels of combined tests. The risk here as well as in other situations is that important deviations from the model are not detected, because the effect of significant components of the deviance is masked by non-significant parts. To prevent this we suggest that the overall test is always combined with an analysis of the individual terms in the partitionings mentioned. This can readily be built into the simulation scheme described, by recording not only the total deviance but the individual components of the deviance as well.

4.4.4 Asymptotic tests in decomposable models

In the decomposable case we have simple formulae for calculating the dimensions of hierarchical model subspaces. More precisely, repeated use of (4.29) on a perfect sequence of cliques gives

Proposition 4.35 *If $\mathcal{C}$ are the cliques of a decomposable graph then the corresponding hierarchical model subspace has dimension*

$$\dim H_\mathcal{C} = \sum_{C \in \mathcal{C}} \prod_{\delta \in C} |\mathcal{I}_\delta| - \sum_{S \in \mathcal{S}} \nu(S) \prod_{\delta \in S} |\mathcal{I}_\delta|, \qquad (4.59)$$

where $\nu(S)$ is the multiplicity of S in any perfect numbering of $\mathcal{C}$.

The degrees of freedom can then be calculated by subtracting the dimensions of the model subspaces of the reduced and full models.

Alternatively the partitioning of tests can be used directly since, clearly, the degrees of freedom are additive over the partitionings described. The

Lancaster components are asymptotically χ^2-distributed with just one degree of freedom. Thus adding these up will provide the approximate distribution of any of the parts of the deviance statistic in the general case.

When some of the estimated mean cell counts are equal to zero (which is equivalent to some of the sufficient marginal counts being zero) it is, in the case of comparing two decomposable models, relatively straightforward to calculate the necessary modified dimensions of the hierarchical model subspaces. Essentially one replaces $\prod_{\delta \in C} |\mathcal{I}_\delta|$ and the similar terms for the separators in (4.59) with the number of positive entries in the relevant marginal tables. In the case of testing one conditional independence, to obtain a reasonable χ^2-approximation to the distribution of the test statistic, one has to use the modified dimension relative to the marginals under the hypothesis. Solving the modified degrees of freedom when testing the removal of one edge $e \in C^*$ between α and β become

$$\sum_{i_b \in \mathcal{I}_b} \left\{ \nu^+\left(\mathcal{I}^{i_b}_{b \cup \{\alpha\}}\right) - 1 \right\} \left\{ \nu^+\left(\mathcal{I}^{i_b}_{b \cup \{\beta\}}\right) - 1 \right\},$$

where $b = C^* \setminus \{\alpha, \beta\}$ and $\nu^+\left(\mathcal{I}^{i_b}_{b \cup \{\alpha\}}\right)$ is the number of positive entries in the $b \cup \{\alpha\}$ marginal of the i_b-slice. In the general case, the modified degrees of freedom should be added over a decomposition of the testing problem according to Proposition 4.32.

We hasten, however, to point out that zero marginal tables most often occur in situations where the general expected number of cell counts is so small that the asymptotic results are far from useful anyway, and the exact simulated methods therefore are preferable.

4.5 Recursive models

Whereas the previous sections dealt with models describing associations between variables in a symmetric fashion, the present section deals with the situation where some variables are responses to others in a recursive response structure.

Such models are appropriate, for example, when – from subject matter knowledge – the variables can be linearly ordered or, equivalently, numbered as

$$\Delta = \{\delta_1, \ldots, \delta_k\}$$

in such a way that the variable δ_j is considered to be a response to variables $\delta_1, \ldots, \delta_{j-1}$, but explanatory to variables $\delta_{j+1}, \ldots, \delta_k$.

In other cases only a partial order can be identified, since some of the variables appear to be on a symmetric footing. Then chain graph models or block-recursive models, to be described in the next section, are typically more appropriate.

We describe the recursive graphical models before the more general recursive hierarchical models. The advantages of doing this are both technical and conceptual.

Throughout this section we consider the single multinomial sampling scheme, i.e. the set of counts is assumed to follow the multinomial distribution (4.5).

4.5.1 Recursive graphical models

A *recursive graphical model* is specified by assuming that the unknown probability distribution p belongs to the set $M(\mathcal{G})$ of distributions that obey the Markov property with respect to a directed acyclic graph $\mathcal{G}$. For the purpose of interpretation, the arrows in the directed graph must respect an ordering of the variables obtained from subject matter knowledge such that arrows point from explanatory variables to responses and not conversely.

As shown in Theorem 3.27, there are several equivalent formulations of the Markov property on a directed acyclic graph, in particular the statement $p \in M(\mathcal{G})$ is equivalent to the existence of a factorization

$$p(i) = \prod_{\delta \in \Delta} p\left(i_\delta \,|\, i_{\mathrm{pa}(\delta)}\right). \tag{4.60}$$

If all variables are connected in the graph, the recursive model is *saturated*. Then the directed Markov property involves no conditional independence restrictions, implying that the saturated recursive models also are saturated models in the sense of Section 4.2.2.

Observe that for a saturated recursive model to make sense, the variables must be linearly ordered, since otherwise there would be cycles in the complete directed graph.

Maximum likelihood estimation

Consider the problem of maximizing the likelihood function in a recursive graphical model $M(\mathcal{G})$ based upon a multinomial sample. From (4.60) we obtain a factorization of the likelihood function as

$$\begin{aligned} L(p) &\propto \prod_{i \in \mathcal{I}} \left\{ \prod_{\delta \in \Delta} p\left(i_\delta \,|\, i_{\mathrm{pa}(\delta)}\right) \right\}^{n(i)} \\ &= \prod_{\delta \in \Delta} \prod_{i_{\mathrm{cl}(\delta)} \in \mathcal{I}_{\mathrm{cl}(\delta)}} p\left(i_\delta \,|\, i_{\mathrm{pa}(\delta)}\right)^{n(i_{\mathrm{cl}(\delta)})} = \prod_{\delta \in \Delta} L_\delta(p), \end{aligned} \tag{4.61}$$

where $\mathrm{cl}(\delta) = \delta \cup \mathrm{pa}(\delta)$ as usual. This factorization displays the likelihood function as a product of likelihood functions L_δ, each being proportional to the likelihood function obtained when sampling the variables in $\mathrm{cl}(\delta)$ with fixed $\mathrm{pa}(\delta)$-marginals.

Since the model sets no further restrictions on the conditional probabilities, the joint likelihood function can be maximized by maximizing each of the factors. Each of these is in turn proportional to the likelihood function for a saturated model involving the variables in cl(δ) and we therefore obtain from Proposition 4.5

Theorem 4.36 *The maximum likelihood estimate in a recursive graphical model $M(\mathcal{G})$ based upon a multinomial sample is given as*

$$\hat{p}(i) = \prod_{\delta \in \Delta} \frac{n\left(i_{\mathrm{cl}(\delta)}\right)}{n\left(i_{\mathrm{pa}(\delta)}\right)}. \qquad (4.62)$$

In the above formula it should be remembered that $n(i_\emptyset) = |n|$. This will appear in the denominator whenever a variable δ has no parents.

Example 4.37 Consider the recursive graphical model for an $R \times S \times T$-table with the three variables $\Delta = \{\alpha, \beta, \gamma\}$ related as

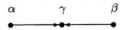

reflecting marginal independence of the variables α and β. From (4.62) it follows that the maximum likelihood estimate of the unknown probability, based upon a multinomial sample, is equal to

$$\hat{p}_{rst} = \frac{n_{rst}}{n_{rs+}} \frac{n_{+s+}}{|n|} \frac{n_{r++}}{|n|} = \frac{n_{rst}\, n_{+s+}\, n_{r++}}{n_{rs+}\, |n|^2}.$$

Observe that there is no sufficient data reduction beyond that of the table of counts $\{n_{rst}\}$ itself. □

The situation in this example, where we have explicit maximum likelihood estimates but only a limited amount of sufficient reduction, is typical for the recursive models.

In the general case there will be no sufficient reduction beyond the reduction to the table of marginal counts for the maximal elements among the sets $\bigl(\mathrm{cl}(\delta)\bigr)_{\delta \in \Delta}$. That these marginal counts are minimally sufficient is seen directly from the expression for the likelihood function in (4.61). We omit the details of this.

Example 4.38 Consider the recursive model for the $R \times S \times T \times U \times V$-table with the five variables $\Delta = \{\alpha, \beta, \gamma, \delta, \varepsilon\}$ related as

From (4.62) we obtain that the maximum likelihood estimate of the unknown probability, based upon a multinomial sample, is equal to

$$\hat{p}_{rstuv} = \frac{n_{r++++}}{|n|} \frac{n_{rs+++}}{n_{r++++}} \frac{n_{++t++}}{|n|} \frac{n_{+s+u+}}{n_{+s+++}} \frac{n_{+st+v}}{n_{+st++}}$$

$$= \frac{n_{rs+++} \, n_{++t++} \, n_{+s+u+} \, n_{+st+v}}{|n|^2 \, n_{+s+++} \, n_{+st++}}.$$

The marginal counts $\{n_{rs+++}, n_{+st+v}, n_{+s+u+}\}$ are minimally sufficient in the model. □

A special case occurs when the directed acyclic graph $\mathcal{G}$ is perfect and therefore $M(\mathcal{G}) = M_F(\mathcal{G}^\sim) = M_E(\mathcal{G}^\sim)$; see Proposition 3.29. Then the recursive graphical model is identical to the graphical interaction model with the decomposable undirected graph $\mathcal{G}^\sim$ as its graph. This is reflected in the expression for the maximum likelihood estimate. The recursive factorization of the probabilities reduces to the factorization (4.32) and the factorization of the maximum likelihood estimate reduces to (4.39). In both cases the reduction is obtained by observing that all terms of the type $p(i_{\text{cl}(\delta)})$ and $n(i_{\text{cl}(\delta)})$, for such variables δ where $\text{cl}(\delta)$ is not maximal (not a clique), will appear both in a numerator and in a denominator of (4.60) and (4.62) respectively, and therefore cancel. Observe that the saturated recursive model is a special case of a recursive graphical model with a perfect graph.

We shall not make much effort to give a detailed description of the sampling distribution of the maximum likelihood estimate. Clearly the conditional distribution of $\{\hat{p}(i_\delta \,|\, i_{\text{pa}(\delta)})\}$ given the counts $\{n(i_{\text{pa}(\delta)})\}$ is a scaled multinomial distribution. It is however also true that

$$\{\hat{p}(i_\delta \,|\, i_{\text{pa}(\delta)})\} \perp\!\!\!\perp \{\hat{p}(i_{\text{nd}(\delta)})\} \,|\, \{\hat{p}(i_{\text{pa}(\delta)})\},$$

which we refer to as the *directed hyper Markov property*. We abstain from showing the latter but refer to Dawid and Lauritzen (1993) for further details.

Deviance tests

Assume that $\mathcal{G}_0$ is a subgraph of the directed acyclic graph $\mathcal{G}$ obtained by deleting one or more edges of $\mathcal{G}$. Then the recursive graphical model $M_0 = \bar{M}(\mathcal{G}_0)$ is a submodel of the recursive graphical model $M = \bar{M}(\mathcal{G})$ and we can – at least from a mathematical point of view – meaningfully consider the likelihood ratio test statistic for the hypothesis that $p \in M_0$ assuming $p \in M$.

If we first consider the case where only one edge is removed from $\mathcal{G}$ to obtain $\mathcal{G}_0$ and this is the edge $\alpha \to \beta$, say, we have for the two graphs that $\mathrm{pa}_0(\delta) = \mathrm{pa}(\delta)$ for $\delta \neq \beta$ whereas $\mathrm{pa}_0(\beta) = \mathrm{pa}(\beta) \setminus \{\alpha\}$ such that the maximum likelihood estimates become from (4.62)

$$\hat{p}(i) = \prod_{\delta \in \Delta} \frac{n\left(i_{\mathrm{cl}(\delta)}\right)}{n\left(i_{\mathrm{pa}(\delta)}\right)}$$

and

$$\hat{p}_0(i) = \frac{n\left(i_{\mathrm{cl}(\beta)\setminus\{\alpha\}}\right)}{n\left(i_{\mathrm{pa}(\beta)\setminus\{\alpha\}}\right)} \prod_{\delta \in \Delta \setminus \{\beta\}} \frac{n\left(i_{\mathrm{cl}(\delta)}\right)}{n\left(i_{\mathrm{pa}(\delta)}\right)},$$

where $\hat{p}_0$ is the estimate of p in the model M_0. In the likelihood ratio most terms cancel and the deviance reduces to

$$\begin{aligned} d = -2\log Q &= -2\log\frac{L_\beta(\hat{p}_0)}{L_\beta(\hat{p})} \\ &= \sum_{i_{\mathrm{cl}(\beta)} \in \mathcal{I}_{\mathrm{cl}(\beta)}} 2\, n(i_{\mathrm{cl}(\beta)}) \log \frac{n\left(i_{\mathrm{pa}(\beta)\setminus\{\alpha\}}\right) n\left(i_{\mathrm{cl}(\beta)}\right)}{n\left(i_{\mathrm{cl}(\beta)\setminus\{\alpha\}}\right) n\left(i_{\mathrm{pa}(\beta)}\right)}, \end{aligned}$$

which we can recognize from (4.58) as the deviance for testing the conditional independence

$$H_0 : \alpha \perp\!\!\!\perp \beta \mid \mathrm{cl}(\beta) \setminus \{\alpha, \beta\}$$

in the saturated model for the marginal table involving variables in $\mathrm{cl}(\beta)$.

The question now arises: in which distribution is this statistic to be judged? In the section on log–affine models all the models were linear exponential families and exact similar tests were obtained by judging the test statistic in the conditional distribution given the minimal (and complete) sufficient statistic under the hypothesis. However, in the case of recursive models the minimally sufficient statistic is typically not complete and such an extensive conditioning could be too violent.

The procedure that we suggest for obtaining a similar test in the case of testing only one conditional independence is first to marginalize the problem to variables only involving $\mathrm{cl}(\beta)$ and then to condition on the marginal tables that are sufficient in this problem. In effect this has the consequence that the simulation methods described in Section 4.4.3 apply without modification.

In the general case we approach the problem of testing as in Section 4.4.3, realizing that here it is trivial that if $\mathcal{G}_0$ is a directed acyclic subgraph of $\mathcal{G}$, obtained by removing edges, then we can find a sequence $\mathcal{G}_0 \subset \mathcal{G}_1 \subset \cdots \subset \mathcal{G}$ of directed acyclic graphs differing by exactly one edge. As in Proposition 4.32, the deviance can be decomposed into deviances for testing single conditional independences. But here it is a more complex problem to find a 'correct' distribution for evaluation of test probabilities.

RECURSIVE MODELS 111

If all edges removed point at the same variable δ, the situation is relatively straightforward. We first marginalize the problem to consider only variables in the set $\mathrm{cl}(\delta)$ and find that the likelihood ratio factorizes into likelihood ratios that are identical to those obtained by removing undirected edges in the saturated log–linear model involving only these variables. Thus the problem reduces to testing decomposable models as discussed in Example 4.33.

In the more general case, there is some intuitive appeal in proceeding in analogy with the decomposable case, but it has to be admitted that this lacks a sound theoretical justification. We illustrate the procedure by an example.

Example 4.39 Consider the problem of testing the model

assuming that the model

holds true. Then we form a sequence of recursive models differing by one edge by inserting the two models

between those compared, and we then proceed as follows in the simulation procedure. Based upon the observed sufficient marginals, which were found to be $\{n_{rs+++}, n_{+st+v}, n_{+s+u+}\}$, we

1. simulate the table $\{n_{r+t++}\}$ from the observed tables $\{n_{r++++}\}$ and $\{n_{++t++}\}$, and calculate the corresponding deviance d_{01} for testing independence of α and γ;

2. simulate the table $\{n_{rst++}\}$ from the observed table $\{n_{rs+++}\}$ and the simulated $\{n_{r+t++}\}$ table, and calculate the deviance d_{12} for conditional independence of β and γ given α;

3. simulate the table $\{n_{+s+uv}\}$ from the observed tables $\{n_{+s+u+}\}$ and $\{n_{+s++v}\}$ and calculate the deviance d_{23} for conditional independence of δ and ε given β;

4. calculate the total simulated deviance d as the sum of these deviances.

A different simulation procedure results from specifying a different sequence, although the test statistic will have the same distribution. □

As in the case of decomposable models, we warn against uncritical use of overall test statistics but rather advocate separate investigation of the deviances corresponding to each of the independences considered. This might be even more critical here, since the procedure described in the example is somewhat *ad hoc*.

4.5.2 Recursive hierarchical models

These models, analogous to hierarchical log–linear models but taking into account a recursive response structure in the variables, are most directly described as submodels of recursive graphical models.

Let a directed acyclic graph $\mathcal{G}$ be given and also for all $\delta \in \Delta$ a generating class $\mathcal{A}_\delta$ of subsets of $\mathrm{cl}(\delta)$ with the extra property that $\mathrm{pa}(\delta) \subseteq a$ for some $a \in \mathcal{A}_\delta$. The *recursive hierarchical model* with generating classes $(\mathcal{A}_\delta)_{\delta \in \Delta}$ is specified by assuming that the unknown probability p is an element of those directed Markov probabilities in $M(\mathcal{G})$ which simultaneously have marginal distributions $p_{\mathrm{cl}(\delta)}$, each of which is a member of the extended hierarchical model $\bar{M}(\mathcal{A}_\delta)$.

The likelihood function based upon a multinomial sample also here factorizes as in (4.61) into a product of likelihood functions L_δ, each being proportional to the likelihood function obtained when sampling the variables in $\mathrm{cl}(\delta)$ with fixed $\mathrm{pa}(\delta)$-marginals.

Because the restrictions on the factors are independent, the joint likelihood function can be maximized by maximizing each factor separately. Each of these is in turn proportional to the likelihood function for a hierarchical log–linear model involving the variables in $\mathrm{cl}(\delta)$. Since these hierarchical models all have generators that include the marginals $\mathrm{pa}(\delta)$ which are fixed under sampling, we obtain from Theorem 4.11:

Theorem 4.40 *The maximum likelihood estimate in a recursive hierarchical model with directed acyclic graph $\mathcal{G}$ and generating classes $(\mathcal{A}_\delta)_{\delta \in \Delta}$ based upon a multinomial sample is given as*

$$\hat{p}(i) = \prod_{\delta \in \Delta} \frac{\hat{m}_\delta\left(i_{\mathrm{cl}(\delta)}\right)}{n\left(i_{\mathrm{pa}(\delta)}\right)}. \tag{4.63}$$

Here $\hat{m}_\delta$ is the mean vector of the unique element of the extended hierarchical model $\bar{M}(\mathcal{A}_\delta)$ that satisfies

$$\hat{m}(i_a) = n(i_a), \quad i_a \in \mathcal{I}_a, a \in \mathcal{A}_\delta. \tag{4.64}$$

Example 4.41 A simple example of a recursive hierarchical model is obtained by modifying the model of marginal independence of variables α and β in Example 4.37 to also have restrictions on the conditional distribution of γ given the other variables. More precisely, let the graph be as in the example and let the generating classes of a hierarchical recursive model be

$$\mathcal{A}_\alpha = \{\{\alpha\}\}, \quad \mathcal{A}_\beta = \{\{\beta\}\}, \quad \mathcal{A}_\gamma = \{\{\alpha,\beta\},\{\alpha,\gamma\},\{\beta,\gamma\}\}.$$

Using the same notation as in the example, the maximum likelihood estimate becomes

$$\hat{p}_{rst} = \frac{\hat{m}_{rst}\, n_{+s+}\, n_{r++}}{n_{rs+}\, |n|^2}.$$

Here $\hat{m}$ has to be calculated iteratively as described in Section 4.3.1. □

4.6 Block-recursive models

We have previously dealt with models describing associations between variables in a symmetric fashion, and also with models where the variables were arranged in a recursive response structure. This section deals with models that combine the two types of association.

Such models are appropriate when there is a partial ordering of variables as response and explanatory variables, but either a complete ordering is not relevant or there is not sufficient subject matter knowledge to establish the ordering (Wermuth and Lauritzen 1990).

We suppose the variables can be partitioned into blocks as

$$\Delta = \Delta(1) \cup \cdots \cup \Delta(T) \tag{4.65}$$

in such a way that the variables in a given block $\Delta(t)$ are considered to be responses to variables in $\Delta(1)\cup\cdots\cup\Delta(t-1)$, but explanatory to variables in $\Delta(t+1) \cup \cdots \cup \Delta(T)$. Variables within the same block are assumed associated on a symmetric footing. This partitioning of the variables is referred to as a *dependence chain* or a *block-recursive structure*.

As in the section on recursive models, we begin with considering the block-recursive graphical models — also named chain graph models — and we also exclusively investigate the single multinomial sampling scheme.

4.6.1 Chain graph models

A *chain graph model* or *graphical block-recursive model* is specified by assuming that the unknown probability distribution p belongs to the set $M_E(\mathcal{G})$ of extended Markov distributions on a chain graph $\mathcal{G}$. Either these are distributions with strictly positive densities that obey any of the chain graph Markov properties, or they are limits of such distributions. For the purpose of interpretation, the arrows in the directed graph must respect a partial ordering of the variables obtained from subject matter knowledge. Thus if the variables are partitioned as in (4.65), a connection between two variables in the same block must be an undirected edge, whereas edges between variables in different blocks are arrows that point from explanatory variables to responses and not conversely.

As mentioned in Corollary 3.37, a probability distribution on a discrete sample space satisfies the extended Markov property with respect to a chain graph $\mathcal{G}$ if and only if it factorizes as

$$p(i) = \prod_{t=1}^{T} \frac{p\left(i_{\Delta(t) \cup B(t)}\right)}{p\left(i_{B(t)}\right)} = \prod_{t=1}^{T} p\left(i_{\Delta(t)} \mid i_{B(t)}\right) \qquad (4.66)$$

and each denumerator of the central expression is extended Markov with respect to the graph $\mathcal{G}_*(t)$. Here $B(t) = \text{pa}\{\Delta(t)\}$, and $\mathcal{G}_*(t)$ is the undirected graph having $\Delta(t) \cup B(t)$ as its vertices and undirected variables between any pair of vertices that are either connected in $\mathcal{G}$ or both in $B(t)$.

If all variables are connected in the graph, the model is *saturated*. Then the chain Markov property involves no conditional independence restrictions, so the saturated chain graph models are also saturated models in the sense of Section 4.2.2.

Maximum likelihood estimation

Consider the problem of maximizing the likelihood function in a block-recursive graphical model $M_E(\mathcal{G})$ based upon a multinomial sample. From (4.66) we obtain a factorization of the likelihood function as

$$\begin{aligned} L(p) &\propto \prod_{i \in \mathcal{I}} \left\{ \prod_{t=1}^{T} p\left(i_{\Delta(t)} \mid i_{B(t)}\right) \right\}^{n(i)} \\ &= \prod_{t=1}^{T} \prod_{i_{\Delta(t) \cup B(t)} \in \mathcal{I}_{\Delta(t) \cup B(t)}} p\left(i_{\Delta(t)} \mid i_{B(t)}\right)^{n(i_{\Delta(t) \cup B(t)})} \\ &= \prod_{t=1}^{T} L_t(p). \end{aligned} \qquad (4.67)$$

This factorization displays the likelihood function as a product of likelihood functions L_t, each being proportional to the likelihood function obtained when sampling the variables in $\Delta(t) \cup B(t)$ with fixed $B(t)$ marginals.

The restrictions specified by the model on each of the factors are independent, since they restrict the mean vector of each of the conditional distributions to be in the graphical model specified by $M_E\{\mathcal{G}_*(t)\}$. Hence also in this case the joint likelihood function can be maximized by maximizing each factor separately.

The generating classes of these graphical models, viewed as hierarchical models, all have a generator including the marginals $B(t)$ that are fixed under sampling. If we let $\mathcal{C}_*(t)$ denote the cliques of $\mathcal{G}_*(t)$ we thus obtain from Theorem 4.11:

Theorem 4.42 *The maximum likelihood estimate in a chain graph model with graph $\mathcal{G}$ and dependence chain $\Delta(1), \ldots, \Delta(T)$ based upon a multinomial sample is*

$$\hat{p}(i) = \prod_{t=1}^{T} \frac{\hat{m}_t^*\left(i_{\Delta(t) \cup B(t)}\right)}{n\left(i_{B(t)}\right)}. \qquad (4.68)$$

Here $\hat{m}_t^*$ is the mean vector of the unique element of the graphical model $M_E\{\mathcal{G}_*(t)\}$ that satisfies

$$\hat{m}_t^*(i_a) = n(i_a), \quad i_a \in \mathcal{I}_a, a \in \mathcal{C}_*(t). \qquad (4.69)$$

So in the case of chain graph models we can reduce the problem of calculating the maximum likelihood estimates to a system of similar problems involving the associated undirected graphical models.

Example 4.43 Consider the chain graph model for the $R \times S \times T \times U$-table with the four variables $\Delta = \{\alpha, \beta, \gamma, \delta\}$ related as

where the dependence chain for example could be given as $\Delta(1) = \{\alpha, \beta\}$ and $\Delta(2) = \{\gamma, \delta\}$. The two associated undirected graphs $\mathcal{G}_*(1)$ and $\mathcal{G}_*(2)$ are

The maximum likelihood estimate of the unknown probability, based upon a multinomial sample, is then first calculated for the first component to be

$$\hat{m}^*_{rs} = \frac{n_{r+++}\, n_{+s++}}{|n|}.$$

In the second component $\hat{m}^*_{rstu}$ must be found by iteration since the corresponding graph is not decomposable. The final estimate then becomes

$$\hat{p}_{rstu} = \frac{n_{r+++}\, n_{+s++}\, \hat{m}^*_{rstu}}{|n|^2\, n_{rs++}}.$$

□

Deviance tests

Assume that $\mathcal{G}_0$ is a subgraph of the chain graph $\mathcal{G}$ obtained by deleting one or more edges of $\mathcal{G}$. Then the chain graph model $M_0 = M_E(\mathcal{G}_0)$ is a submodel of the chain graph model $M = M_E(\mathcal{G})$ and we consider the likelihood ratio test statistic for the hypothesis that $p \in M_0$ assuming $p \in M$.

Initially we consider the case where only one edge is removed from $\mathcal{G}$ to obtain $\mathcal{G}_0$. This can then either be a directed edge or an undirected edge. If $\alpha \to \beta$ is a directed edge to be removed and t^* is the smallest t containing β, we have for the two graphs that $B_0(t) = B(t)$ for $t \neq t^*$ whereas $B_0(t^*) = B(t^*) \setminus \{\alpha\}$ such that the maximum likelihood estimates become from (4.68) in the model M

$$\hat{p}(i) = \prod_{t=1}^{T} \frac{\hat{m}^*_t\left(i_{\Delta(t) \cup B(t)}\right)}{n\left(i_{B(t)}\right)}$$

and in the model M_0

$$\hat{p}_0(i) = \frac{\hat{m}^{00}_{t^*}\left(i_{\Delta(t^*) \cup B_0(t^*)}\right)}{n\left(i_{B_0(t^*)}\right)} \prod_{t \neq t^*} \frac{\hat{m}^*_t\left(i_{\Delta(t) \cup B(t)}\right)}{n\left(i_{B(t)}\right)}.$$

Here $\hat{m}^{00}_{t^*}$ is the estimate of m in the graphical model $M_E(\mathcal{G}_{00*}(t^*))$. The graph $\mathcal{G}_{00*}(t^*)$ has vertices $\Delta(t^*) \cup B_0(t^*)$ and undirected edges between any pair of variables that are connected in $\mathcal{G}_0$ or both in $B_0(t^*)$. In the likelihood ratio most terms cancel and the deviance reduces to

$$\begin{aligned} d &= -2 \log Q = -2 \log \frac{L_{t^*}(\hat{p}_0)}{L_{t^*}(\hat{p})} \\ &= \sum_{i_{\Delta(t^*) \cup B(t^*)}} 2\, n(i_{\Delta(t^*) \cup B(t^*)}) \log \frac{n\left(i_{B_0(t^*)}\right) \hat{m}^*_{t^*}\left(i_{\Delta(t^*) \cup B(t^*)}\right)}{\hat{m}^{00}_{t^*}\left(i_{\Delta(t^*) \cup B_0(t^*)}\right) n\left(i_{B(t^*)}\right)}. \end{aligned}$$

If we let $\hat{m}_{t^*}^0$ be the estimate of m in the model $M_E(\mathcal{G}_{0*}(t^*))$, where $\mathcal{G}_{0*}(t^*)$ has vertices $\Delta(t^*) \cup B(t^*)$ and undirected edges between any pair of variables that are connected in $\mathcal{G}_0$ or both in $B(t^*)$, it holds that

$$\hat{m}_{t^*}^0\left(i_{\Delta(t^*)\cup B(t^*)}\right) = \frac{\hat{m}_{t^*}^{00}\left(i_{\Delta(t^*)\cup B_0(t^*)}\right)\, n\left(i_{B(t^*)}\right)}{n\left(i_{B_0(t^*)}\right)}.$$

This follows by recalling that $B_0(t^*) = B(t^*) \setminus \{\alpha\}$ and marginalizing both sides to cliques in the graph $\mathcal{G}_{0*}(t^*)$. Hence

$$d = \sum_{i_{\Delta(t^*)\cup B(t^*)}} 2n\left(i_{\Delta(t^*)\cup B(t^*)}\right) \log \frac{\hat{m}_{t^*}^*\left(i_{\Delta(t^*)\cup B(t^*)}\right)}{\hat{m}_{t^*}^0\left(i_{\Delta(t^*)\cup B(t^*)}\right)},$$

which we recognize as the deviance for testing the conditional independence

$$H_0 : \alpha \perp\!\!\!\perp \beta \mid \left(\Delta(t^*) \cup B(t^*)\right) \setminus \{\alpha, \beta\}$$

in the model $M_E(\mathcal{G}^*(t^*))$ for the corresponding marginal table.

The question again arises: in which distribution is this statistic to be judged? As in the case of recursive models, the minimally sufficient statistic is not complete in general, so conditioning on the sufficient statistic could well be too violent.

The procedure that we suggest for obtaining a similar test in the case of testing only one conditional independence is first to marginalize the problem to variables only involving $\Delta(t^*) \cup B(t^*)$ and then to condition on the marginal tables that are sufficient in this problem.

In general there are only asymptotic results available describing the distribution of the deviance, so whether we marginalize or not makes little difference. It leads to the deviance being approximately χ^2-distributed with degrees of freedom equal to those obtained in the marginalized test of the corresponding undirected graphical model.

In the case where an undirected edge between α and β is to be removed, the story is not much different. Again the likelihood ratio is only affected in the factor L_{t^*} containing both α and β in the $\Delta(t^*)$ part. So here it can be directly seen that the deviance reduces to the deviance for removing one edge from the corresponding undirected model.

Example 4.44 Consider the chain graph model for the $R \times S \times T \times U$-table with the four variables $\Delta = \{\alpha, \beta, \gamma, \delta\}$ differing from the one studied in Example 4.43 by the edge between α and β, having dependence chain $\Delta(1) = \{\alpha, \beta\}$ and $\Delta(2) = \{\gamma, \delta\}$. The two associated undirected graphs $\mathcal{G}^*(1)$ and $\mathcal{G}^*(2)$ are

The deviance for removing the undirected edge between α and β is calculated in the $\Delta(1)$-marginal table and is equal to

$$d = \sum_{rs} 2 n_{rs++} \log \frac{n_{rs++}|n|}{n_{r+++} n_{+s++}},$$

which is the deviance for independence in the undirected graphical model $M_E(\mathcal{G}_*(1))$. The deviance for removing the directed edge between α and δ is equal to

$$d = \sum_{rstu} 2 n_{rstu} \log \frac{n_{+s++} n_{++t+} \hat{m}^*_{rstu}}{n_{rs++} n_{+st+} n_{++tu}},$$

where iteration is needed to calculate $\hat{m}^*_{rstu}$. This deviance is the deviance for testing conditional independence between α and δ given β and γ in the undirected model with graph $\mathcal{G}_*(2)$. □

When more than one edge is removed from $\mathcal{G}$ to obtain $\mathcal{G}_0$, the deviance factorizes in the obvious way. We omit the details.

4.6.2 Block-recursive hierarchical models

The models, analogous to hierarchical log–linear models but taking into account a block-recursive response structure in the variables, are most easily described as submodels of chain graph models.

Hence, let a chain graph $\mathcal{G}$ with chain components $\Delta(1), \ldots, \Delta(T)$ be given, and also for all t a generating class $\mathcal{A}_t$ of subsets of $\Delta(t) \cup B(t)$ with the extra property that for all t, $B(t) \subseteq a$ for some $a \in \mathcal{A}_t$. The *recursive hierarchical model* with generating classes $\mathcal{A}_t, t = 1, \ldots, T$ is specified by assuming that the unknown probability p is an extended chain Markov probability which simultaneously has marginal distributions $p_{\Delta(t) \cup B(t)}$, each of which is a member of the extended hierarchical model $\bar{M}(\mathcal{A}_t)$.

The likelihood function based upon a multinomial sample also here factorizes as in (4.67) into a product of likelihood functions L_t, each being proportional to the likelihood function obtained when sampling the variables in $\Delta(t) \cup B(t)$ with fixed $B(t)$-marginals.

Since the restrictions on the factors are independent, the joint likelihood function can be maximized by maximizing each factor separately. Each of these is in turn proportional to the likelihood function for a hierarchical log–linear model involving the variables in $\Delta(t)\cup B(t)$. Since these hierarchical models all have generators including the marginals $B(t)$ that are fixed under sampling, we obtain from Theorem 4.11:

Theorem 4.45 *The maximum likelihood estimate in a block-recursive hierarchical model with graph $\mathcal{G}$, dependence chain $\Delta(1), \ldots, \Delta(T)$, and generating classes $\mathcal{A}_1, \ldots, \mathcal{A}_T$ based upon a multinomial sample is*

$$\hat{p}(i) = \prod_{t=1}^{T} \frac{\hat{m}_t^*\left(i_{\Delta(t)\cup B(t)}\right)}{n\left(i_{B(t)}\right)}. \qquad (4.70)$$

Here $\hat{m}_t^*$ is the mean vector of the unique element of the extended hierarchical log–linear model with generating class $\mathcal{A}_t$ that satisfies

$$\hat{m}_t^*(i_a) = n(i_a), \quad i_a \in \mathcal{I}_a, a \in \mathcal{A}_t. \qquad (4.71)$$

Example 4.46 A simple example of a block-recursive hierarchical model is obtained by modifying the model in Example 4.43. More precisely, let the graph be

with the dependence chain being $\Delta(1) = \{\alpha, \beta\}$ and $\Delta(2) = \{\gamma, \delta\}$. Let the generating classes of a hierarchical recursive model be

$$\mathcal{A}_1 = \{\{\alpha\}, \{\beta\}\}, \quad \mathcal{A}_2 = \{\{\alpha, \beta, \delta\}, \{\beta, \gamma\}, \{\gamma, \delta\}\}.$$

Using a notation similar to that used in the quoted example, the maximum likelihood estimate becomes

$$\hat{p}_{rstu} = \frac{n_{r+++}\, n_{+s++}\, \hat{m}^*_{rstu}}{|n|^2\, n_{rs++}} = \frac{n_{r+++}\, n_{+s++}\, n_{rst+}\, \hat{m}_{stu}}{|n|^2\, n_{rs++}\, n_{+st+}}.$$

Here $\hat{m}_{stu}$ is the maximum likelihood estimate in the hierarchical model for the $\{\beta, \gamma, \delta\}$-marginal table including all two-factor interactions. This has to be calculated iteratively as described in Section 4.3.1. □

4.6.3 Decomposable block-recursive models

A particular interesting class of block-recursive models appears when all the hierarchical models involved in the description of the model are decomposable models. Since this in particular implies that the models must be graphical, all decomposable block-recursive models are chain graph models, but chain graph models whose associated undirected graphs $\mathcal{G}_*(t)$ are decomposable graphs.

It is worth noting that all recursive graphical models are decomposable in this sense, since the associated undirected graphs in this case are complete subgraphs involving the variables $\mathrm{cl}(\delta), \delta \in \Delta$.

The chain graph models in Example 4.43 and in Example 4.44 are both non-decomposable because the associated undirected graph $\mathcal{G}_*(2)$ is the 4-cycle in both cases. The chain graph model having the same graph as the hierarchical model in Example 4.46 is decomposable, but the model discussed in the example is not, because it is not graphical. Below is an example of a decomposable block-recursive model.

Example 4.47 Consider the chain graph model for the $R \times S \times T \times U \times V$-table with the five variables $\Delta = \{\alpha, \beta, \gamma, \delta, \varepsilon\}$ related as

where the chain components are $\Delta(1) = \{\alpha, \beta, \gamma\}$ and $\Delta(2) = \{\delta, \varepsilon\}$. The two associated undirected graphs $\mathcal{G}_*(1)$ and $\mathcal{G}_*(2)$ are

and each of the associated graphs is decomposable. Hence the model is decomposable and the maximum likelihood estimate of the unknown probability, based upon a multinomial sample, can be explicitly calculated to be

$$\hat{p}_{rstuv} = \frac{n_{rs+++}\, n_{+st++}}{|n|\, n_{+s+++}} \frac{n_{+s+u+}\, n_{+st+v}}{n_{+s+++}} \frac{1}{n_{+st++}}$$
$$= \frac{n_{rs+++}\, n_{+s+u+}\, n_{+st+v}}{|n|\, n_{+s+++}^2}.$$

□

There seems to be no need to perform a detailed study of the decomposable block-recursive models *per se* since the general results for chain graph models imply that the analysis can be composed from analyses of the undirected models which we have described earlier in sufficient detail.

4.7 Notes

4.7.1 Collapsibility

An important notion for the practical and conceptual analysis of contingency tables is that of *collapsibility*. There are two fundamental but distinct concepts. One is concerned with conditions for absence of phenomena like the Yule–Simpson paradox. In other words one studies conditions that ensure various measures of association between variables to be stable under marginalization of tables. This concept does not refer directly to the statistical model, but rather to a particular element of the corresponding subset of distributions. Important articles concerned with this type of *parametric collapsibility* are Bishop (1971), Whittemore (1978), Wermuth (1987), Geng (1992).

A different type of collapsibility was introduced in Lauritzen (1989a) and studied in detail by Asmussen and Edwards (1983). A graphical model $M(\mathcal{G})$ is said to be *collapsible* onto a subset $A \subseteq \Delta$ if it holds that

$$\hat{p}_A(i_A) = \hat{p}(i_A),$$

where $\hat{p}_A(i_A)$ is the maximum likelihood estimate of the marginal distribution $p(i_A)$ based on the marginal data and the marginal model $M(\mathcal{G}_A)$. A similar notion for hierarchical log–linear models is also studied in detail in Asmussen and Edwards (1983).

In words this condition says that estimation of relations among criteria in A can be performed in the A-marginal table. This can be a useful conceptual and computational tool in reducing the dimensionality of any given problem. This type of collapsibility could conveniently be termed *model collapsibility*.

Asmussen and Edwards (1983) show that a graphical model is collapsible onto a subset A if and only if $\mathrm{bd}(B)$ is complete for any connected component B of A^c. Algorithms for identification of collapsible subsets have, for example, been given by Geng (1989), Madigan and Mosurski (1990) and Badsberg (1995).

4.7.2 Bibliographical notes

There is an extensive general literature on the statistical theory of contingency tables and it will lead too far astray to to give a complete survey of that literature here. But a few selected references seem to be in order. The study of log–linear models and even recursive models in higher dimension began in a formal sense with the classic paper by Birch (1963) on three-dimensional contingency tables although there were fundamental precursors by Fisher (1925) and Bartlett (1935). The theory of log–linear models was studied in high generality and consequence by Goodman (1970, 1971) who among many other things also initiated the study of directed models (Goodman 1973). The formal theory of log–linear models was developed in rigour and impressive detail by Haberman (1974), with a compact version of some of the most important results given in A.H. Andersen (1974). Bishop *et al.* (1975) also had an enormous influence on the development of the subject.

The explicit regocnition of the full analogy between the notion of interaction as used in the theory of Markov random fields and the corresponding notion for contingency tables is of later date, and Darroch *et al.* (1980) appears to be the earliest reference. An elementary version of the theory of graphical models is given in Lauritzen (1989a). Wermuth and Lauritzen (1983) study recursive graphical models and their relation to decomposable, undirected graphical models and Edwards and Kreiner (1983) is an early study of the application potential of graphical models for contingency tables.

Some modern treatments of log–linear models have to some extent used the graphical ideas; see for example R. Christensen (1990).

5
Multivariate normal models

5.1 Basic facts and concepts

5.1.1 Notation

The graphical models in this chapter have a particularly simple interpretation and a rather detailed statistical theory. The models assume that the variables observed follow a regular multivariate normal distribution. Conditional independence restrictions in the multivariate normal distribution can be expressed in a simple fashion through zero restrictions on the inverse covariance matrix.

We will need some special notation for vectors in $\mathcal{R}^{|\Gamma|}$ and matrices with entries indexed by Γ. An arbitrary element of $\mathcal{R}^{|\Gamma|}$ is denoted as any of

$$y = y_\Gamma = (y_\gamma)_{\gamma \in \Gamma},$$

and for an arbitrary subset $d \subseteq \Gamma$ we let

$$y_d = (y_\gamma)_{\gamma \in d}$$

denote a $|d|$-dimensional subvector of y.

A $|\Gamma| \times |\Gamma|$ matrix with entries indexed by Γ is written as any of

$$A = A_\Gamma = A_{\Gamma\Gamma} = \{a_{\gamma\mu}\}_{\gamma,\mu \in \Gamma},$$

whereas for two arbitrary subsets d and e of Γ we let

$$A_{de} = \{a_{\gamma\mu}\}_{\gamma \in d, \mu \in e}$$

denote a $|d| \times |e|$ submatrix of A. For a partitioning $\Gamma = d \cup e$ with $d \cap e = \emptyset$ we can then use any of the block matrix notations

$$A = \begin{pmatrix} A_d & A_{de} \\ A_{ed} & A_e \end{pmatrix} = \begin{pmatrix} A_{dd} & A_{de} \\ A_{ed} & A_{ee} \end{pmatrix}.$$

For a $|d| \times |e|$ matrix $A = \{a_{\gamma\mu}\}_{\gamma \in d, \mu \in e}$ we let $[A]^\Gamma$ denote the matrix obtained from A by filling up with zero entries to obtain full dimension

$|\Gamma| \times |\Gamma|$, i.e.

$$([A]^\Gamma)_{\gamma\mu} = \begin{cases} a_{\gamma\mu} & \text{if } \gamma \in d, \mu \in e \\ 0 & \text{otherwise.} \end{cases}$$

When matrix operations are combined with forming submatrices, we use the convention that the matrix operation is performed first, i.e.

$$A_d^{-1} = (A^{-1})_d.$$

5.1.2 The saturated model

The model where no conditional independence restrictions are assumed to hold is called *the saturated model* as in the previous chapter. This model is concerned with a sample $(y^1, \ldots, y^n)$ of independent random vectors from a multivariate normal distribution $\mathcal{N}_{|\Gamma|}(\xi, \Sigma)$, where ξ and Σ are unknown and arbitrary apart from the restriction that Σ is assumed to be positive definite.

Exact results

Using (C.1), we get the likelihood function, expressed in the parameters (ξ, K) as

$$L(\xi, K) = (2\pi)^{-n|\Gamma|/2} (\det K)^{n/2} \prod_{\nu=1}^{n} \exp\left\{-\langle y^\nu - \xi, K(y^\nu - \xi)\rangle/2\right\}$$

$$\propto (\det K)^{n/2} \exp\left\{-\sum_{\nu=1}^{n}(y^\nu - \xi)^\top K(y^\nu - \xi)/2\right\}$$

$$= (\det K)^{n/2} \exp\left[-\operatorname{tr}\{K(y - e\xi^\top)^\top(y - e\xi^\top)\}/2\right]$$

$$= (\det K)^{n/2} \exp\left\{-\operatorname{tr}(Ky^\top y/2) + \operatorname{tr}(Ky^\top e\xi^\top) - n\operatorname{tr}(K\xi\xi^\top)/2\right\}$$

$$= (\det K)^{n/2} \exp\left\{-\operatorname{tr}(Ky^\top y/2) + \xi^\top Ky^\top e - n\xi^\top K\xi/2\right\}. \quad (5.1)$$

As in (C.6) we have let y be the $n \times |\Gamma|$ matrix with $(y^\nu)^\top$ as rows and e is the vector in $\mathcal{R}^n$ with all entries equal to one. Note that $y^\top y$ is the matrix of sums of squares and products of the coordinates of y^ν, and $y^\top e$ is the sum of the vectors y^ν.

To maximize the likelihood function, we choose to take advantage of the theory of exponential models as described in Appendix D. Although it is unnecessary in this particular case, it turns out to be convenient when we later discuss graphical models with conditional independence restrictions.

The expression (5.1) identifies the statistical model determined by the family of multivariate normal distributions with unknown ξ and unknown

BASIC FACTS AND CONCEPTS

concentration matrix K as an exponential model. To see this, we first recall from (B.21) that

$$\langle A, B \rangle = \text{tr}(A^\top B)$$

defines an inner product on the vector space of matrices of any fixed dimension, in particular also on the subspace $\mathcal{S}_{|\Gamma|}$ of symmetric $|\Gamma| \times |\Gamma|$ matrices.

Comparing with the definition of an exponential model in (D.1), we define the canonical parameter as $\theta = (K, h)$ where $h = K\xi$, the base measure μ as Lebesgue measure on $\mathcal{R}^{n \times |\Gamma|}$, and the canonical statistic as $t(y) = (-y^\top y/2, y^\top e)$. If we let the inner product on $\mathcal{S}_{|\Gamma|} \times \mathcal{R}^{|\Gamma|}$ be

$$\langle (S_1, s_1), (S_2, s_2) \rangle = \text{tr}(S_1 S_2) + s_1^\top s_2, \tag{5.2}$$

where $S_1, S_2 \in \mathcal{S}_{|\Gamma|}$ and $s_1, s_2 \in \mathcal{R}^{|\Gamma|}$, then the exponent in (5.1) can be written as

$$-\text{tr}(Ky^\top y/2) + \xi^\top K y^\top e - n\xi^\top K\xi/2 = \langle \theta, t(y) \rangle - n\xi^\top K\xi/2.$$

The cumulant function is found using that $\xi = K^{-1}h$:

$$\begin{aligned}\psi(K, h) &= \log\{(2\pi)^{n|\Gamma|/2}(\det K)^{-n/2}\} + nh^\top K^{-1} h/2 \\ &= (n|\Gamma|/2)\log(2\pi) - (n/2)\log\det K + nh^\top K^{-1} h/2. \end{aligned} \tag{5.3}$$

Since the integral

$$\int_{\mathcal{R}^{|\Gamma|}} e^{-(y-\xi)^\top K(y-\xi)/2} \, dy$$

is finite if and only if K is positive definite, it follows that the model is a regular exponential model. The closed convex support C of the canonical statistic is the set of pairs (A, b), where A is a symmetric $|\Gamma| \times |\Gamma|$ matrix, $b \in \mathcal{R}^{|\Gamma|}$ and further $A - bb^\top/n$ is non-negative definite, i.e.

$$C = \left\{ (A, b) \in \mathcal{S}_{|\Gamma|} \times \mathcal{R}^{|\Gamma|} \mid A - bb^\top/n \geq 0 \right\}. \tag{5.4}$$

The interior of C consists of those pairs (A, b) where $A - bb^\top/n$ is positive definite. The correctness of this is most easily established using the identity

$$\text{ssd} = \sum_{i=1}^{n} (y^i - \bar{y})(y^i - \bar{y})^\top = y^\top y - y^\top ee^\top y/n.$$

The details are left to the reader. We are then ready to give the basic estimation result for the saturated model.

Theorem 5.1 *In the saturated multivariate normal model, the maximum likelihood estimates of the unknown mean and covariance matrix exist if and only if*

$$\text{ssd} = y^\top y - y^\top ee^\top y/n$$

is positive definite. This happens with probability one if $n > |\Gamma|$ and never when $n \leq |\Gamma|$. When the estimates exist they are given as

$$\hat{\xi} = \bar{y} = y^\top e/n, \quad \hat{\Sigma} = ssd/n = y^\top y/n - y^\top ee^\top y/n^2,$$

and independently distributed as

$$\hat{\xi} \sim \mathcal{N}_{|\Gamma|}(\xi, \Sigma/n), \quad \hat{\Sigma} \sim \mathcal{W}_{|\Gamma|}(n-1, \Sigma/n).$$

Proof: Under the stated assumptions, the random variable SSD follows a Wishart distribution with $n-1$ degrees of freedom and parameter Σ, see (C.8).

The rank of SSD is at most $n-1$. Corollary C.16 then implies that SSD is positive definite with probability one when $n > |\Gamma|$, and singular when $n \leq |\Gamma|$.

The main result about estimation in exponential models, Theorem D.1, asserts that the maximum likelihood estimate exists if and only if the statistic is in the interior of its convex support, in this case if and only if ssd is positive definite. Then it is determined by the equations

$$\mathbf{E}(-Y^\top Y/2) = -y^\top y/2, \quad \mathbf{E}(Y^\top e) = y^\top e.$$

Expressing the left hand sides in terms of ξ and Σ leads to the equations

$$-(n\Sigma + n\xi\xi^\top)/2 = -y^\top y/2, \quad n\xi = y^\top e.$$

Solving for ξ and Σ leads to the expressions for the estimates. The distributional result follows directly from Proposition C.14. □

Asymptotic results

It is of interest to compare the exact distributional results for the saturated model with the asymptotic results obtained from the theory of exponential models.

Asymptotically, the estimates $(\hat{K}, \hat{h})$ are unbiased. However, if we exploit Proposition C.10, we find for $n > |\Gamma| + 2$

$$\mathbf{E}(\hat{K}) = \mathbf{E}(\hat{\Sigma}^{-1}) = \frac{n}{n - |\Gamma| - 2} K$$

and also

$$\mathbf{E}(\hat{h}) = \mathbf{E}(\hat{K}\hat{\xi}) = \mathbf{E}(\hat{K})\mathbf{E}(\hat{\xi}) = \frac{n}{n - |\Gamma| - 2} K\xi = \frac{n}{n - |\Gamma| - 2} h.$$

The bias can be considerable unless n is very large compared to $|\Gamma|$. From Proposition D.3 we know that the asymptotic covariance of $(\hat{K}, \hat{h})$ is given as the inverse of the covariance of the canonical statistic. The latter can be determined as the Hessian of the logarithm to the cumulant function in (5.3). We choose, however, to determine the covariance directly from its distribution. We must for $(A, b) \in \mathcal{S}_{|\Gamma|} \times \mathcal{R}^{|\Gamma|}$ find

$$\mathbf{V}\{\langle (A,b), t(Y) \rangle\} = \mathbf{V}\{(-\operatorname{tr}(AY^\top Y/2) + b^\top Y^\top e\}.$$

Since
$$Y^\top Y = SSD + n\bar{Y}\bar{Y}^\top,$$
this amounts to finding

$$\mathbf{V}\{\langle (A,b), t(Y) \rangle\} - \mathbf{V}\{-\operatorname{tr}(A\,SSD)/2 + n(b^\top \bar{Y} - \bar{Y}^\top A\bar{Y}/2)\}.$$

Since SSD and $\bar{Y}$ are independent, we can add the variances of the terms. The variance of the first term is given in Proposition C.10 as

$$\mathbf{V}\{\operatorname{tr}(ASSD)/2\} = \frac{1}{4}2(n-1)\langle A, (\Sigma \otimes \Sigma)A\rangle = \frac{n-1}{2}\operatorname{tr}(A\Sigma A\Sigma). \quad (5.5)$$

To find the variance of the second term we let $Z = \bar{Y} - \xi$ and note that then
$$\mathbf{V}(b^\top \bar{Y} - \bar{Y}^\top A\bar{Y}/2) = \mathbf{V}\{(b^\top - \xi^\top A)Z - Z^\top AZ/2\}.$$

Letting $U = (b^\top - \xi^\top A)Z - Z^\top AZ/2$, we find

$$\mathbf{E}(U^2) = (b^\top - \xi^\top A)\Sigma(b - A\xi)/n + \mathbf{E}\{(Z^\top AZ)^2\}/4$$

and
$$\mathbf{E}(U) = \mathbf{E}(Z^\top AZ)/2,$$
whereby
$$\begin{aligned}\mathbf{V}(U) &= (b^\top - \xi^\top A)\Sigma(b - A\xi)/n + \mathbf{V}(Z^\top AZ)/4 \\ &= (b^\top - \xi^\top A)\Sigma(b - A\xi)/n + \operatorname{tr}(A\Sigma A\Sigma)/(2n^2).\end{aligned}$$

Combining with (5.5) we then have

$$\mathbf{V}\{\langle (A,b), t(Y)\rangle\} = n\operatorname{tr}(A\Sigma A\Sigma)/2 + n(b^\top - \xi^\top A)\Sigma(b - A\xi). \quad (5.6)$$

The covariance operator can be identified as

$$v(\theta) = H = n\begin{pmatrix} \frac{1}{2}(\Sigma \otimes \Sigma + \Sigma \otimes \xi\xi^\top + \xi\xi^\top \otimes \Sigma) & F^\top \\ F & \Sigma \end{pmatrix}, \quad (5.7)$$

where
$$F(A) = -\Sigma A\xi, \quad F^\top(b) = -\frac{1}{2}(\Sigma b\xi^\top + \xi b^\top \Sigma).$$

That $F^\top$ is the transpose of F follows from the identity

$$\begin{aligned}\langle A, F^\top(b)\rangle &= -\frac{1}{2}\operatorname{tr}(A\Sigma b\xi^\top + A\xi b^\top \Sigma)\\ &= -\frac{1}{2}\{\operatorname{tr}(A\Sigma b\xi^\top) + \operatorname{tr}(\Sigma b\xi^\top A)\}\\ &= -\operatorname{tr}(A\Sigma b\xi^\top) = -\xi^\top A\Sigma b = \langle F(A), b\rangle.\end{aligned}$$

To verify that the expression for the covariance operator is correct, we find

$$\langle (A,b), H(A,b)\rangle$$
$$= n\left[\operatorname{tr}(A\Sigma A\Sigma)/2 + \operatorname{tr}(A\Sigma A\xi\xi^\top) + \operatorname{tr}\{AF^\top(b)\} + b^\top F(A) + b^\top \Sigma b\right],$$

which reduces to (5.6).

The asymptotic covariance of $(\hat{K}, \hat{h})$ is given by the inverse operator. From (B.2) its upper left corner can be found as

$$\begin{aligned}\left\{\frac{n}{2}(\Sigma\otimes\Sigma + \Sigma\otimes\xi\xi^\top + \xi\xi^\top\otimes\Sigma) - F^\top\Sigma^{-1}F\right\}^{-1} &= \left(\frac{n}{2}\Sigma\otimes\Sigma\right)^{-1}\\ &= \frac{2}{n}K\otimes K,\end{aligned}$$

where we have used

$$\begin{aligned}F^\top\Sigma^{-1}F(A) &= \frac{1}{2}(\Sigma A\xi\xi^\top + \xi\xi^\top A\Sigma)\\ &= \frac{1}{2}\{(\Sigma\otimes\xi\xi^\top)(A) + (\xi\xi^\top\otimes\Sigma)(A)\}. \quad (5.8)\end{aligned}$$

Further calculation exploiting (B.2) gives the full expression for the asymptotic covariance operator

$$\overset{a}{\mathbf{V}}(\hat{K}, \hat{h}) = v(\theta)^{-1} = H^{-1} = n^{-1}\begin{pmatrix}2K\otimes K & L^\top\\ L & M\end{pmatrix}, \quad (5.9)$$

where

$$\begin{aligned}L^\top(b) &= Kb\xi^\top K + K\xi b^\top K\\ L(A) &= 2KAK\xi\\ M(b) &= (1+\xi^\top K\xi)Kb + (b^\top K\xi)K\xi.\end{aligned}$$

The asymptotic covariance of $\hat{K}$ is thus $\overset{a}{\mathbf{V}}(\hat{K}) = 2(K\otimes K)/n$, which should be contrasted with the exact expression from Proposition C.10

$$\mathbf{V}(\hat{K}) = \frac{2n^2\left(K\otimes K + \frac{1}{n-|\Gamma|-2}K\odot K\right)}{(n-|\Gamma|-1)(n-|\Gamma|-2)(n-|\Gamma|-3)}.$$

Again, this can be quite different unless n is much larger than the number of variables $|\Gamma|$.

5.1.3 Conditional independence

Before the graphical models are described in detail, it seems appropriate to clarify the connection between conditional independence and the multivariate normal distribution.

Let $Y = (Y_\gamma)_{\gamma \in \Gamma}$ be a random vector in $\mathcal{R}^{|\Gamma|}$ following a multivariate normal distribution with mean ξ and covariance matrix Σ. Assume the covariance to be regular such that the concentration matrix $K = \Sigma^{-1}$ is well defined. Conditional independence in the multivariate normal distribution is simply reflected in the concentration matrix of the distribution through zero entries. This fact is formalized below.

Proposition 5.2 *Assume that $Y \sim \mathcal{N}_{|\Gamma|}(\xi, \Sigma)$, where Σ is regular. Then it holds for $\gamma, \mu \in \Gamma$ with $\gamma \neq \mu$ that*

$$Y_\gamma \perp\!\!\!\perp Y_\mu \mid Y_{\Gamma \setminus \{\gamma, \mu\}} \iff k_{\gamma\mu} = 0,$$

where $K = \{k_{\alpha\beta}\}_{\alpha, \beta \in \Gamma} = \Sigma^{-1}$ is the concentration matrix of the distribution.

Proof: This is almost a direct consequence of Proposition C.5 and its proof. Comparing (C.2) and (C.3) identifies the matrix

$$K_{\{\gamma,\mu\}} = \begin{pmatrix} k_{\gamma\gamma} & k_{\gamma\mu} \\ k_{\mu\gamma} & k_{\mu\mu} \end{pmatrix} \tag{5.10}$$

as the concentration matrix of the conditional distribution of $Y_{\{\gamma,\mu\}}$ given $Y_{\Gamma \setminus \{\gamma,\mu\}}$. The covariance matrix of this conditional distribution is therefore equal to

$$\Sigma_{\gamma,\mu \mid \Gamma \setminus \{\gamma,\mu\}} = \frac{1}{\det K_{\{\gamma,\mu\}}} \begin{pmatrix} k_{\mu\mu} & -k_{\gamma\mu} \\ -k_{\mu\gamma} & k_{\gamma\gamma} \end{pmatrix}. \tag{5.11}$$

The desired independence now follows from Corollary C.6. □

This fundamental relation forms the basis for all models treated in this chapter. Corresponding to the different Markov properties studied in Chapter 3, we have multivariate normal models defined through restricting particular elements in suitable concentration matrices to be equal to zero.

The entries in the concentration matrix K have a simple interpretation. It follows from (C.2) and (C.3) that the diagonal elements $k_{\gamma\gamma}$ are reciprocals of the conditional variances, given the remaining variables, i.e.

$$k_{\gamma\gamma} = \mathbf{V}(Y_\gamma \mid Y_{\Gamma \setminus \{\gamma\}})^{-1}$$

for all $\gamma \in \Gamma$. Let further $C = \{c_{\alpha\beta}\}_{\alpha,\beta\in\Gamma}$ be the matrix obtained by scaling K to have all diagonal elements equal to one,

$$c_{\gamma\mu} = \frac{k_{\gamma\mu}}{\sqrt{k_{\gamma\gamma}k_{\mu\mu}}}.$$

Then $c_{\gamma\mu}$, the off-diagonal elements in C, are equal to the negative *partial correlation coefficients*

$$\rho_{\gamma\mu\,|\,\Gamma\setminus\{\gamma,\mu\}} = \frac{\mathrm{Cov}(Y_\gamma, Y_\mu\,|\,Y_{\Gamma\setminus\{\gamma,\mu\}})}{\mathbf{V}(Y_\gamma\,|\,Y_{\Gamma\setminus\{\gamma,\mu\}})^{1/2}\mathbf{V}(Y_\mu\,|\,Y_{\Gamma\setminus\{\gamma,\mu\}})^{1/2}} = -c_{\gamma\mu}.$$

This follows from (5.11) since

$$\mathbf{V}(Y_\gamma\,|\,Y_{\Gamma\setminus\{\gamma,\mu\}}) = \frac{k_{\mu\mu}}{k_{\gamma\gamma}k_{\mu\mu} - k_{\gamma\mu}^2}$$

and

$$\mathrm{Cov}(Y_\gamma, Y_\mu\,|\,Y_{\Gamma\setminus\{\gamma,\mu\}}) = \frac{-k_{\gamma\mu}}{k_{\gamma\gamma}k_{\mu\mu} - k_{\gamma\mu}^2}.$$

Note that it also holds that

$$\left(\rho_{\gamma\mu\,|\,\Gamma\setminus\{\gamma,\mu\}}\right)^2 = (c_{\gamma\mu})^2 = 1 - \frac{\det\Sigma\,\det\Sigma_{\Gamma\setminus\{\gamma,\mu\}}}{\det\Sigma_{\Gamma\setminus\{\gamma\}}\det\Sigma_{\Gamma\setminus\{\mu\}}}. \qquad (5.12)$$

This follows from the relations

$$\det\Sigma = \det\Sigma_{\Gamma\setminus\{\gamma\}}\mathbf{V}(Y_\gamma\,|\,Y_{\Gamma\setminus\{\gamma\}}) = \det\Sigma_{\Gamma\setminus\{\gamma\}}k_{\gamma\gamma}^{-1}$$

and

$$\det\Sigma_{\Gamma\setminus\{\mu\}} = \det\Sigma_{\Gamma\setminus\{\gamma,\mu\}}\mathbf{V}(Y_\gamma\,|\,Y_{\Gamma\setminus\{\gamma,\mu\}}) = \det\Sigma_{\Gamma\setminus\{\gamma,\mu\}}\frac{k_{\mu\mu}}{k_{\gamma\gamma}k_{\mu\mu} - k_{\gamma\mu}^2},$$

which both are easy consequences of (C.5).

From Proposition C.5 we have that the conditional distribution of Y_γ given $Y_{\Gamma\setminus\{\gamma\}} = y_{\Gamma\setminus\{\gamma\}}$ is univariate normal. Writing the conditional expectation as

$$\xi_\gamma + \sum_{\mu\in\Gamma\setminus\{\gamma\}} \beta_{\gamma\mu\,|\,\Gamma\setminus\{\gamma\}}(y_\mu - \xi_\mu)$$

and using (C.4) we find the *partial regression coefficient* as

$$\beta_{\gamma\mu\,|\,\Gamma\setminus\{\gamma\}} = -k_{\gamma\mu}/k_{\gamma\gamma}.$$

5.1.4 Interaction

It is illuminating to investigate the additive terms in the logarithm of the normal density, thereby highlighting the analogy to interaction expansions of the discrete models, described in Section B.2 and Section 4.3.

Using the expression (C.1) for the multivariate normal density we get

$$\begin{aligned}
\log f(y) &= c - \langle y - \xi, K(y - \xi)\rangle/2 \\
&= c_1 - \frac{1}{2}\sum_{\gamma \in \Gamma} k_{\gamma\gamma} y_\gamma^2 + \sum_{\gamma \in \Gamma} y_\gamma \left(\sum_{\mu \in \Gamma} k_{\gamma\mu}\xi_\mu\right) - \sum_{\{\gamma,\mu\}} k_{\gamma\mu} y_\gamma y_\mu \\
&= c_1 - \frac{1}{2}\sum_{\gamma \in \Gamma} k_{\gamma\gamma} y_\gamma^2 + \sum_{\gamma \in \Gamma} h_\gamma y_\gamma - \sum_{\{\gamma,\mu\}} k_{\gamma\mu} y_\gamma y_\mu,
\end{aligned} \quad (5.13)$$

where $\{\gamma, \mu\}$ in the sums above represent all unordered pairs of elements of Γ, c and c_1 are constants, and we have let $h = K\xi$.

The expansion shows that the logarithm of the density is additively composed of *quadratic main effects* with coefficients $-k_{\gamma\gamma}/2$, *linear main effects* with coefficients h_γ, and *quadratic interactions* with coefficients $-k_{\gamma\mu}$. We will sometimes use the terms interactions and main effects referring directly to the coefficients and also omit the negative signs and the division by two. So, for example, we will refer to $k_{\gamma\gamma}$ as the quadratic main effect of the variable γ, although this, strictly speaking, should refer to $-k_{\gamma\gamma} y_\gamma^2/2$.

We emphasize that the interaction terms of highest order in (5.13) involve pairs of variables, and there are no terms involving groups of variables with three or more elements. This is in contrast to the discrete case and it follows in particular that within the normal distribution there are no hierarchical interaction models which are not graphical.

5.2 Covariance selection models

The graphical interaction models for the multivariate normal distribution are called covariance selection models. They are determined by assuming conditional independence of selected pairs of variables, given the remaining ones.

Thus, if $\mathcal{G} = (\Gamma, E)$ is an undirected graph and $Y = Y_\Gamma$ is a random variable taking values in $\mathcal{R}^{|\Gamma|}$, the *Gaussian graphical model* or *covariance selection model* for Y with graph $\mathcal{G}$ is given by assuming that Y follows a multivariate normal distribution which obeys the undirected pairwise Markov property with respect to $\mathcal{G}$. Since the density is positive and continuous, this implies the global and local Markov properties and the density factorizes.

It follows from Proposition 5.2 that this is equivalent to assuming the quadratic interactions $k_{\gamma\mu}$ to be equal to zero for all pairs γ, μ which are not adjacent in $\mathcal{G}$. The expression (5.13) for the normal density then reduces to

$$\log f(y) = c_1 - \frac{1}{2}\sum_{\gamma \in \Gamma} k_{\gamma\gamma} y_\gamma^2 + \sum_{\gamma \in \Gamma} h_\gamma y_\gamma - \sum_{\{\gamma,\mu\} \in E} k_{\gamma\mu} y_\gamma y_\mu,$$

where the linear interaction coefficients are given as

$$h_\gamma = k_{\gamma\gamma}\xi_\gamma + \sum_{\mu \in \mathrm{bd}(\gamma)} k_{\gamma\mu}\xi_\mu.$$

Let $\mathcal{S}(\mathcal{G})$ denote the set of symmetric matrices A satisfying for all $\gamma, \mu \in \Gamma$ that

$$\gamma \not\sim \mu \implies a_{\gamma\mu} = 0$$

and $\mathcal{S}^+(\mathcal{G})$ those elements of $\mathcal{S}(\mathcal{G})$ that are positive definite. Then the covariance selection model for Y can be compactly described as

$$Y \sim \mathcal{N}_{|\Gamma|}(\xi, \Sigma), \quad \Sigma^{-1} \in \mathcal{S}^+(\mathcal{G}).$$

5.2.1 Maximum likelihood estimation

The likelihood equations

Consider a sample $(y^1, \ldots, y^n)$ from a covariance selection model. The likelihood function is obtained from (5.1):

$$L(\xi, K) \propto (\det K)^{n/2} \exp\left\{-\operatorname{tr}(Ky^\top y/2) + \xi^\top K y^\top e - n\xi^\top K\xi/2\right\}.$$

For an arbitrary matrix A, we let $A(\mathcal{G})$ denote the matrix with entries

$$A(\mathcal{G})_{\gamma\mu} = \begin{cases} 0 & \text{if } \gamma \not\sim \mu \\ a_{\gamma\mu} & \text{otherwise.} \end{cases}$$

Exploiting that $K \in \mathcal{S}(\mathcal{G})$ we find that

$$\operatorname{tr}(Ky^\top y) = \operatorname{tr}\{Kss(\mathcal{G})\},$$

where we have let $ss = y^\top y$. Letting as before $h = K\xi$, the likelihood function reduces to

$$L(K, \xi) \propto (\det K)^{n/2} \exp\left[-\operatorname{tr}\{Kss(\mathcal{G})/2\} + h^\top y^\top e - n\xi^\top K\xi/2\right]. \tag{5.14}$$

The restriction which is imposed on the distribution of Y by the model is linear in the canonical parameter K. Hence the hypothesis $K \in \mathcal{S}^+(\mathcal{G})$ is an affine hypothesis, cf. Section D.1, and it follows that a covariance selection model is itself a regular exponential model with canonical statistic equal to $(-ss(\mathcal{G})/2, y^\top e)$. The following result about maximum likelihood estimation then follows directly.

Theorem 5.3 *In the covariance selection model, the maximum likelihood estimates of the unknown mean and covariance matrix exist if*

$$ssd = y^\top y - y^\top ee^\top y/n$$

is positive definite. If $n > |\Gamma|$ this happens with probability one. When the estimates exist, the estimate of the mean is

$$\hat{\xi} = \bar{y} = y^\top e/n,$$

whereas the estimate of the unknown covariance matrix Σ is determined as the unique solution to the system of equations

$$n\hat{\sigma}_{\rho\rho} = ssd_{\rho\rho}, \quad n\hat{\sigma}_{\gamma\mu} = ssd_{\gamma\mu}, \quad \rho \in \Gamma, \{\gamma, \mu\} \in E,$$

which also satisfies the model restriction $\Sigma^{-1} \in \mathcal{S}^+(\mathcal{G})$.

Proof: The maximum likelihood estimates are obtained directly as for the saturated model:

$$\mathbf{E}\{-SS(\mathcal{G})/2\} = -ss(\mathcal{G})/2, \quad \mathbf{E}(Y^\top e) = y^\top e.$$

The first of these equations is a restriction of the corresponding equation from the saturated model. The result then follows directly from Theorem D.1 and Theorem 5.1 by expressing the expectations in terms of ξ and Σ, and eliminating ξ from the first equation. □

Note that the condition $n > |\Gamma|$ for existence of the maximum likelihood estimate in this case is only sufficient, not necessary. From the likelihood equations it follows that a necessary condition is that $n > \max_{C \in \mathcal{C}} |C|$, as otherwise ssd_C would not all be positive definite. However, this condition is not sufficient for the existence. The problem has been studied in some detail by Buhl (1993). We shall return to a discussion of this issue later in Section 5.3.2.

An alternative way of writing the part of the estimating equations involving Σ is

$$n\hat{\Sigma}(\mathcal{G}) = ssd(\mathcal{G}).$$

To make the similarity between these equations and the maximum likelihood equations (4.22) for hierarchical log–linear models apparent, we could also write

$$n\hat{\Sigma}_{cc} = ssd_{cc} \quad \text{for all } c \in \mathcal{C}, \tag{5.15}$$

where $\mathcal{C}$ is the set of cliques of $\mathcal{G}$.

In a general covariance selection model no exact distributional results concerning the estimates of the covariance matrix are available. Clearly, since the equations for the mean and covariance are separate, it follows from the saturated case, Theorem 5.1, that $\hat{\xi}$ and $\hat{\Sigma}$ are independent and that $\hat{\xi} \sim \mathcal{N}_{|\Gamma|}(\xi, \Sigma/n)$ as usual.

The asymptotic distribution of the maximum likelihood estimate is multivariate normal from standard exponential family theory. The asymptotic covariance in the case of a general covariance selection model is less straightforward. One can proceed as follows. Let $N_{\mathcal{G}}$ denote the orthogonal projection onto $\mathcal{S}(\mathcal{G})$, i.e. the linear map that sends a symmetric matrix A into $A(\mathcal{G})$. The covariance operator of the sufficient statistic then becomes from (5.7) and Proposition B.8

$$v_{\mathcal{G}}(\theta) = H_{\mathcal{G}} = n \begin{pmatrix} \frac{1}{2}N_{\mathcal{G}}(\Sigma \otimes \Sigma + \Sigma \otimes \xi\xi^{\top} + \xi\xi^{\top} \otimes \Sigma)N_{\mathcal{G}} & N_{\mathcal{G}}F^{\top} \\ FN_{\mathcal{G}} & \Sigma \end{pmatrix}.$$

The asymptotic covariance operator of the maximum likelihood estimate is obtained by inversion of this operator; see also Dempster (1973).

Iterative proportional scaling

Theorem 5.3 identifies which equations to solve in order to maximize the likelihood function, but it gives no advice on doing so. In general the equations concerning the estimates for the covariance matrix have to be solved by iterative methods. Below we describe one of these, which is based upon the method of iterative partial maximization described in Section A.4 and also used for hierarchical log–affine models in Section 4.3.1. It consists of iteratively and successively adjusting the covariance matrices for the clique marginals appearing in (5.15). The algorithm was discussed in detail by Speed and Kiiveri (1986) along with other algorithms.

Let ssd from a sample $(y^1, \ldots, y^n)$ be given, and consider a covariance selection model with graph $\mathcal{G}$. For $K \in \mathcal{S}^{+}(\mathcal{G})$ and $c \in \mathcal{C}$, define the operation of 'adjusting the c-marginal' by

$$T_c K = K + \left[n(ssd_{cc})^{-1} - (K_{cc}^{-1})^{-1}\right]^{\Gamma}. \tag{5.16}$$

This operation is clearly well defined if ssd_{cc} is positive definite. If we let $a = \Gamma \setminus c$ and exploit (B.2) we find

$$K_{cc}^{-1} = \Sigma_{cc} = \left\{K_{cc} - K_{ca}(K_{aa})^{-1}K_{ac}\right\}^{-1}, \tag{5.17}$$

giving the alternative expression

$$T_c K = \begin{pmatrix} n(ssd_{cc})^{-1} + K_{ca}(K_{aa})^{-1}K_{ac} & K_{ca} \\ K_{ac} & K_{aa} \end{pmatrix}. \tag{5.18}$$

Using this expression for $T_c K$, we find the covariance $\tilde{\Sigma}_{cc}$ corresponding to the adjusted concentration matrix

$$\begin{aligned}\tilde{\Sigma}_{cc} &= (T_c K)^{-1}_{cc} \\ &= \{n(ssd_{cc})^{-1} + K_{ca}(K_{aa})^{-1}K_{ac} - K_{ca}(K_{aa})^{-1}K_{ac}\}^{-1} \\ &= ssd_{cc}/n, \end{aligned} \qquad (5.19)$$

hence $T_c K$ does indeed adjust the marginals. From (5.16) it is seen that the pattern of zeros in K is preserved, i.e. $T_c K$ is in $\mathcal{S}(\mathcal{G})$ if K is, and applying Lemma B.1 to (5.18) shows that it stays positive definite. Hence the adjusted concentration matrix $T_c K$ is in $\mathcal{S}^+(\mathcal{G})$ if K is.

In fact, it is not difficult to see that the operation T_c also scales proportionally in the sense that

$$f\{y \mid \hat{\xi}, (T_c K)^{-1}\} = f(y \mid \hat{\xi}, K^{-1}) \frac{f(y_c \mid \hat{\xi}_c, ssd_{cc}/n)}{f(y_c \mid \hat{\xi}_c, \Sigma_{cc})}.$$

This clearly demonstrates the analogy to the procedure used for hierarchical log–affine models.

Next we choose any ordering $(c_1, \ldots, c_k)$ of the cliques in $\mathcal{G}$. Choose further an arbitrary starting value $K_0 \in \mathcal{S}^+(\mathcal{G})$ and define recursively for $r = 0, 1, \ldots$

$$K_{r+1} = (T_{c_1} \cdots T_{c_k}) K_r. \qquad (5.20)$$

Then we have

Theorem 5.4 *Consider a sample from a covariance selection model with graph $\mathcal{G}$ and assume that ssd is such that the maximum likelihood estimate $\hat{K}$ of K exists. Then*

$$\hat{K} = \lim_{r \to \infty} K_r.$$

Proof: We must realize that this is a special instance of iterative partial maximization, discussed in Section A.4. To do this, we let

$$\Theta = \left\{ (K, \xi) \in \mathcal{S}^+(\mathcal{G}) \times \mathcal{R}^{|\Gamma|} \mid L(\xi, K) \geq L(\hat{\xi}, K_0) \right\},$$

where $K_0 \in \mathcal{S}^+(\mathcal{G})$ is chosen arbitrarily. Since a covariance selection model is a regular exponential model and the maximum likelihood estimate is assumed to exist, Θ is compact by Lemma D.2.

It is obvious that the transformation T_c is continuous for all $c \in \mathcal{C}$ and, as mentioned, also that it maps $\mathcal{S}^+(\mathcal{G})$ into itself.

Next we establish that $(T_c K, \hat{\xi})$ maximizes the likelihood function over the section

$$\Theta_c = \Theta_c(K, \hat{\xi}) = \left\{ (A, b) \in \mathcal{S}^+(\mathcal{G}) \times \mathcal{R}^{|\Gamma|} \mid A_{aa} = K_{aa}, A_{ac} = K_{ac}, b = \hat{\xi} \right\},$$

where we have let $a = \Gamma \setminus c$. In Θ_c it holds that

$$\begin{aligned}&\operatorname{tr}\{K ssd^\xi(\mathcal{G})/2\}\\ &= \operatorname{tr}\{K_{cc} ssd^\xi_{cc}/2\} + \operatorname{tr}\{A_{aa} ssd^\xi(\mathcal{G})_{aa}/2\} + \operatorname{tr}\{A_{ac} ssd^\xi(\mathcal{G})_{ca}\},\end{aligned}$$

where

$$ssd^\xi = \sum_{\nu=1}^{n}(y^\nu - \xi)(y^\nu - \xi)^\top.$$

Using the expression (5.41) for the likelihood function, we identify the subfamily determined by the section as an exponential family with canonical statistic $-ssd^\xi_{cc}/2$, leading to the likelihood equations

$$-n\Sigma_{cc}/2 = -ssd^\xi_{cc}/2.$$

Using (5.19) and the identity $ssd^{\hat{\xi}}_{cc} = ssd_{cc}$ identifies $T_c(K)$ as maximizer.

Since we know already that the global maximum of L is uniquely determined, the theorem now follows from Proposition A.3. □

Decomposition of covariance selection models

Consider a covariance selection model with graph $\mathcal{G} = (\Gamma, E)$ and assume the graph to be decomposed by (A, B, C). From Definition 2.1 we recall that this means that the sets A, B, and C are disjoint, $\Gamma = A \cup B \cup C$, C separates A from B, and C is complete in $\mathcal{G}$. As in Section 4.3.1, we say that the covariance selection model is a *direct join* of the covariance selection models for the marginals $Y_{A \cup C}$ and $Y_{B \cup C}$, with graphs $\mathcal{G}_{A \cup C}$ and $\mathcal{G}_{B \cup C}$ respectively. First a useful lemma.

Lemma 5.5 *Let $K \in \mathcal{S}(\mathcal{G})$, and let (A, B, C) be a disjoint partitioning of $\mathcal{G}$ with C separating A from B. Then*

$$K = [K_{A \cup C}]^\Gamma + [K_{B \cup C}]^\Gamma - [K_C]^\Gamma \qquad (5.21)$$

and for any symmetric $|\Gamma| \times |\Gamma|$-matrix L, we have

$$\operatorname{tr}(KL) = \operatorname{tr}(K_{A \cup C} L_{A \cup C}) + \operatorname{tr}(K_{B \cup C} L_{B \cup C}) - \operatorname{tr}(K_C L_C). \qquad (5.22)$$

Further, if K is regular then

$$K = \left[(K^{-1}_{A \cup C})^{-1}\right]^\Gamma + \left[(K^{-1}_{B \cup C})^{-1}\right]^\Gamma - \left[(K^{-1}_C)^{-1}\right]^\Gamma \qquad (5.23)$$

and the determinant satisfies

$$\det(K) = \frac{\det(K^{-1}_C)}{\det(K^{-1}_{A \cup C})\det(K^{-1}_{B \cup C})}. \qquad (5.24)$$

Proof: Since C separates A from B and $K \in \mathcal{S}(\mathcal{G})$, we have

$$K = \begin{pmatrix} K_{AA} & K_{AC} & 0 \\ K_{CA} & K_{CC} & K_{CB} \\ 0 & K_{BC} & K_{BB} \end{pmatrix}.$$

The relation (5.21) is immediate and the identity (5.22) follows by direct calculation.

If K is regular we can use (B.2) to get

$$\left(K_{A\cup C}^{-1}\right)^{-1} = \begin{pmatrix} K_{AA} & K_{AC} \\ K_{CA} & K_{CC} - K_{CB}\left(K_{BB}\right)^{-1}K_{BC} \end{pmatrix}, \quad (5.25)$$

and similarly for $\left(K_{B\cup C}^{-1}\right)^{-1}$. Further, by the same formula,

$$\left(K_C^{-1}\right)^{-1} = K_{CC} - (K_{CA}, K_{CB}) \begin{pmatrix} K_{AA} & 0 \\ 0 & K_{BB} \end{pmatrix}^{-1} \begin{pmatrix} K_{AC} \\ K_{BC} \end{pmatrix}$$

$$= K_{CC} - K_{CA}\left(K_{AA}\right)^{-1}K_{AC} - K_{CB}\left(K_{BB}\right)^{-1}K_{BC}. \quad (5.26)$$

Combining these relations gives (5.23). For the final identity involving the determinant, we use (B.1) to get

$$\det(K) = \det(K_{AA})/\det\left(K_{B\cup C}^{-1}\right). \quad (5.27)$$

Exploiting (5.25) and (5.26) we also find

$$\det\left(K_{A\cup C}^{-1}\right) = \det\left(K_C^{-1}\right)/\det K_{AA}. \quad (5.28)$$

Solving for $\det K_{AA}$ in (5.28) and inserting the result into (5.27) gives the equation (5.24). $\square$

In Lemma 5.5 K may have negative eigenvalues. If we assume that K is also positive definite, both (5.23) and (5.24) follow immediately from the conditional independence factorization (3.17) of the normal density with K as concentration matrix and mean $\xi = 0$:

$$\begin{aligned}
f(y) &= (2\pi)^{-|\Gamma|/2}(\det K)^{1/2}\exp\left\{-\langle y, Ky\rangle/2\right\} \\
&= \frac{f(y_{A\cup C})f(y_{B\cup C})}{f(y_C)} \\
&= (2\pi)^{-|\Gamma|/2}\frac{\left(\det K_{A\cup C}^{-1}\right)^{-1/2}\left(\det K_{B\cup C}^{-1}\right)^{-1/2}}{\left(\det K_C^{-1}\right)^{-1/2}} \\
&\quad \times \exp\left\{-\langle y_{A\cup C}, (K_{A\cup C}^{-1})^{-1}y_{A\cup C}\rangle/2\right\} \\
&\quad \times \frac{\exp\left\{-\langle y_{B\cup C}, (K_{B\cup C}^{-1})^{-1}y_{B\cup C}\rangle/2\right\}}{\exp\left\{-\langle y_C, (K_C^{-1})^{-1}y_C\rangle/2\right\}}.
\end{aligned}$$

When K is just an arbitrary regular matrix, we must work a bit more to get the result.

For a sample $(y^1, \ldots, y^n)$ of size n we may consider the marginal samples $(y^1_{A\cup C}, \ldots, y^n_{A\cup C})$ and $(y^1_{B\cup C}, \ldots, y^n_{B\cup C})$ that only concern variables in the two subsets. Let $\hat{K}_{[A\cup C]}$ and $\hat{K}_{[B\cup C]}$ denote the maximum likelihood estimates of the concentration matrices in the two marginal models with graphs $\mathcal{G}_{A\cup C}$ and $\mathcal{G}_{B\cup C}$, based on the data in the marginal samples only. With a similar notation for other parameters we then have

$$\hat{\Sigma}_{[A\cup C]} = \left(\hat{K}_{[A\cup C]}\right)^{-1}, \quad \hat{h}_{[A\cup C]} = \hat{K}_{[A\cup C]} \hat{\xi}_{[A\cup C]}$$

and analogously for the entities involving $B \cup C$ and C.

Proposition 5.6 *Consider a sample from a covariance selection model with graph $\mathcal{G} = (\Gamma, E)$ decomposed by (A, B, C). The maximum likelihood estimate $\hat{K}$ exists if and only if the estimates $\hat{K}_{[A\cup C]}$ and $\hat{K}_{[B\cup C]}$ exist. If the estimates exist, they satisfy*

$$\hat{K} = \left[\hat{K}_{[A\cup C]}\right]^{\Gamma} + \left[\hat{K}_{[B\cup C]}\right]^{\Gamma} - n\left[(ssd_{CC})^{-1}\right]^{\Gamma}, \qquad (5.29)$$

and

$$\det \hat{K} = \left(\det \hat{K}_{[A\cup C]}\right)\left(\det \hat{K}_{[B\cup C]}\right) n^{-|C|} (\det ssd_{CC}). \qquad (5.30)$$

Further, it holds that

$$\hat{\Sigma}_{[A\cup C]} = \hat{\Sigma}_{A\cup C}, \quad \hat{\Sigma}_{[B\cup C]} = \hat{\Sigma}_{B\cup C} \qquad (5.31)$$

and

$$\hat{h} = \left[\hat{h}_{[A\cup C]}\right]^{\Gamma} + \left[\hat{h}_{[B\cup C]}\right]^{\Gamma} - \left[\hat{h}_{[C]}\right]^{\Gamma}. \qquad (5.32)$$

Proof: Assume first that the maximum likelihood estimates exist in the two models determined by the subgraphs. Since C is complete in all involved graphs, ssd_{CC} is the maximum likelihood estimate of $n\Sigma_{CC}$ in all the models, by (5.15), and hence it must be positive definite. Define now

$$\tilde{K} = \left[\hat{K}_{[A\cup C]}\right]^{\Gamma} + \left[\hat{K}_{[B\cup C]}\right]^{\Gamma} - n\left[(ssd_{CC})^{-1}\right]^{\Gamma}$$

and let $K^1 = \hat{K}_{[A\cup C]}$ and $K^2 = \hat{K}_{[B\cup C]}$. Writing $\tilde{K}$ in block form gives

$$\tilde{K} = \begin{pmatrix} K^1_{AA} & K^1_{AC} & 0 \\ K^1_{CA} & K^1_{CC} + K^2_{CC} - n(ssd_{CC})^{-1} & K^2_{CB} \\ 0 & K^2_{BC} & K^2_{BB} \end{pmatrix}. \qquad (5.33)$$

From (B.2) we then find

$$\tilde{K}_{A\cup C}^{-1} =$$

$$\begin{pmatrix} K_{AA}^1 & K_{AC}^1 \\ K_{CA}^1 & K_{CC}^1 + K_{CC}^2 - n\left(ssd_{CC}\right)^{-1} - K_{CB}^2 \left(K_{BB}^2\right)^{-1} K_{BC}^2 \end{pmatrix}^{-1},$$

provided the inverses exist. But because C is complete in $\mathcal{G}_{B\cup C}$,

$$\left(K^2\right)_{CC}^{-1} = ssd_{CC}/n = \left\{K_{CC}^2 - K_{CB}^2 \left(K_{BB}^2\right)^{-1} K_{BC}^2\right\}^{-1}. \quad (5.34)$$

Inserting this into the previous expression shows that we have

$$\tilde{K}_{A\cup C}^{-1} = \left(K^1\right)^{-1} = \left(\hat{K}_{[A\cup C]}\right)^{-1} \quad (5.35)$$

and $\tilde{K}$ is positive definite. By symmetry, we also have

$$\tilde{K}_{B\cup C}^{-1} = \left(K^2\right)^{-1} = \left(\hat{K}_{[B\cup C]}\right)^{-1}. \quad (5.36)$$

The relation (5.31) now follows from (5.35) and (5.36).

Next let $a \subseteq \Gamma$ be complete. Then either $a \subseteq (A \cup C)$ or $a \subseteq (B \cup C)$, since (A, B, C) is a decomposition of $\mathcal{G}$. Suppose the former. Then

$$\tilde{K}_{aa}^{-1} = \left(\tilde{K}_{A\cup C}^{-1}\right)_{aa} = \left(K^1\right)_{aa}^{-1} = ssd_{aa}/n,$$

where the last equality follows from the fact that $K^1 = \hat{K}_{[A\cup C]}$. If $a \subseteq (B \cup C)$ we argue analogously. Hence $\tilde{K}^{-1}$ satisfies (5.15) for all complete subsets and $\tilde{K}$ is therefore equal to $\hat{K}$, the unique maximum likelihood estimate of K in the model with graph $\mathcal{G}$.

The determinantial relation (5.30) follows from (5.24) using the identities (5.34), (5.35), and (5.36). Finally, since it is obvious that

$$\hat{\xi}_{[D]} = \hat{\xi}_D = \bar{y}_D \text{ for all } D \subseteq \Gamma,$$

we get for D equal to either of $A \cup C$, $B \cup C$ or C that

$$\left[\hat{h}_{[D]}\right]^\Gamma = \left[\hat{K}_{[D]}\hat{\xi}_{[D]}\right]^\Gamma = \left[\hat{K}_{[D]}\right]^\Gamma \hat{\xi},$$

whereby (5.32) is a consequence of (5.29). The proof is complete. $\square$

Note that we have $\hat{\Sigma}_{[A\cup C]} = \hat{\Sigma}_{A\cup C}$ but

$$\hat{K}_{[A\cup C]} = \hat{K}_{A\cup C} - \left[\hat{K}_{CB}\left(\hat{K}_{BB}\right)^{-1}\hat{K}_{BC}\right]^{A\cup C}. \quad (5.37)$$

Using (B.2) we find

$$\begin{pmatrix} \hat{\Sigma}_{AB} \\ \hat{\Sigma}_{CB} \end{pmatrix} = -\hat{K}_{A\cup C}^{-1} \begin{pmatrix} 0 \\ \hat{K}_{CB} \end{pmatrix} \left(\hat{K}_{BB}\right)^{-1}$$

$$= -\hat{\Sigma}_{A\cup C} \begin{pmatrix} 0 \\ \hat{K}_{CB} \end{pmatrix} \left(\hat{K}_{BB}\right)^{-1},$$

whereby

$$\hat{\Sigma}_{AB} = -\hat{\Sigma}_{AC}\hat{K}_{CB}\left(\hat{K}_{BB}\right)^{-1}.$$

From (5.37) it follows that

$$\hat{K}_{AC} = \left(\hat{\Sigma}_{[A\cup C]}\right)^{-1}_{AC}, \quad \hat{K}_{CB} = \left(\hat{\Sigma}_{[B\cup C]}\right)^{-1}_{CB}.$$

Combining the last two relations with (5.31) and (C.4) we obtain

$$\hat{\Sigma}_{AB} = (K_{AA})^{-1} \hat{K}_{AC}\hat{\Sigma}_{CC}\hat{K}_{CB}\left(\hat{K}_{BB}\right)^{-1}$$

$$= \hat{\Sigma}_{AC}\left(\hat{\Sigma}_{CC}\right)^{-1}\hat{\Sigma}_{CB}$$

$$= n\hat{\Sigma}_{AC}\left(ssd_{CC}\right)^{-1}\hat{\Sigma}_{CB}.$$

In the special case where $\mathcal{G}_{A\cup C}$ and $\mathcal{G}_{B\cup C}$ are both complete, the last of these expressions reduces further to

$$\hat{\Sigma}_{AB} = \frac{1}{n} ssd_{AC} \left(ssd_{CC}\right)^{-1} ssd_{CB}.$$

Decomposition of the graph also results in a decomposition formula for the asymptotic covariance. To show this, we first realize that a decomposition induces a split as defined in Section A.5.

Lemma 5.7 *Consider a covariance selection model with graph $\mathcal{G}$ decomposed by (A, B, C). Then Y_C is a split of $Y = (Y_A, Y_B, Y_C)$ with respect to the parameters*

$$\kappa = (K_{AA}, K_{AC}, h_A), \quad \lambda = (K_{BB}, K_{BC}, h_B), \quad \mu = (\Sigma_C, \xi_C).$$

Proof: It is immediate from writing out the density that κ parametrizes the conditional distributions of Y_A given $Y_C = y_C$ and similarly with λ and μ. The variation independence follows from the fact that all zero restrictions in the model are placed independently in κ and λ, combined with Lemma B.1 which ensures the equivalence of positive definiteness of K with positive definiteness of K_{AA}, K_{BB}, and Σ_C. □

This lemma ensures the asymptotic independence of $\hat{\kappa}$, $\hat{\lambda}$ and $\hat{\mu}$ which is to be exploited below.

Proposition 5.8 *Consider a covariance selection model with graph $\mathcal{G}$ decomposed by (A, B, C). Then the asymptotic covariance of $\hat{K}$ satisfies*

$$\overset{a}{V}\left(\hat{K}\right) = \overset{a}{V}\left(\hat{K}_{[A \cup C]}\right) + \overset{a}{V}\left(\hat{K}_{[B \cup C]}\right) - \frac{2}{n}(\Sigma_C)^{-1} \otimes (\Sigma_C)^{-1}, \quad (5.38)$$

expressed alternatively as

$$\overset{a}{V}\left\{\operatorname{tr}\left(D\hat{K}\right)\right\} = \overset{a}{V}\left\{\operatorname{tr}\left(D_{A \cup C}\hat{K}_{[A \cup C]}\right)\right\} + \overset{a}{V}\left\{\operatorname{tr}\left(D_{B \cup C}\hat{K}_{[B \cup C]}\right)\right\}$$
$$- \frac{2}{n}\operatorname{tr}\left\{D_C(\Sigma_C)^{-1}D_C(\Sigma_C)^{-1}\right\}, \quad (5.39)$$

for an arbitrary symmetric matrix D.

Proof: Define the matrices

$$\hat{K}_{A \setminus C} = \hat{K}_{A \cup C} - \hat{K}_C, \quad H_{CA} = \hat{K}_{CA}\left(\hat{K}_{AA}\right)^{-1}\hat{K}_{AC},$$

and similarly with B replacing A. Here and in the remaining part of the proof we have suppressed the zero extension $[\cdot]^\Gamma$. From (5.37) we find

$$\hat{K}_{[A \cup C]} = \hat{K}_{A \setminus C} + \hat{K}_C - H_{CB}, \quad \hat{K}_{[B \cup C]} = \hat{K}_{B \setminus C} + \hat{K}_C - H_{CA},$$

and (5.26) shows that

$$\hat{K}_{[C]} = \hat{K}_C - H_{CA} - H_{CB}. \quad (5.40)$$

Combining these we obtain

$$\hat{K}_{[A \cup C]} = \hat{K}_{A \setminus C} + H_{CA} + \hat{K}_{[C]}, \quad \hat{K}_{[B \cup C]} = \hat{K}_{B \setminus C} + H_{CB} + \hat{K}_{[C]}.$$

Exploiting Lemma 5.7 we find that $\hat{K}_{A \setminus C} + H_{CA}$, $\hat{K}_{[C]}$ and $\hat{K}_{B \setminus C} + H_{CB}$ are mutually independent, whereby

$$\overset{a}{V}\left(\hat{K}_{[A \cup C]}\right) = \overset{a}{V}\left(\hat{K}_{A \setminus C} + H_{CA}\right) + \overset{a}{V}\left(\hat{K}_{[C]}\right)$$

and similarly with $\overset{a}{V}\left(\hat{K}_{[B \cup C]}\right)$. We then find

$$\overset{a}{V}\left(\hat{K}_{[A \cup C]}\right) + \overset{a}{V}\left(\hat{K}_{[B \cup C]}\right) - \overset{a}{V}\left(\hat{K}_{[C]}\right)$$
$$= \overset{a}{V}\left(\hat{K}_{A \setminus C} + H_{CA}\right) + \overset{a}{V}\left(\hat{K}_{B \setminus C} + H_{CB}\right) + \overset{a}{V}\left(\hat{K}_{[C]}\right)$$
$$= \overset{a}{V}\left(\hat{K}_{A \setminus C} + \hat{K}_{B \setminus C} + \hat{K}_C\right) = \overset{a}{V}\left(\hat{K}\right),$$

where we have exploited the asymptotic independence again as well as (5.40). From (5.9) we have that

$$\overset{a}{V}\left(\hat{K}_{[C]}\right) = \frac{2}{n}(\Sigma_C)^{-1} \otimes (\Sigma_C)^{-1}$$

and (5.38) follows. Using the definition of covariance for a symmetric matrix gives (5.39). □

A relation similar to (5.38) can be derived for the joint asymptotic covariance of the estimate $(\hat{K}, \hat{h})$. We omit the details.

5.2.2 Deviance tests

Consider the problem of testing the hypothesis that $K \in \mathcal{S}^+(\mathcal{G}_0)$ under the assumption that $K \in \mathcal{S}^+(\mathcal{G})$, where $\mathcal{G}_0 = (\Gamma, E_0)$ is a subgraph of $\mathcal{G}$ obtained by removing a certain number of edges. This corresponds to testing an additional list of statements concerning the conditional independence of some pairs of variables, given the remaining.

The likelihood equations imply that for $A \in \mathcal{S}(\mathcal{G})$ we have

$$\operatorname{tr}\left\{A\hat{K}^{-1}\right\} = \operatorname{tr}\left\{A\hat{\Sigma}(\mathcal{G})\right\} = \operatorname{tr}\left\{A\,ssd(\mathcal{G})\right\}/n,$$

whereby

$$\operatorname{tr}\left\{\hat{K}\,ssd(\mathcal{G})\right\} = n\operatorname{tr}\left\{\hat{K}\hat{K}^{-1}\right\} = n|\Gamma|.$$

After simplifications analogous to those on page 124 we can rewrite (5.14) as

$$L(K, \xi) \propto (\det K)^{n/2} \exp\left[-\operatorname{tr}\{K\,ssd^\xi(\mathcal{G})/2\}\right], \qquad (5.41)$$

where as before

$$ssd^\xi = \sum_{\nu=1}^n (y^\nu - \xi)(y^\nu - \xi)^\top.$$

Using that $ssd^{\hat{\xi}} = ssd$, it follows that

$$L(\hat{K}, \hat{\xi}) \propto$$
$$(\det \hat{K})^{n/2} \exp\left[-\operatorname{tr}\{\hat{K}\,ssd(\mathcal{G})/2\}\right] = (\det \hat{K})^{n/2} e^{-n|\Gamma|/2}. \quad (5.42)$$

From (5.42) we calculate the deviance as

$$d = -2\log \frac{L(\hat{K}_0, \hat{\xi}_0)}{L(\hat{K}, \hat{\xi})} = -2\log \frac{L(\hat{K}_0, \hat{\xi})}{L(\hat{K}, \hat{\xi})} = n\left(\log \det \hat{K} - \log \det \hat{K}_0\right),$$

where $(\hat{K}_0, \hat{\xi}_0) = (\hat{K}_0, \hat{\xi})$ denote the maximum likelihood estimates under the model with graph $\mathcal{G}_0$.

From the general theory of exponential models the asymptotic distribution of the deviance D, as n tends to infinity, is a χ^2 distribution with degrees of freedom equal to $|E| - |E_0|$, the number of edges that are in $\mathcal{G}$ but not in $\mathcal{G}_0$.

The exact distribution is inaccessible in general, but more accurate approximations than the one just described are available.

Consider first the special case, where E_0 differs from E by one edge only, $\{\gamma, \mu\}$ say, so that we are testing the conditional independence of Y_γ and Y_μ, given the remaining variables. It may then be shown (Eriksen 1996) that the likelihood ratio test statistic, raised to the power $2/n$, approximately follows a beta distribution. More precisely, if we let

$$b = \left(\frac{L(\hat{K}_0, \hat{\xi})}{L(\hat{K}, \hat{\xi})}\right)^{2/n} = \frac{\det \hat{K}_0}{\det \hat{K}} = \exp(-d/n),$$

we have that, approximately,

$$B \sim \mathcal{B}\{(n - a(\gamma, \mu) - 2)/2, 1/2\},$$

where $a(\gamma, \mu) = |\operatorname{bd}(\gamma) \cap \operatorname{bd}(\mu)|$ is the number of common neighbours for the variables under consideration. Remember that the *ssd* matrix is based on n observations and has therefore $n - 1$ degrees of freedom. The above approximation is considerably more accurate than the χ^2 approximation of the deviance, and in a number of cases the distributional result is even exact, as described later in Section 5.3.3.

This special instance of the beta distribution can be transformed into a Student's t distribution. In fact, we can perform the conditional independence test by rejecting for large values of

$$|t| = \left[\{n - a(\gamma, \mu) - 2\}\left(b^{-1} - 1\right)\right]^{1/2} = \left[\{n - a(\gamma, \mu) - 2\}\left(e^{d/n} - 1\right)\right]^{1/2}$$

and $|T|$ is distributed as the absolute value of Student's t with degrees of freedom equal to $n - a(\gamma, \mu) - 2$. The χ^2 approximation to the distribution of the deviance D appears by first approximating the t-distribution with a standard normal distribution and then approximating further as

$$t^2 = \{n - a(\gamma, \mu) - 2\}\left(e^{d/n} - 1\right) \approx d.$$

In the general case where more than one edge is removed at a time, i.e. when E_0 differs from E with k edges and $k > 1$, a better approximation of the distribution of the deviance statistic can be obtained as follows. First find a sequence $\mathcal{G}_0, \ldots, \mathcal{G}_k$ such that $\mathcal{G} = \mathcal{G}_k$ and $\mathcal{G}_i$ and $\mathcal{G}_{i-1}$ differ by exactly one edge $\{\gamma_i, \mu_i\}$ for $i = 1, \ldots, k$. Let $a(\gamma_i, \mu_i) = |\operatorname{bd}_i(\gamma_i) \cap \operatorname{bd}_i(\mu_i)|$ be the number of common neighbours of (γ_i, μ_i) in $\mathcal{G}_i$. Then, clearly, the deviance is d equal to the sum of deviances d_i for conditional independence of γ_i and μ_i under the assumption of the model determined by $\mathcal{G}_i$. Further, it holds that the deviance D is approximately distributed as

$$D \stackrel{\mathcal{D}}{\approx} -n \sum_{i=1}^{k} \log B_i, \qquad (5.43)$$

where B_i are independent and distributed as $\mathcal{B}\{(n - a(\gamma_i, \mu_i) - 2)/2, 1/2\}$. The deviance can then, for example, be evaluated in this distribution by the methods in J.L. Jensen (1991, 1995) or Monte Carlo methods. As in the case of testing the absence of one edge, the approximation (5.43) is considerably more accurate than the usual χ^2 approximation. And, as it was true for the removal of just one edge, it is exact in certain cases; see Section 5.3.3.

Alternatively, the χ^2 approximation can be improved by dividing the deviance with a Bartlett correction factor. Using that the expectation of the logarithm of a χ^2-distribution with f degrees of freedom is equal to

$$\mathbf{E}\{\log \chi^2(f)\} = \log f - f^{-1} - f^{-2}/3 + O(f^{-4}),$$

see Porteous (1985a), and letting $c(f) = \log f - f^{-1} - f^{-2}/3$, we get that the Bartlett corrected statistic $D^* = kD/\beta$ is approximately $\chi^2(k)$-distributed, where

$$\beta = \sum_{i=1}^{k} c\{n - a(\gamma_i, \mu_i) - 1\} - \sum_{i=1}^{k} c\{n - a(\gamma_i, \mu_i) - 2\}.$$

Use of the χ^2-distribution for evaluating the uncorrected test statistic can be very misleading and is strongly advised against (Porteous 1989).

5.3 Decomposable models

In analogy with Section 4.4 we here study the special features of covariance selection models whose interaction graphs are decomposable. Theorem 2.25 implies that these models are built up from saturated models by successive direct joins. This structure makes it possible to break down the statistical analysis of a decomposable model into small analyses of saturated submodels in an elegant way.

5.3.1 Basic factorizations

In Lemma 5.5 we gave formulae that essentially described how the Markov property was reflected in additive properties of concentration matrices as well as multiplicative properties of covariance matrices across decompositions of the graph. A similar formula (3.17) holds for the density itself.

As shown in Proposition 2.17, we can number the cliques of a decomposable graph $\mathcal{G}$ to form a perfect sequence, i.e. a sequence $C_1, \ldots, C_k$ where each combination of subgraphs induced by $H_{j-1} = C_1 \cup \cdots \cup C_{j-1}$ and C_j is a decomposition. Repeated use of (3.17) gives

$$f(y) = \frac{\prod_{j=1}^{k} f(y_{C_j})}{\prod_{j=2}^{k} f(y_{S_j})} = \frac{\prod_{C \in \mathcal{C}} f(y_C)}{\prod_{S \in \mathcal{S}} f(y_S)^{\nu(S)}}, \qquad (5.44)$$

where $S_j = H_{j-1} \cap C_j$ is the sequence of separators and $\nu(S)$ is an index counting the number of times a given separator S occurs in a perfect sequence (possibly including the empty set).

From Lemma 5.5 we further find

$$K = \Sigma^{-1} = \sum_{C \in \mathcal{C}} [K_C]^\Gamma - \sum_{S \in \mathcal{S}} \nu(S) [K_S]^\Gamma$$

$$= \sum_{C \in \mathcal{C}} \left[(\Sigma_C)^{-1} \right]^\Gamma - \sum_{S \in \mathcal{S}} \nu(S) \left[(\Sigma_S)^{-1} \right]^\Gamma$$

as well as

$$\det \Sigma = \frac{\prod_{C \in \mathcal{C}} \det \Sigma_C}{\prod_{S \subset S} (\det \Sigma_S)^{\nu(S)}}.$$

The factorizations induce a similar factorization of the joint density of the matrix

$$SSD = Y^\top Y - Y^\top e e^\top Y / n,$$

which is distributed as $\mathcal{W}(n-1, \Sigma)$. Combining the above relations with (C.9) yields

$$w_{p,n-1,\Sigma}(s) =$$

$$c(p, n-1)^{-1} \left(\frac{\prod_{C \in \mathcal{C}} \det \Sigma_C}{\prod_{S \in \mathcal{S}} (\det \Sigma_S)^{\nu(S)}} \right)^{-(n-1)/2} (\det s)^{(n-p-2)/2}$$

$$\times \exp \left\{ - \left(\sum_{C \in \mathcal{C}} \text{tr}(K_C s_C) - \sum_{S \in \mathcal{S}} \nu(S) \text{tr}(K_S s_S) \right) / 2 \right\}, \quad (5.45)$$

which should be compared with (4.37).

5.3.2 Maximum likelihood estimation

Exact results

In Proposition 5.6 we derived a formula for combining maximum likelihood estimates of concentration matrices in two covariance selection models to find the estimate in the model formed by their direct join.

Combining this with the usual simple estimates in the saturated models, explicit formulae for the maximum likelihood estimate in a decomposable covariance selection model can be derived. More precisely we find by repeated use of (5.29) that

$$\hat{K} = n \left\{ \sum_{j=1}^{k} \left[(ssd_{C_j})^{-1} \right]^\Gamma - \sum_{j=2}^{k} \left[(ssd_{S_j})^{-1} \right]^\Gamma \right\}. \quad (5.46)$$

Using the alternative expression where the separators and cliques are not numbered and the expression for the determinant (5.30), we also get

Fig. 5.1. Graph of a decomposable covariance selection model for a 6-dimensional random variable.

Proposition 5.9 *In a decomposable covariance selection model with graph $\mathcal{G}$, the maximum likelihood estimate of the mean vector and concentration matrix exists with probability one if and only if $n > \max_{C \in \mathcal{C}} |C|$. It is then given as*

$$\hat{\xi} = \bar{y}, \quad \hat{K} = n \left\{ \sum_{C \in \mathcal{C}} \left[(ssd_C)^{-1} \right]^{\Gamma} - \sum_{S \in \mathcal{S}} \nu(S) \left[(ssd_S)^{-1} \right]^{\Gamma} \right\}, \quad (5.47)$$

where $\mathcal{C}$ is the set of cliques of $\mathcal{G}$ and $\mathcal{S}$ the set of separators with multiplicities ν in any perfect sequence. The determinant of the estimate can be calculated as

$$\det \hat{K} = n^{|\Gamma|} \frac{\prod_{S \in \mathcal{S}} (\det ssd_S)^{\nu(S)}}{\prod_{C \in \mathcal{C}} \det ssd_C}. \quad (5.48)$$

Example 5.10 As an example, consider the covariance selection model for a 6-dimensional random variable with the graph given in Fig. 5.1. We have $\Gamma = \{1, 2, 3, 4, 5, 6\}$ and the cliques are

$$C_1 = \{1, 2, 4\}, \; C_2 = \{1, 3, 4\}, C_3 = \{4, 5\}, \; C_4 = \{4, 6\},$$

where the numbering is perfect with separators

$$S_2 = \{1, 4\}, \; S_3 = S_4 = \{4\}.$$

Thus the multiplicities are

$$\nu(\{1, 4\}) = 1, \; \nu(\{4\}) = 2.$$

The determinant of the concentration matrix factorizes as

$$\det K = \frac{\det K_{\{1,2,4\}} \, \det K_{\{1,3,4\}} \, \det K_{\{4,5\}} \, \det K_{\{4,6\}}}{\det K_{\{1,4\}} \, (k_{44})^2},$$

and its estimate can be calculated as

$$\det \hat{K} = n^6 \frac{\det ssd_{\{1,4\}} \, (ssd_{44})^2}{\det ssd_{\{1,2,4\}} \, \det ssd_{\{1,3,4\}} \, \det ssd_{\{4,5\}} \, \det ssd_{\{4,6\}}}.$$

The estimate $\hat{K}$ itself decomposes as

$$\hat{K} = \begin{pmatrix} \hat{k}^{11}_{\{1,2,4\}} & \hat{k}^{12}_{\{1,2,4\}} & 0 & \hat{k}^{14}_{\{1,2,4\}} & 0 & 0 \\ \hat{k}^{21}_{\{1,2,4\}} & \hat{k}^{22}_{\{1,2,4\}} & 0 & \hat{k}^{24}_{\{1,2,4\}} & 0 & 0 \\ 0 & 0 & 0 & 0 & 0 & 0 \\ \hat{k}^{41}_{\{1,2,4\}} & \hat{k}^{42}_{\{1,2,4\}} & 0 & \hat{k}^{44}_{\{1,2,4\}} & 0 & 0 \\ 0 & 0 & 0 & 0 & 0 & 0 \\ 0 & 0 & 0 & 0 & 0 & 0 \end{pmatrix}$$

$$+ \begin{pmatrix} \hat{k}^{11}_{\{1,3,4\}} & 0 & \hat{k}^{13}_{\{1,3,4\}} & \hat{k}^{14}_{\{1,3,4\}} & 0 & 0 \\ 0 & 0 & 0 & 0 & 0 & 0 \\ \hat{k}^{31}_{\{1,3,4\}} & 0 & \hat{k}^{33}_{\{1,3,4\}} & \hat{k}^{34}_{\{1,3,4\}} & 0 & 0 \\ \hat{k}^{41}_{\{1,3,4\}} & 0 & \hat{k}^{43}_{\{1,3,4\}} & \hat{k}^{44}_{\{1,3,4\}} & 0 & 0 \\ 0 & 0 & 0 & 0 & 0 & 0 \\ 0 & 0 & 0 & 0 & 0 & 0 \end{pmatrix}$$

$$+ \begin{pmatrix} 0 & 0 & 0 & 0 & 0 & 0 \\ 0 & 0 & 0 & 0 & 0 & 0 \\ 0 & 0 & 0 & 0 & 0 & 0 \\ 0 & 0 & 0 & \hat{k}^{44}_{\{4,5\}} & \hat{k}^{45}_{\{4,5\}} & 0 \\ 0 & 0 & 0 & \hat{k}^{54}_{\{4,5\}} & \hat{k}^{55}_{\{4,5\}} & 0 \\ 0 & 0 & 0 & 0 & 0 & 0 \end{pmatrix}$$

$$+ \begin{pmatrix} 0 & 0 & 0 & 0 & 0 & 0 \\ 0 & 0 & 0 & 0 & 0 & 0 \\ 0 & 0 & 0 & 0 & 0 & 0 \\ 0 & 0 & 0 & \hat{k}^{44}_{\{4,6\}} & 0 & \hat{k}^{46}_{\{4,6\}} \\ 0 & 0 & 0 & 0 & 0 & 0 \\ 0 & 0 & 0 & \hat{k}^{64}_{\{4,6\}} & 0 & \hat{k}^{66}_{\{4,6\}} \end{pmatrix}$$

$$- \begin{pmatrix} \hat{k}^{11}_{\{1,4\}} & 0 & 0 & \hat{k}^{14}_{\{1,4\}} & 0 & 0 \\ 0 & 0 & 0 & 0 & 0 & 0 \\ 0 & 0 & 0 & 0 & 0 & 0 \\ \hat{k}^{41}_{\{1,4\}} & 0 & 0 & \hat{k}^{44}_{\{1,4\}} + 2\frac{n}{ssd_{44}} & 0 & 0 \\ 0 & 0 & 0 & 0 & 0 & 0 \\ 0 & 0 & 0 & 0 & 0 & 0 \\ 0 & 0 & 0 & 0 & 0 & 0 \end{pmatrix}.$$

Here $\hat{k}_A^{ij}$ is the ijth element in $\hat{K}_{[A]} = n\,(ssd_A)^{-1}$, where we have exploited that A is complete. □

As in the discrete case, the distribution of the maximum likelihood estimate obeys fundamental conditional independences. In fact, whenever S separates A from B in $\mathcal{G}$ we have

$$\hat{\Sigma}_{A \cup S} \perp\!\!\!\perp \hat{\Sigma}_{B \cup S} \mid \hat{\Sigma}_S. \tag{5.49}$$

This property of the distribution of the maximum likelihood estimate is the so-called hyper Markov property (Dawid and Lauritzen 1993).

The general question of existence of maximum likelihood estimates in non-decomposable covariance selection models has been studied by Buhl (1993). For a given undirected graph $\mathcal{G}$ we may look for a decomposable cover of $\mathcal{G}$, i.e. a decomposable graph $\mathcal{G}'$ of which $\mathcal{G}$ is a subgraph. From Proposition 5.9 we find that the maximum likelihood estimate exists in the model determined by $\mathcal{G}'$ if and only if $n > n' = \max_{C \in \mathcal{C}'} |C|$, where $\mathcal{C}'$ is the set of cliques in $\mathcal{G}'$. As the model determined by $\mathcal{G}$ is an affine submodel of the model determined by $\mathcal{G}'$, this condition would then imply the existence also of the maximum likelihood estimate in the model determined by $\mathcal{G}$. In general there are many different decomposable covers. If we let n^+ denote the smallest possible value of n', we thus have that the condition $n > n^+$ is sufficient for the existence of the maximum likelihood estimate. Even determining the number n^+ is a difficult combinatorial problem (Kjærulff 1992).

On the other hand, as noted just after the proof of Theorem 5.3, if we let n^- denote the maximal clique size of $\mathcal{G}$, a necessary condition is that $n > n^-$. If we have $n^- < n \leq n^+$ it is unclear what happens.

Buhl (1993) shows that in the case of a p-cycle, we have $n^- = 2$ and $n^+ = 3$. If now $n = 3$, the probability that the maximum likelihood estimate exists is strictly between 0 and 1. In fact, if the true covariance is the identity, we have that

$$P\{\text{MLE exists} \mid \Sigma = I\} = 1 - \frac{2}{(p-1)!}.$$

A similar result holds for the graph obtained from the cycle by adding a single vertex and letting this be adjacent to all other vertices. But the general case is unclear.

Asymptotic results

As for decomposable models in the discrete case, simple and explicit results for the asymptotic distribution of $\hat{K}$ can be derived. In fact, from (5.38) it follows that, with the usual notation, we have

$$\overset{a}{V}\left(\hat{K}\right) = \frac{2}{n}\left\{\sum_{C\in\mathcal{C}}(\Sigma_C)^{-1}\otimes(\Sigma_C)^{-1} - \sum_{S\in\mathcal{S}}\nu(S)(\Sigma_S)^{-1}\otimes(\Sigma_S)^{-1}\right\} \quad (5.50)$$

or, alternatively, for an arbitrary symmetric matrix D,

$$\overset{a}{V}\left\{\mathrm{tr}\left(D\hat{K}\right)\right\} = \frac{2}{n}\sum_{C\in\mathcal{C}}\mathrm{tr}\left\{D_C(\Sigma_C)^{-1}D_C(\Sigma_C)^{-1}\right\}$$
$$-\frac{2}{n}\sum_{S\in\mathcal{S}}\nu(S)\,\mathrm{tr}\left\{D_S(\Sigma_S)^{-1}D_S(\Sigma_S)^{-1}\right\}.(5.51)$$

Example 5.11 If we consider the problem discussed in Example 5.10, we get

$$\overset{a}{V}\left(\hat{k}^{24}\right) = \overset{a}{V}\left\{\mathrm{tr}\left(\tilde{E}_{24}\hat{K}\right)/\sqrt{2}\right\} = \left\{\left(k^{24}_{\{1,2,4\}}\right)^2 + k^{22}_{\{1,2,4\}}k^{44}_{\{1,2,4\}}\right\}/n,$$

where $\tilde{E}_{24}$ is given in (B.22), and

$$\overset{a}{V}\left(\hat{k}^{11}\right) = \overset{a}{V}\left\{\mathrm{tr}\left(\tilde{E}_{11}\hat{K}\right)\right\}$$
$$= \frac{2}{n}\left\{\left(k^{11}_{\{1,2,4\}}\right)^2 + \left(k^{11}_{\{1,3,4\}}\right)^2 - \left(k^{11}_{\{1,4\}}\right)^2\right\},$$

whereas

$$\overset{a}{V}\left(\hat{k}^{44}\right) = \frac{2}{n}\left(k^{44}_{\{1,2,4\}}\right)^2 + \frac{2}{n}\left(k^{44}_{\{1,3,4\}}\right)^2 + \frac{2}{n}\left(k^{44}_{\{4,5\}}\right)^2 + \frac{2}{n}\left(k^{44}_{\{4,6\}}\right)^2$$
$$-\frac{2}{n}\left(k^{44}_{\{1,4\}}\right)^2 - \frac{4}{n}(\sigma_{44})^2.$$

Note for example also that $\hat{k}^{12}$ and $\hat{k}^{46}$ are uncorrelated. □

5.3.3 Exact tests in decomposable models

Test for conditional independence

Before we discuss the general case, we deal with the problem of testing conditional independence of two variables γ and μ under the saturated normal model. As in the discrete case, the general testing problem in nested decomposable models can be reduced to a combination of testing problems of this kind.

From Section 5.2.2 we find that the deviance test can be performed by rejecting for small values of

$$b = \exp(-d/n) = \frac{\det\hat{K}_0}{\det\hat{K}},$$

where $\hat{K}$ and $\hat{K}_0$ are the estimates under the model and the hypothesis. Since both the saturated model and the saturated model involving only one conditional independence restriction are decomposable, we further find from (5.48) combined with (5.12) that

$$b = \frac{\det ssd_{\Gamma\setminus\{\gamma,\mu\}} \det ssd}{\det ssd_{\Gamma\setminus\{\gamma\}} \det ssd_{\Gamma\setminus\{\mu\}}} = 1 - \left(\hat{\rho}_{\gamma\mu\,|\,\Gamma\setminus\{\gamma,\mu\}}\right)^2,$$

where $\hat{\rho}_{\gamma\mu\,|\,\Gamma\setminus\{\gamma,\mu\}}$ is the empirical partial correlation coefficient.

An exact test is formed by judging the test statistic in its conditional distribution, given the statistic which is complete and sufficient under the hypothesis. Since SSD and $\hat{\xi}$ are independent, this amounts to the conditional distribution of B given $(SSD_{\Gamma\setminus\{\gamma\}}, SSD_{\Gamma\setminus\{\mu\}})$.

Now, conditionally on $Y^1_{\Gamma\setminus\{\gamma,\mu\}}, \ldots, Y^n_{\Gamma\setminus\{\gamma,\mu\}}$, the squared empirical partial correlation coefficient is distributed as $\mathcal{B}\{1/2, (n - |\Gamma|)/2\}$, and hence B is distributed as $\mathcal{B}\{(n - |\Gamma|)/2, 1/2\}$. But the conditional distribution does not depend on the condition and is therefore also the marginal distribution of B.

Finally, by appealing to Basu's theorem (Theorem A.4), B is independent of the complete and sufficient statistic under the hypothesis, which is equal to $(SSD_{\Gamma\setminus\{\gamma\}}, SSD_{\Gamma\setminus\{\mu\}})$, and the exact test can be performed by comparing the observed value of B to the quantiles of the beta distribution $\mathcal{B}\{(n - |\Gamma|)/2, 1/2\}$. Hence what we found in Section 5.2.2 to be approximately correct in the general case is exact in the present case. Also, as shown in that section, the test can alternatively be performed by rejecting for large values of

$$|t| = \left\{(n - |\Gamma|)\left(b^{-1} - 1\right)\right\}^{1/2}.$$

In the given case we find that

$$t = \sqrt{n - |\Gamma|}\frac{\hat{\rho}_{\gamma\mu\,|\,\Gamma\setminus\{\gamma,\mu\}}}{\left\{1 - \left(\hat{\rho}_{\gamma\mu\,|\,\Gamma\setminus\{\gamma,\mu\}}\right)^2\right\}^{1/2}},$$

which is distributed as Student's t with $n - |\Gamma|$ degrees of freedom. It can be recognized as the t-statistic for the hypothesis that the partial regression coefficient $\hat{\beta}_{\gamma\mu\,|\,\Gamma\setminus\{\gamma\}}$ is equal to zero in the model for linear regression of Y_γ on $Y_{\Gamma\setminus\{\gamma\}}$.

Consider next the case where $\mathcal{G}$ is a decomposable graph and we wish to investigate the hypothesis that $\Sigma \in \mathcal{S}^+(\mathcal{G}_0)$, assuming that $\Sigma \in \mathcal{S}^+(\mathcal{G})$, where $\mathcal{G}_0 = (V_0, E_0)$ is a decomposable subgraph of $\mathcal{G}$ with the same vertices, i.e. $V = V_0 = \Gamma$. First we assume that $\mathcal{G}_0$ has exactly one edge $e = \{\gamma, \mu\}$ less than $\mathcal{G}$. This means that we test the hypothesis of the additional conditional independence

$$H_0 : \gamma \perp\!\!\!\perp \mu \mid \Gamma \setminus \{\gamma, \mu\},$$

assuming that the conditional independences already given by $\mathcal{G}$ all hold.

We have previously considered the case where the model $\mathcal{S}(\mathcal{G})$ is saturated, i.e. the graph $\mathcal{G}$ is complete and has only one clique which is Γ itself.

In the case of a general decomposable graph $\mathcal{G}$, Lemma 2.19 ensures that e is a member of one clique C^* only. Moreover we have

Proposition 5.12 *The exact deviance test for the hypothesis H_0 of conditional independence is identical to the exact deviance test of*

$$H_0' : \gamma \perp\!\!\!\perp \mu \mid C^* \setminus \{\gamma, \mu\}$$

in the saturated model for the C^*-marginal. Here C^* is the unique clique of $\mathcal{G}$ containing e. In particular the test can be performed by rejecting for small values of

$$b = \exp(-d/n) = \frac{\det \hat{K}_0}{\det \hat{K}}$$

which is distributed as $\mathcal{B}\{(n - |C^*|)/2, 1/2\}$.

Proof: We have to show that the deviance statistics in the two problems are identical, but also that their conditional distributions given the sufficient statistics agree.

First number the cliques of $\mathcal{G}$ to form a perfect sequence in $\mathcal{G}$ with $C_1 = C^*$. This can be done by Lemma 2.18. The graph $\mathcal{G}_0$ has cliques equal to

$$\mathcal{C}_0 = \text{red}\{C_0^*, C_1^*, C_2, \ldots, C_k\},$$

where without loss of generality we can assume $S^* = S_2 \subseteq C_1^*$ such that the sequence of sets above is perfect in $\mathcal{G}_0$. Using (5.30) we obtain

$$\det \hat{K} = \det \hat{K}_{[C^*]} \det \hat{K}_{[C_2 \cup \cdots \cup C_k]} n^{-|S^*|} (\det ssd_{S^*})$$

and similarly for $\det \hat{K}_0$. In particular only the first factor is different in the two cases and the remaining factors all cancel. Moreover, using (5.48) we find

$$b = \frac{\det \hat{K}_0}{\det \hat{K}} = \frac{\det ssd_{C^* \setminus \{\gamma, \mu\}} \det ssd_{C^*}}{\det ssd_{C_0^*} \det ssd_{C_1^*}}.$$

We recognize this as the test statistic for the corresponding saturated model and its distribution is as stated. Basu's theorem identifies the distribution as identical to the conditional distribution given the sufficient statistic, and the proof is complete. □

Example 5.13 Continuation of Example 5.10. Let $\mathcal{G}$ be the graph of Fig. 5.1. Removing the edge $\{1, 4\}$ destroys the decomposability of the graph.

All other single edges are members of only one clique. Thus the deviance test for removing the edge $\{1,2\}$ can be performed in the $\{1,2,4\}$ marginal and the test statistic in Proposition 5.12 becomes

$$b = \frac{ssd_{44} \det ssd_{\{1,2,4\}}}{\det ssd_{\{1,4\}} \det ssd_{\{2,4\}}},$$

which is distributed as $\mathcal{B}\{(n-3)/2, 1/2\}$. □

Next we consider the situation where the decomposable submodel $\mathcal{G}_0$ has k edges less than $\mathcal{G}$. By Lemma 2.21 there is a sequence $\mathcal{G}_0 \subset \cdots \subset \mathcal{G}_k = \mathcal{G}$ of graphs that are decomposable and differ by one edge only. If we let e_i denote the edge that is in $\mathcal{G}_i$ but not in $\mathcal{G}_{i-1}$, we obtain the following

Proposition 5.14 *The exact deviance test for testing a decomposable submodel $\Sigma \in \mathcal{S}^+(\mathcal{G}_0)$, assuming a decomposable model $\Sigma \in \mathcal{S}^+(\mathcal{G})$, can be performed by rejecting for small values of*

$$b = \exp(-d/n) = \frac{\det \hat{K}_0}{\det \hat{K}},$$

which is distributed as the product

$$B \stackrel{\mathcal{D}}{=} B_1 \cdots B_k$$

of independent beta random variables $\mathcal{B}\{(n-|C_i^*|)/2, 1/2\}$, where C_i^* is the unique clique of $\mathcal{G}_i$ that contains e_i.

Proof: Let $\hat{K}_i$ denote the maximum likelihood estimate of the concentration matrix under the model with graph $\mathcal{G}_i$. We then get for the likelihood ratios

$$b = \frac{\det \hat{K}_0}{\det \hat{K}} = \frac{\det \hat{K}_0}{\det \hat{K}_1} \frac{\det \hat{K}_1}{\det \hat{K}_2} \cdots \frac{\det \hat{K}_{k-1}}{\det \hat{K}_k} = \prod_{j=1}^{k} b_j.$$

Each of the terms in the product is by Proposition 5.12 a test statistic for conditional independence in the marginal saturated model given by the single clique of which the missing edge is a member, and each of these is beta-distributed as stated in the proposition. The components of the deviance are also independent. This follows from repeated use of Basu's theorem: B_k is independent of the sufficient statistic under the model with graph $\mathcal{G}_{k-1}$. But $(B_1, \ldots, B_{k-1})$ is a function of this statistic and hence B_k and $(B_1, \ldots, B_{k-1})$ are independent. Continuing this argument eventually yields the desired independence. □

This distributional result was used as an approximation in Section 5.2.2.

It seems appropriate to warn against misuse of the procedure of calculating overall significance levels of combined tests as we did at the end of

Section 4.4.3. We suggest that the overall test always should be combined with an analysis of the individual factors.

If overall levels are to be calculated, the approximations of J.L. Jensen (1991) can be used with very high accuracy since B has a Box-type distribution.

Proposition 5.14 can be extended to the case where $\mathcal{G}_0$ is a regular subgraph of $\mathcal{G}$, meaning that $\mathcal{G}$ and $\mathcal{G}_0$ have the same non-decomposable atoms (Eriksen 1996).

Example 5.15 Continuation of Example 5.10. Suppose the additional conditional independences to be tested are specified by the absence of edges $\{1,2\}$, $\{1,4\}$ and $\{3,4\}$, i.e. we investigate whether the group of variables $\{1,3\}$ are independent of the variables $\{2,4,5,6\}$, assuming the model with graph in Fig. 5.1.

The test statistic can then be factored into beta statistics as follows. One factor is the test statistic for removing $\{1,2\}$ in the $\{1,2,4\}$ marginal as calculated in the previous example:

$$b_3 = \frac{ssd_{44} \det ssd_{\{1,2,4\}}}{\det ssd_{\{1,4\}} \det ssd_{\{2,4\}}},$$

which is distributed as $\mathcal{B}\{(n-3)/2, 1/2\}$; the next is the statistic for removing $\{1,4\}$, supposing that $\{1,2\}$ has been removed already:

$$b_2 = \frac{ssd_{33} \det ssd_{\{1,3,4\}}}{\det ssd_{\{1,3\}} \det ssd_{\{3,4\}}},$$

and this is also distributed as $\mathcal{B}\{(n-3)/2, 1/2\}$; the last factor is the statistic for testing independence between the variables 3 and 4 in the $\{3,4\}$ marginal:

$$b_1 = \frac{\det ssd_{\{3,4\}}}{ssd_{33} ssd_{44}},$$

which is $\mathcal{B}\{(n-2)/2, 1/2\}$. Multiplying the factors together finally yields

$$b = \frac{\det ssd_{\{1,2,4\}} \det ssd_{\{1,3,4\}}}{\det ssd_{\{1,4\}} \det ssd_{\{2,4\}} \det ssd_{\{1,3\}}},$$

as all other factors cancel. □

5.4 Notes

5.4.1 Chain graph models

As in the discrete case, there is a natural extension of the theory that deals with models that have response structure in the variables, corresponding to the Markov properties associated with a directed acyclic graph or, more generally, a chain graph.

Results that are very analogous to the discrete case are obtained. For example, as an analogue of (4.68) we find the general formula for the estimate of the inverse covariance matrix as

$$\hat{K} = \sum_{\tau \in \mathcal{T}} \left[\hat{K}^*_{\tau_*}\right]^{|\Gamma|} - n \left[(ssd_{\mathrm{pa}(\tau)})^{-1}\right]^{|\Gamma|}. \qquad (5.52)$$

Here τ is a chain component, and τ_* denotes the undirected graph that has $\tau \cup \mathrm{pa}(\tau)$ as vertices and undirected edges between a pair (α, β) if either both of these are in $\mathrm{pa}(\tau)$ or there is an edge, directed or undirected, between them in the chain graph $\mathcal{G}$. The symbol $\hat{K}^*_{\tau_*}$ denotes the maximum likelihood estimate of the inverse covariance in the covariance selection model with graph τ_*. In the particular case where the chain components are singletons, we have univariate recursive regressions and the estimate $\hat{K}^*$ is explicitly given as

$$\hat{K}^* = n\,(ssd_{\tau_*})^{-1}.$$

As in the discrete case, explicit formulae are also available if the chain graph is decomposable, i.e. if all the graphs τ_* are decomposable.

In the Gaussian case, all models can alternatively be defined through a particular type of linear structural equation, as shown by Wermuth (1992). More precisely, define a *recursive path analysis system* as a system of linear structural equations

$$\Lambda Y = U,$$

where $Y = Y_V$ is the vector of variables, partitioned into groups corresponding to the dependence chain $V(1), \ldots, V(T)$, Λ — partitioned accordingly — is an upper block-triangular matrix of coefficients with positive definite symmetric matrices in the blocks along the diagonal, and U is a vector of residuals. We assume further that the covariance matrix Φ of the residuals is block-diagonal and with the blocks in its diagonal *equal to the corresponding blocks in* Λ. With this additional restriction, Λ and the covariance matrix Σ of Y are in one-to-one correspondence.

Wermuth (1992) shows that if these conditions are satisfied, then the elements of Λ can be interpreted as appropriate particular partial concentrations, whereby the conditional independence restrictions of a graphical chain model amount to specifying the elements in Λ to be zero for all corresponding missing edges in the chain graph. As examples we consider the models shown in Fig. 5.2.

Example 5.16 The model on the right in Fig. 5.2 is equivalent to its undirected counterpart and has the conditional independence restriction

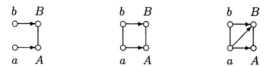

Fig. 5.2. Three chain models for continuous variables. The model on the far right has an explicit solution whereas the two other models need iteration to maximize the joint likelihood function.

$A \perp\!\!\!\perp b \mid \{a, B\}$. Thus the covariance matrix can be estimated explicitly by using equation (5.46).

The representation of the model as a recursive path analysis system is given by

$$\begin{pmatrix} \lambda_{AA} & \lambda_{AB} & \lambda_{Aa} & 0 \\ \lambda_{BA} & \lambda_{BB} & \lambda_{Ba} & \lambda_{Bb} \\ 0 & 0 & \lambda_{aa} & \lambda_{ab} \\ 0 & 0 & \lambda_{ba} & \lambda_{bb} \end{pmatrix} \begin{pmatrix} Y_A \\ Y_B \\ Y_a \\ Y_b \end{pmatrix} = \begin{pmatrix} U_A \\ U_B \\ U_a \\ U_b \end{pmatrix}, \qquad (5.53)$$

where $\lambda_{AB} = \lambda_{BA}, \lambda_{ab} = \lambda_{ba}$ and (U_A, U_B) and (U_a, U_b) are independent with covariance matrices given by the blocks along the diagonal of Λ.

The model in the centre of Fig. 5.2 has a representation in structural equations as

$$\begin{pmatrix} \lambda_{AA} & \lambda_{AB} & \lambda_{Aa} & 0 \\ \lambda_{BA} & \lambda_{BB} & 0 & \lambda_{Bb} \\ 0 & 0 & \lambda_{aa} & \lambda_{ab} \\ 0 & 0 & \lambda_{ba} & \lambda_{bb} \end{pmatrix} \begin{pmatrix} Y_A \\ Y_B \\ Y_a \\ Y_b \end{pmatrix} = \begin{pmatrix} U_A \\ U_B \\ U_a \\ U_b \end{pmatrix},$$

i.e. the equation (5.53) modified by also assuming $\lambda_{Ba} = 0$. The model is equivalent to the corresponding undirected non-decomposable model where no arrows are present. Thus iteration is needed to calculate maximum likelihood estimates.

Finally we consider the model on the left in Fig. 5.2. As structural equations it is given as

$$\begin{pmatrix} \lambda_{AA} & \lambda_{AB} & \lambda_{Aa} & 0 \\ \lambda_{BA} & \lambda_{BB} & 0 & \lambda_{Bb} \\ 0 & 0 & \lambda_{aa} & 0 \\ 0 & 0 & 0 & \lambda_{bb} \end{pmatrix} \begin{pmatrix} Y_A \\ Y_B \\ Y_a \\ Y_b \end{pmatrix} = \begin{pmatrix} U_A \\ U_B \\ U_a \\ U_b \end{pmatrix}.$$

Thus here the coefficient $\lambda_{ab}(=\lambda_{ba})$ has also been set to zero, reflecting the marginal independence of a and b. This model is not equivalent to its undirected counterpart. Iteration, combined with the formula (5.52), is needed to calculate the estimates. The joint concentration matrix is estimated by

$$\hat{K} = \hat{K}^* - n \begin{pmatrix} 0 & 0 & 0 & 0 \\ 0 & 0 & 0 & 0 \\ 0 & 0 & ssd^{aa}_{\{ab\}} & ssd^{ab}_{\{ab\}} \\ 0 & 0 & ssd^{ba}_{\{ab\}} & ssd^{bb}_{\{ab\}} \end{pmatrix} + \begin{pmatrix} 0 & 0 & 0 & 0 \\ 0 & 0 & 0 & 0 \\ 0 & 0 & n/ssd_a & 0 \\ 0 & 0 & 0 & n/ssd_b \end{pmatrix},$$

where $\hat{K}^*$ is the estimate of K in the second model. But note that the likelihood ratio for comparing this model with the previous model reduces to that of marginal independence between a and b and leads to a standard test. □

5.4.2 Lattice models

Andersson and Perlman (1993) study conditional independence models for the multivariate normal distribution that are determined through distributive lattices of subspaces. By choosing a suitable coordinate system, such models correspond to univariate recursive models given by directed acyclic graphs that are transitive, i.e. they satisfy that $\alpha \mapsto \beta$ implies $\alpha \to \beta$ (Andersson et al. 1995b).

5.4.3 Collapsibility

There are notions of collapsibility which are completely analogous to the discrete case. Wermuth (1989a) discusses notions related to parametric collapsibility and Frydenberg (1990b) studies stability of models under marginalization. The graphical condition for model collapsibility is identical in the discrete and pure continuous cases.

5.4.4 Bibliographical notes

Covariance selection models were originally introduced by Dempster (1972) and the relation to models for contingency tables was explicitly discovered by Wermuth (1976a). Many of the results given in the present exposition appeared for the first time in Porteous (1985b), who also conjectured the asymptotic formula (5.38).

The univariate recursive case was also discussed in some detail in Wermuth (1980), and Kiiveri and Speed (1982) give a description of the relation between Gaussian recursive graphical models and some traditional structural equation models such as dealt with for example in LISREL (Jöreskog 1973, 1981).

The case of complex random variables has recently been studied in some detail by H.H. Andersen *et al.* (1995).

6
Models for mixed data

6.1 Basic facts and concepts

In this chapter we unify the graphical models developed for contingency tables and for the multivariate normal distribution. At the same time we extend the theory and methodology to deal with variables of which some are qualitative – as discussed in the chapter on contingency tables – and some are quantitative – as in the chapter on multivariate normal models. The results in this chapter will eventually contain many of the results already described.

The building blocks for the class of models to be developed are multivariate distributions of a type that can capture covariation of discrete and continuous random variables. The distributions have the property that the conditional distribution of the continuous random variables given the discrete variables is multivariate normal. Before we move any further, we spend some time introducing this class of distributions and studying its properties.

6.1.1 CG distributions

Notation and terminology

We begin the formal developments by introducing some notation. It will hopefully be apparent that the notation conforms with that used in earlier chapters.

The set of variables V is partitioned as $V = \Delta \cup \Gamma$ into variables of *discrete* (Δ) and *continuous* (Γ) type. A typical element of the joint state space is denoted as in one of the possibilities below:

$$x = (x_\alpha)_{\alpha \in V} = (i, y) = \{(i_\delta)_{\delta \in \Delta}, (y_\gamma)_{\gamma \in \Gamma}\},$$

where i_δ are qualitative and y_γ are real-valued. A particular combination $i = (i_\delta)_{\delta \in \Delta}$ is referred to as a *cell* and the set of cells is denoted by $\mathcal{I}$. The joint distribution of the variables is supposed to have a density f with

$$\log f(x) = \log f(i, y) = g(i) + h(i)^\top y - y^\top K(i) y / 2, \quad (6.1)$$

BASIC FACTS AND CONCEPTS   159

in which case we say that X follows a *CG distribution*. For each i, $g(i)$ is a real number, $h(i) = \{h(i)_\gamma\}_{\gamma \in \Gamma}$ a vector in $\mathcal{R}^\mathcal{I}$, and $K(i) = \{k(i)_{\gamma\mu}\}_{\gamma,\mu \in \Gamma}$ a positive definite $|\Gamma| \times |\Gamma|$ matrix. The triple (g, h, K) is the *canonical characteristics* of the distribution. That X follows a CG distribution is equivalent to the statements

$$p(i) = P(X_\Delta = i) > 0 \quad \text{and} \quad \mathcal{L}(X_\Gamma \mid X_\Delta = i) = \mathcal{N}_{|\Gamma|}\{\xi(i), \Sigma(i)\},$$

where $X_A = (X_\alpha)_{\alpha \in A}$ and so on, as in the previous chapters, and

$$\Sigma(i) = K(i)^{-1}, \quad \xi(i) = K(i)^{-1} h(i).$$

The triple (p, ξ, Σ) are the *moment characteristics* of the distribution. The probabilities $p(i)$ can be calculated from the canonical characteristics as

$$p(i) = (2\pi)^{|\Gamma|/2} \{\det K(i)\}^{-1/2} \exp\{g(i) + h(i)^\top K(i)^{-1} h(i)/2\}. \quad (6.2)$$

This can be seen by first writing

$$f(i, y) = \exp\left[g^*(i) - \{y - \xi(i)\}^\top K(i)\{y - \xi(i)\}/2\right],$$

where we have let

$$g^*(i) = g(i) + h(i)^\top K(i)^{-1} h(i)/2,$$

letting $z = y - \xi(i)$, and finally integrating over z using that if $z \in \mathcal{R}^p$ and K is a positive definite matrix, then

$$\int_{\mathcal{R}^p} e^{-z^\top K z/2} \, dz = (2\pi)^{\frac{p}{2}} (\det K)^{-\frac{1}{2}}. \quad (6.3)$$

In words, a CG distribution has positive cell probabilities and the conditional distribution of the continuous variables, given the discrete, is regular multivariate normal. There is a case for allowing zero probabilities as well as singular conditional distributions, but we choose to avoid this complication here even though this is not in line with our treatment of the discrete case.

Inverting the relation (6.2) we can calculate the discrete component of the canonical characteristics from the probabilities and the linear and quadratic terms as in either of the two alternatives below:

$$\begin{aligned} g(i) &= \log p(i) - \frac{|\Gamma|}{2} \log(2\pi) \\ &\quad + \frac{1}{2} \log \det K(i) - h(i)^\top K(i)^{-1} h(i)/2 \quad (6.4) \\ &= \log p(i) - \frac{|\Gamma|}{2} \log(2\pi) \\ &\quad - \frac{1}{2} \log \det \Sigma(i) - \xi(i)^\top \Sigma(i)^{-1} \xi(i)/2. \quad (6.5) \end{aligned}$$

Note that any combination of the elements of these triples of characteristics can be used to parametrize the distributions and for each $i \in \mathcal{I}$, the triple varies freely in the product space of $\mathcal{R}$, $\mathcal{R}^{|\Gamma|}$ and $\mathcal{S}^+_{|\Gamma|}$. The *mixed characteristics* (p, h, K) are often convenient.

We say that the distribution is *homogeneous* if the covariance is independent of i, i.e. if $\Sigma(i) \equiv \Sigma$ or, equivalently, if $K(i) \equiv K$.

Marginal and conditional distributions

Next we investigate the behaviour of marginal and conditional distributions of variables that follow CG distributions. For a subset A of V we denote the marginal density of X_A by f_A and for $B = V \setminus A$ the conditional density of X_B given $X_A = x_A$ as $f^{B|A}$. Also we sometimes abbreviate and write i_A for $i_{A \cap \Delta}$, $\mathcal{I}_A$ for $\mathcal{I}_{A \cap \Delta}$, and y_A for $y_{A \cap \Gamma}$.

The marginal of a CG distribution is not always a CG distribution. However, we have

Proposition 6.1 *If f is a CG density and $B \subseteq \Gamma$, then the marginal f_A is a CG density with canonical characteristics (g_A, h_A, K_A) given as*

$$\begin{aligned}
g_A(i) &= g(i) \\
&\quad + \left[|B| \log(2\pi) - \log \det K(i)_B + h(i)_B^\top \{K(i)_B\}^{-1} h(i)_B\right]/2 \\
h_A(i) &= h(i)_A - K(i)_{AB} \{K(i)_B\}^{-1} h(i)_B \\
K_A(i) &= K(i)_A - K(i)_{AB} \{K(i)_B\}^{-1} K(i)_{BA}.
\end{aligned}$$

Proof: Let

$$\mu(i) = -\{K(i)_B\}^{-1} K(i)_{BA} y_A + \{K(i)_B\}^{-1} h(i)_B.$$

Then we find by direct calculation that

$$\begin{aligned}
f(i, y) &= \exp\left[-\{y_B - \mu(i)\}^\top K(i)_B \{y_B - \mu(i)\}/2\right] \\
&\quad \times \exp\left[y_A^\top \{h(i)_A - K(i)_{AB}\{K(i)_B\}^{-1} h(i)_B\}\right] \\
&\quad \times \exp\left(-y_A^\top \left[K(i)_A - K(i)_{AB}\{K(i)_B\}^{-1} K(i)_{BA}\right] y_A/2\right) \\
&\quad \times \exp\left[g(i) + h(i)_B^\top \{K(i)_B\}^{-1} h(i)_B/2\right].
\end{aligned}$$

Now y_B only appears in the first factor. This can be integrated by letting $z = y_B - \mu(i)$ and using (6.3). The result follows from inspection of the relevant terms. □

Note that the relation (6.2) comes out as a special case of this proposition when $B = \Gamma$.

Alternatively, the marginal distribution could have been identified by its moment characteristics, expressed by the moment characteristics of the full distribution. This would lead to the formulae

$$p_A(i) = p(i), \quad \xi_A(i) = \xi(i)_A, \quad \Sigma_A(i) = \Sigma(i)_A.$$

These formulae are much simpler than those just given above and reflect a general phenomenon: the canonical characteristics give simple expressions for conditional distributions, whereas the moment characteristics are simple to use for marginalization.

The situation is different if the variables to be integrated out are discrete.

Proposition 6.2 *If f is a CG density, $B \subseteq \Delta$, and $B \perp\!\!\!\perp \Gamma \mid \Delta \setminus B$, then the marginal f_A is a CG density with canonical characteristics (g_A, h_A, K_A) given as*

$$g_A(i_{\Delta \setminus B}) = \log \sum_{j: j_{\Delta \setminus B} = i_{\Delta \setminus B}} e^{g(j)}, \quad h_A(i_{\Delta \setminus B}) = h(i), \quad K_A(i_{\Delta \setminus B}) = K(i).$$

Proof: That f_A is a CG density is most elegantly seen as follows. We must show that the conditional distribution of the continuous variables X_Γ given the remaining discrete variables $X_{\Delta \setminus B}$ is regular multivariate normal. But, by the conditional independence,

$$\mathcal{L}(X_\Gamma \mid X_{\Delta \setminus B} = i_{\Delta \setminus B}) = \mathcal{L}(X_\Gamma \mid X_\Delta = i_\Delta) \tag{6.6}$$

and the distribution on the right-hand side is regular multivariate normal by assumption.

Since $K(i)$ is the conditional concentration matrix, by (6.6) it can only depend on i through $i_{\Delta \setminus B}$, which gives the third identity concerning the canonical characteristics. A similar argument gives the second identity. The final identity comes from summing in the expression (6.1) for a general CG density, using the constancy of $h(j)$ and $K(j)$:

$$\begin{aligned} f(i_{\Delta \setminus B}, y) &= \sum_{j: j_{\Delta \setminus B} = i_{\Delta \setminus B}} \exp\{g(j) + h(j)^\top y - y^\top K(j) y/2\} \\ &= \sum_{j: j_{\Delta \setminus B} = i_{\Delta \setminus B}} \exp\{g(j) + h(i_{\Delta \setminus B})^\top y - y^\top K(i_{\Delta \setminus B}) y/2\} \\ &= \exp\{h(i_{\Delta \setminus B})^\top y - y^\top K(i_{\Delta \setminus B}) y/2\} \sum_{j: j_{\Delta \setminus B} = i_{\Delta \setminus B}} \exp g(j). \end{aligned}$$

This completes the proof. □

The moment characteristics of the marginal distribution are also simple here:
$$p_A(i_{\Delta\setminus B}) = p(i_{\Delta\setminus B}) = \sum_{j:j_{\Delta\setminus B}=i_{\Delta\setminus B}} p(j)$$

and
$$\xi_A(i_{\Delta\setminus B}) = \xi(i), \quad \Sigma_A(i_{\Delta\setminus B}) = \Sigma(i).$$

Marginalization of CG distributions over discrete variables leads in general to complicated mixture distributions. In fact, the converse to Proposition 6.2 holds, i.e. if the marginal f_A of a CG density over a set B of discrete variables is again a CG density, then $B \perp\!\!\!\perp \Gamma \mid \Delta \setminus B$ (Frydenberg 1990b). But even though the correct marginal of a CG distribution may not be CG itself, there is an alternative way of marginalizing which has interesting properties. For a general subset A with $B = V \setminus A$ as above, we define the *weak marginal* $f_{[A]}$ of f as the CG distribution with moment characteristics $(p_{[A]}, \xi_{[A]}, \Sigma_{[A]})$ where, using abbreviated notation $i_A = i_{A\cap\Delta}$ and so on,

$$p_{[A]}(i_A) = P(I_A = i_A) = \sum_{j:j_A=i_A} p(j) \tag{6.7}$$

$$\xi_{[A]}(i_A) = \mathbf{E}(Y_A \mid I_A = i_A) = \sum_{j:j_A=i_A} \frac{p(j)}{p_{[A]}(i_A)} \xi(j)_A \tag{6.8}$$

$$\Sigma_{[A]}(i_A) = \mathbf{V}(Y_A \mid I_A = i_A)$$
$$= \sum_{j:j_A=i_A} \frac{p(j)}{p_{[A]}(i_A)} \{\xi(j)_A - \xi_{[A]}(i_A)\} \{\xi(j)_A - \xi_{[A]}(i_A)\}^\top$$
$$+ \sum_{j:j_A=i_A} \frac{p(j)}{p_{[A]}(i_A)} \Sigma(j)_A, \tag{6.9}$$

i.e. the moment characteristics of the weak marginal are the correct (conditional) moments of the full joint distribution. It follows from this remark and Propositions 6.1 and 6.2 that we have

Proposition 6.3 *If $B \subseteq \Gamma$ or $(B \cap \Delta) \perp\!\!\!\perp (A \cap \Gamma) \mid (A \cap \Delta)$ then the weak marginal to A is the true marginal, i.e. $f_A = f_{[A]}$. In particular this holds for marginals to discrete sets $A \subseteq \Delta$.*

In general the weak marginal $f_{[A]}$ of f is the CG density which is closest in Kullback–Leibler divergence to the true marginal f_A. This is a consequence of the following lemma.

Lemma 6.4 *Let $f(i, y)$ be a density with finite conditional moments*

$$p(i) = P(I = i), \quad \xi(i) = \mathbf{E}(Y \mid I = i), \quad \Sigma(i) = \mathbf{V}(Y \mid I = i),$$

where $\Sigma(i)$ is positive definite for all $i \in \mathcal{I}$, and let $[f]$ denote the CG density with moment characteristics (p, ξ, Σ). Then $[f]$ minimizes the Kullback–Leibler divergence $D(f, f^*)$ among all CG distributions f^*.

Proof: If f^* is an arbitrary CG density we have

$$\begin{aligned} D(f, f^*) &= \sum_i \int \log \frac{f(i,y)}{f^*(i,y)} f(i,y) \, dy \\ &= \sum_i \int \log \left\{ \frac{f(i,y)}{[f](i,y)} \frac{[f](i,y)}{f^*(i,y)} \right\} f(i,y) \, dy \\ &= D(f, [f]) + \sum_i \int \log \frac{[f](i,y)}{f^*(i,y)} f(i,y) \, dy. \end{aligned}$$

If $([g], [h], [K])$ and (g^*, h^*, K^*) denote the canonical characteristics of $[f]$ and f^*, we have that

$$\log \frac{[f](i,y)}{f^*(i,y)} = [g](i) - g^*(i) + \{[h](i) - h^*(i)\}^\top y - y^\top \{[K](i) - K^*(i)\} y/2.$$

As this for each i is a quadratic function in y, its integral with respect to a density depends only on its second order (conditional) moments and therefore

$$\sum_i \int \log \frac{[f](i,y)}{f^*(i,y)} f(i,y) \, dy = \sum_i \int \log \frac{[f](i,y)}{f^*(i,y)} [f](i,y) \, dy = D([f], f^*).$$

Thus we find

$$D(f, f^*) = D(f, [f]) + D([f], f^*) \geq D(f, [f])$$

and there is only equality if $[f] = f^*$. This completes the proof. □

The weak marginal behaves in many respects like an ordinary marginal, for example when marginals are iterated as in the next proposition.

Proposition 6.5 *Weak marginalization of CG distributions satisfies for $C \subseteq A$ that $\{f_{[A]}\}_{[C]} = f_{[C]}$.*

Proof: The result is a consequence of the fact that weak marginals have correct (conditional) moments, and the proof we give is making a direct calculation. We get for the discrete part

$$\begin{aligned} \{p_{[A]}\}_{[C]}(i_C) &= \sum_{j_A: j_C = i_C} p_{[A]}(j_A) = \sum_{j_A: j_C = i_C} \sum_{k: k_A = j_A} p(k) \\ &= \sum_{k: k_C = i_C} p(k) = p_{[C]}(i_C). \end{aligned}$$

For the conditional mean we have

$$\{\xi_{[A]}\}_{[C]}(i_C) = \sum_{j_A:j_C=i_C} \xi_{[A]}(j_A)_C \frac{p_{[A]}(j_A)}{p_{[C]}(i_C)}$$

$$= \sum_{j_A:j_C=i_C} \sum_{k:k_A=j_A} \xi(k)_C \frac{p(k)}{p_{[A]}(j_A)} \frac{p_{[A]}(j_A)}{p_{[C]}(i_C)}$$

$$= \sum_{k:k_C=i_C} \xi(k)_C \frac{p(k)}{p_{[C]}(i_C)} = \xi_{[C]}(i_C).$$

Similar expressions hold for the variances; we abstain from the details here.
□

In contrast to the situation concerning marginals, conditioning with any subset of variables preserves the CG distribution.

Proposition 6.6 *If f is a CG density, $B \subseteq V$ and $A = V \setminus B$, then for all values of x_A the conditional density $f^{B \mid A}$ of X_B given $X_A = x_A$ is a CG density. Its canonical characteristics $(g^{B \mid A}, h^{B \mid A}, K^{B \mid A})$ are*

$$g^{B \mid A}(i_B) = g(i) + h(i)_A^\top y_A - y_A^\top K(i)_A y_A/2 - \log \kappa(i_A, y_A)$$
$$h^{B \mid A}(i_B) = h(i)_B - K(i)_{BA} y_A$$
$$K^{B \mid A}(i_B) = K(i)_B,$$

where $\kappa(i_A, y_A)$ is a normalizing constant

$$\kappa(i_A, y_A) = (2\pi)^{\frac{|B \cap \Gamma|}{2}} \sum_{i_B \in \mathcal{I}_B} \exp\left\{g(i) + h(i)_A^\top y_A - y_A^\top K(i)_A y_A/2\right\}$$
$$\times \{\det K(i)_B\}^{-\frac{1}{2}} \exp\left\{h^{B \mid A}(i_B)^\top (K(i)_B)^{-1} h^{B \mid A}(i_B)/2\right\}.$$

Proof: The conditional distribution is found from the unconditional by fixing the value of the conditioning variable and normalizing. We write the expression (6.1) in a little more detail as

$$\log f(i, y) = g(i) + h(i)_A^\top y_A + h(i)_B^\top y_B$$
$$- \left\{y_A^\top K(i)_A y_A + 2 y_A^\top K(i)_{AB} y_B + y_B^\top K(i)_B y_B\right\}/2.$$

Identifying linear and quadratic terms in y_B, recalling that y_A is fixed, gives the expressions above. The normalizing constant is calculated as

$$\kappa(i_A, y_A) = \sum_{i_B \in \mathcal{I}_B} \exp\left\{g(i) + h(i)_A^\top y_A - y_A^\top K(i)_A y_A/2\right\}$$
$$\times \int_{\mathcal{R}^{|B \cap \Gamma|}} \exp\left\{h^{B \mid A}(i_B)^\top y_B - y_B^\top K(i)_B y_B/2\right\} dy_B.$$

Exploiting (6.3) again we find

$$\int_{\mathcal{R}^{|B \cap \Gamma|}} \exp\left\{h^{B \mid A}(i_B)^\top y_B - y_B^\top K(i)_B y_B/2\right\} dy_B$$
$$= (2\pi)^{\frac{|B \cap \Gamma|}{2}} \{\det K(i)_B\}^{-\frac{1}{2}} \exp\left\{h^{B \mid A}(i_B)^\top (K(i)_B)^{-1} h^{B \mid A}(i_B)/2\right\}$$

and the proof is complete. □

The moment characteristics of the conditional distribution are more complicated. Some calculation yields that they have the form

$$\log p^{B \mid A}(i_B) = u(i_B \mid i_A) + v(i_B \mid i_A)^\top y_A$$
$$\qquad\qquad\qquad - y_A^\top W(i_B \mid i_A) y_A - \log \kappa(i_A, y_A)$$
$$\xi^{B \mid A}(i_B) = c(i_B \mid i_A) + C(i_B \mid i_A) y_A$$
$$\Sigma^{B \mid A}(i_B) = D(i_B \mid i_A),$$

where

$$D(i_B \mid i_A) = \{K(i)_B\}^{-1} = \Sigma(i)_B - \Sigma(i)_{BA}\{\Sigma(i)_A\}^{-1}\Sigma(i)_{AB}$$
$$C(i_B \mid i_A) = -\{K(i)_B\}^{-1} K(i)_{BA} = \Sigma(i)_{BA}\{\Sigma(i)_A\}^{-1}$$
$$c(i_B \mid i_A) = \xi(i)_B - C(i_B \mid i_A)\xi(i)_A$$
$$W(i_B \mid i_A) = \{\Sigma(i)_A\}^{-1}/2$$
$$v(i_B \mid i_A) = \{\Sigma(i)_A\}^{-1}\xi(i)_A$$
$$u(i_B \mid i_A) = \log p(i) - \frac{|\Gamma \cap A|}{2} \log(2\pi)$$
$$\qquad\qquad\qquad - \left[\log \det \Sigma(i)_A + \xi(i)_A^\top \{\Sigma(i)_A\}^{-1}\xi(i)_A\right]/2.$$

The densities $f^{B \mid A}(\cdot \mid x_A)$ are called *CG regressions*, and *homogeneous CG regressions* if the original density f is homogeneous.

There is an interesting factorization result relating conditional independence to weak marginals. Assume that $V = A \cup B \cup C$ with A, B, C being disjoint. Then we have

Proposition 6.7 *If f is a CG distribution such that $A \perp\!\!\!\perp B \mid C$ and either $B \subseteq \Gamma$ or $C \subseteq \Delta$, then the following relations hold for the weak marginals:*

$$f_{[A \cup C]} = f_{A \cup C}, \quad f = \frac{f_{[A \cup C]} f_{[B \cup C]}}{f_{[C]}}, \quad f^{B \mid C} = \frac{f_{B \cup C}}{f_C} = \frac{f_{[B \cup C]}}{f_{[C]}}.$$

Proof: The first relation follows from Proposition 6.3 since either we have $B \subseteq \Gamma$ in which case this follows directly, or if $C \subseteq \Delta$ we can use properties of conditional independence so that from $B \perp\!\!\!\perp A \mid C$, $C = C \cap \Delta$, and $A \cap \Gamma = (A \cup C) \cap \Gamma$ we obtain

$$(B \cap \Delta) \perp\!\!\!\perp \{(A \cup C) \cap \Gamma\} \mid \{(A \cup C) \cap \Delta)\},$$

whereby Proposition 6.3 applies.

If $C \subseteq \Delta$ we may substitute B for A in the first relation to get the identity $f_{[B \cup C]} = f_{B \cup C}$. We also have that $f_{[C]} = f_C$ since this is always true for a set of discrete variables C. The third identity thus follows trivially. The second identity now follows by combining the previous findings with the conditional independence that implies the factorization

$$f = \frac{f_{A \cup C} f_{B \cup C}}{f_C}. \tag{6.10}$$

We finally consider the case $B \subseteq \Gamma$. Let $\tilde{f} = f_{[C]} f^{B \mid C}$ where now $f^{B \mid C}$ is a multivariate normal distribution. The density $\tilde{f}$ is clearly of CG type. Below we find its moment characteristics $(\tilde{p}, \tilde{\xi}, \tilde{\Sigma})$. We get

$$\tilde{p}(i_{B \cup C}) = p_{[C]}(i_C) = p(i_C) = p_{[B \cup C]}(i_{B \cup C})$$

and

$$\begin{aligned}
\tilde{\xi}(i_{B \cup C}) &= \tilde{\mathbf{E}}\left\{\tilde{\mathbf{E}}(Y_{B \cup C} \mid Y_C, I_C = i_C)\right\} = \mathbf{E}\left\{\mathbf{E}(Y_{B \cup C} \mid Y_C, I_C = i_C)\right\} \\
&= \mathbf{E}(Y_{B \cup C} \mid I_C = i_C) = \xi_{[B \cup C]}(i_{B \cup C}).
\end{aligned}$$

Here $\tilde{\mathbf{E}}$ denotes expectation with respect to $\tilde{f}$ and $\mathbf{E}$ expectation with respect to f. We have exploited that $\mathbf{E}(Y_{B \cup C} \mid Y_C, I_C = i_C)$ depends linearly on Y_C and that $f_{[C]}$ has correct conditional first and second moments. Similarly we get

$$\begin{aligned}
\tilde{\Sigma}(i_{B \cup C}) &= \tilde{\mathbf{E}}\left\{\tilde{\mathbf{E}}(Y_{B \cup C} Y_{B \cup C}^\top \mid Y_C, I_C = i_C)\right\} - \tilde{\xi}(i_{B \cup C})\tilde{\xi}(i_{B \cup C})^\top \\
&= \mathbf{E}(Y_{B \cup C} Y_{B \cup C}^\top \mid I_C = i_C) - \tilde{\xi}(i_{B \cup C})\tilde{\xi}(i_{B \cup C})^\top \\
&= \Sigma_{[B \cup C]}(i_{B \cup C}).
\end{aligned}$$

Hence $\tilde{f}$ and $f_{[B \cup C]}$ have the same moment characteristics and must therefore be equal. The third identity of the proposition follows. The second identity again follows from combining (6.10) with the identities which have already been established. □

Sampling statistics

Throughout this chapter we consider a sample of independent and identically distributed observations $(x^1, \ldots, x^{|n|})$ where x^ν have both discrete and continuous components $x^\nu = (i^\nu, y^\nu)$. Sometimes we sort the observations according to the observed value of the discrete variable. We then have

BASIC FACTS AND CONCEPTS

a contingency table as in Chapter 4, but here each cell has $n(i)$ observations of the continuous variables and not just the counts $n(i)$.

It is convenient to introduce the following notation for standard sampling statistics, where A is an arbitrary subset of V. As before, we abbreviate to $i_A = i_{A \cap \Delta}$, $\mathcal{I}_A = \mathcal{I}_{A \cap \Delta}$, $y_A = y_{A \cap \Gamma}$.

$$
\begin{aligned}
d(i_A) &= \{\nu \mid i_A^\nu = i_A\} \\
n(i_A) &= |d(i_A)| = \text{the number of observations in cell } i_A \\
s(i_A) &= \sum_{\nu \in d(i_A)} y^\nu = \text{the sum of the } y\text{-values in cell } i_A \\
\bar{y}(i_A) &= s(i_A)/n(i_A) \\
ss(i_A) &= \sum_{\nu \in d(i_A)} y^\nu (y^\nu)^\top \\
&= \text{the sum of squares of the } y\text{-values in cell } i_A \\
ssd(i_A) &= ss(i_A) - s(i_A)s(i_A)^\top / n(i_A) \\
&= \sum_{\nu \in d(i_A)} \{y^\nu - \bar{y}(i_A)\}\{y^\nu - \bar{y}(i_A)\}^\top \\
&= \text{the sum of squares of deviations from the mean in cell } i_A \\
s &= \sum_{i \in \mathcal{I}} s(i) = \sum_{\nu=1}^{|n|} y^\nu = \text{the total sum} \\
ss &= \sum_{i \in \mathcal{I}} ss(i) = \sum_{\nu=1}^{n} y^\nu (y^\nu)^\top \\
&= \text{the total sum of squares} \\
ssd(A) &= \sum_{i_A \in \mathcal{I}_A} ssd(i_A) \\
&= \text{the sum of squares of deviations from the mean} \\
&\quad \text{within the cells of the marginal table } \mathcal{I}_A \\
ssd &= ssd(V).
\end{aligned}
$$

Note that then $ssd(\emptyset)$ is the total sum of squares of deviations.

Strictly speaking, the quantities $\bar{y}(i_A)$ and hence also $ssd(i_A)$ are only defined for $n(i_A) > 0$, but they can be assigned any value – 0, say – in the case $n(i_A) = 0$. We use capital letters for the random variables corresponding to all the quantities above. An expression for $B \subset V$ such as $ssd(i_A)_B$ denotes as usual the submatrix $\{ssd(i_A)_{\gamma\mu}\}_{\gamma,\mu \in B \cap \Gamma}$, and similarly for the other quantities.

6.1.2 The saturated models

We investigate two types of saturated models. Both types are unrestricted as regards conditional independence properties, and the only restrictions pertain to distributional type. The *saturated mixed model* lets X have an arbitrary CG distribution. The *homogeneous saturated mixed model* restricts the distribution of X to a CG distribution which is homogeneous, i.e. the conditional covariance given the discrete variables does not depend on the condition. The sampling distributions of the sufficient statistics are described below in the two cases.

Proposition 6.8 *The set of statistics $\{N(i), \bar{Y}(i), SSD(i)\}_{i \in \mathcal{I}}$ is minimal sufficient and has a sampling distribution which can be described as follows:*

(i) *given the table of counts $\{N(i)\}_{i \in \mathcal{I}}$, all components of $\{\bar{Y}(i)\}_{i \in \mathcal{I}}$ and $\{SSD(i)\}_{i \in \mathcal{I}}$ are independent and also independent of $(I^1, \ldots, I^{|n|})$;*

(ii) *for all $i \in \mathcal{I}$ we have $\{\bar{Y}(i), SSD(i)\} \perp\!\!\!\perp \{N(j)\}_{j \neq i} \mid N(i)$;*

(iii) *the set of counts $\{N(i)\}_{i \in \mathcal{I}}$ has a multinomial distribution with parameters $|n|$ and $p = \{p(i)\}_{i \in \mathcal{I}}$;*

(iv) $\mathcal{L}\{\bar{Y}(i) \mid N(i) = n(i)\} = \mathcal{N}_{|\Gamma|}\{\xi(i), n(i)^{-1}\Sigma(i)\}$, *when $n(i) > 0$;*

(v) $\mathcal{L}\{SSD(i) \mid N(i) = n(i)\} = \mathcal{W}_{|\Gamma|}\{\Sigma(i), n(i) - 1\}$, *when $n(i) > 1$.*

In the homogeneous case, i.e. when $\Sigma(i) \equiv \Sigma$, the sufficient statistic can be reduced to $[\{N(i), \bar{Y}(i)\}_{i \in \mathcal{I}}, SSD]$. Then it holds further that SSD is independent of $[\{N(i), \bar{Y}(i)\}_{i \in \mathcal{I}}]$ and distributed as $\mathcal{W}_{|\Gamma|}(\Sigma, |n| - |\mathcal{I}|)$.

Proof: Consider the joint density of $(X^1, \ldots, X^{|n|})$:

$$\prod_{\nu=1}^{|n|} f\{x^\nu; (p, \xi, \Sigma)\} = \prod_{\nu=1}^{|n|} p(i^\nu) \phi\{y^\nu; \xi(i^\nu), \Sigma(i^\nu)\}$$

$$= \prod_{i \in \mathcal{I}} p(i)^{n(i)} \prod_{i \in \mathcal{I}} \prod_{\nu \in d(i)} \phi\{y^\nu; \xi(i), \Sigma(i)\}$$

$$\propto \prod_{i \in \mathcal{I}} p(i)^{n(i)} \prod_{i \in \mathcal{I}} \prod_{\nu \in d(i)} \det\{\Sigma(i)\}^{-1/2} e^{-\{y^\nu - \xi(i)\}^\top \Sigma(i)^{-1} \{y^\nu - \xi(i)\}/2}$$

$$= \prod_{i \in \mathcal{I}} \left(\frac{p(i)}{\det\{\Sigma(i)\}^{1/2}}\right)^{n(i)} e^{-\operatorname{tr}\left[\Sigma(i)^{-1}\left\{ssd(i) + (\bar{y}(i) - \xi(i))(\bar{y}(i) - \xi(i))^\top\right\}\right]/2}.$$

The results involving multinomial distributions are obvious. The remaining facts follow from standard results about sampling distributions of the multivariate normal distribution, cf. Appendix C, by conditioning on the random variables $(I^1, \ldots, I^{|n|})$. □

Note that in general both $\bar{Y}(i)$ and $SSD(i)$ have non-standard distributions, being mixtures of normal and Wishart distributions with different covariance matrices or degrees of freedom depending on the random quantities $N(i)$. When it comes to maximizing the likelihood function, we have

Proposition 6.9 *The likelihood function for the saturated model attains its maximum if and only if $ssd(i)$ is positive definite for all $i \in \mathcal{I}$, which is almost surely equal to the event that $n(i) > |\Gamma|$ for all $i \in \mathcal{I}$. If the maximum likelihood estimate exists, it is given as having moment characteristics equal to the empirical moments, i.e..*

$$\hat{p}(i) = n(i)/|n|, \quad \hat{\xi}(i) = \bar{y}(i), \quad \hat{\Sigma}(i) = ssd(i)/n(i).$$

Proof: Continuing from the expression in the proof of Proposition 6.8, we get for the likelihood function

$$\begin{aligned} L(f) &= L(p, \xi, \Sigma) \\ &= L_1\left\{(p(i))_{i \in \mathcal{I}}; (n(i))_{i \in \mathcal{I}}\right\} \prod_{i \in \mathcal{I}} L_{i,n(i)}\{\xi(i), \Sigma(i); \bar{y}(i), ssd(i)\}. \end{aligned}$$

From this expresssion we see that the likelihood function factorizes into a product of likelihood functions as in (4.11) and (5.1) with parameters that are variation-independent between factors. The joint likelihood function has therefore a maximum if and only if each of the factors has, and the maximization can be performed by maximizing factors separately. From Proposition 4.5 and Theorem 5.1 we then obtain the desired result. □

Note that the number of observations in any given cell is random and that the critical events $\{n(i) \leq |\Gamma|\}$ thus have positive probability. The practical consequence of this is that saturated models with many cells require large data sets. For any given set of data one can of course check the condition $n(i) > |\Gamma|$ just as easily as if it had been non-random.

In the homogeneous case we obtain a slightly better result.

Proposition 6.10 *The likelihood function for the homogeneous, saturated model attains its maximum if and only if $n(i) > 0$ for all $i \in \mathcal{I}$ and ssd is positive definite. If $|n| < |\Gamma| + |\mathcal{I}|$ this never happens. If $|n| \geq |\Gamma| + |\mathcal{I}|$ then this is almost surely equal to the event that $n(i) > 0$ for all $i \in \mathcal{I}$. If it exists, the maximum likelihood estimate is given as*

$$\hat{p}(i) = n(i)/|n|, \quad \hat{\xi}(i) = \bar{y}(i), \quad \hat{\Sigma} = ssd/|n|.$$

Proof: Analogous to that of Proposition 6.9. We shall abstain from giving full details but just mention that $\hat{p}(i)$ exists if and only if $n(i) > 0$, and $\hat{\Sigma}$ exists if and only if ssd is positive definite. In this case the positive definiteness of ssd is almost surely equivalent to the non-random condition $|n| \geq |\Gamma| + |\mathcal{I}|$ because SSD has a Wishart distribution with $|n| - |\mathcal{I}|$ degrees of freedom; cf. Proposition 6.8. □

In the non-homogeneous case, the maximized value of the likelihood function is equal to

$$\begin{aligned}
L(\hat{p}, \hat{\xi}, \hat{\Sigma}) &= \prod_{i \in \mathcal{I}} \{n(i)/|n|\}^{n(i)} \prod_{i \in \mathcal{I}} \left\{ \det \hat{\Sigma}(i) \right\}^{-n(i)/2} e^{-\operatorname{tr}\{\hat{\Sigma}(i)^{-1} ssd(i)\}/2} \\
&= \prod_{i \in \mathcal{I}} \{n(i)/|n|\}^{n(i)} \prod_{i \in \mathcal{I}} n(i)^{n(i)|\Gamma|/2} \{\det ssd(i)\}^{-n(i)/2} \\
&\quad \times \prod_{i \in \mathcal{I}} e^{-\operatorname{tr}\{n(i) ssd(i)^{-1} ssd(i)\}/2} \\
&= |n|^{-|n|} e^{-|n||\Gamma|/2} \\
&\quad \times \prod_{i \in \mathcal{I}} n(i)^{n(i)(|\Gamma|/2+1)} \{\det ssd(i)\}^{-n(i)/2}, \quad (6.11)
\end{aligned}$$

where we have ignored the constant $(2\pi)^{-|n||\Gamma|/2}$. In the homogeneous case we similarly find

$$L(\hat{\hat{p}}, \hat{\hat{\xi}}, \hat{\hat{\Sigma}}) = |n|^{|n|(|\Gamma|/2-1)} \{\det ssd\}^{-|n|/2} e^{-|n||\Gamma|/2} \prod_{i \in \mathcal{I}} n(i)^{n(i)}, \quad (6.12)$$

ignoring the same constant.

Next we investigate the deviance test for homogeneity in the saturated model, i.e. the deviance test for the hypothesis that f is a homogeneous CG density under the asssumption that f is a CG density or, equivalently, the deviance test for variance homogeneity $\Sigma(i) = \Sigma$. From the above formulae we find the deviance as

$$d = |n| \log \det ssd - \sum_i n(i) \log \det ssd(i) - |\Gamma||n| \log |n| + |\Gamma| \sum_i n(i) \log n(i).$$

It is common to use not this deviance but instead the minor modification suggested by Bartlett (1937):

$$d^* = f \log \det ssd - \sum_i f_i \log \det ssd(i) - |\Gamma| f \log f + |\Gamma| \sum_i f_i \log f_i,$$

where $f_i = n(i) - 1$ and $f = |n| - |\mathcal{I}|$. The modified deviance test is evaluated in the conditional distribution given the statistic which is sufficient

BASIC FACTS AND CONCEPTS

under the hypothesis. As the conditional distribution of $SSD(i)$ given $n(i)$ is Wishart with f_i degrees of freedom, the distribution of both D and D^* is considered in Section C.3. They are of Box type and the approximations described in that section therefore apply.

Mixed models as exponential models

For later developments it is practical to establish that the saturated mixed model is an exponential model in both the homogeneous and general cases. To do this, we rewrite the logarithm of the density (6.1) as

$$\begin{aligned}
\log f(x) &= g(i) + h(i)^\top y - y^\top K(i)y/2 \\
&= \sum_{j \in \mathcal{I}} \chi^j(i) \left\{ g(j) + h(j)^\top y - y^\top K(j)y/2 \right\} \\
&= \sum_{j \in \mathcal{I}} \chi^j(i) g(j) + \sum_{j \in \mathcal{I}} \chi^j(i) h(j)^\top y - \sum_{j \in \mathcal{I}} \operatorname{tr}\{K(j)\chi^j(i) yy^\top/2\},
\end{aligned}$$

where as usual

$$\chi^j(i) = \begin{cases} 1 & \text{if } j = i \\ 0 & \text{otherwise.} \end{cases}$$

Next we introduce an inner product between two vectors (u, v, W) and (a, b, C) in the vector space $(\mathcal{R} \times \mathcal{R}^{|\Gamma|} \times \mathcal{S}_\Gamma)^{\mathcal{I}}$ which contains the canonical and moment characteristics

$$\langle (u, v, W), (a, b, C) \rangle = \sum_{i \in \mathcal{I}} u(i) a(i) + v(i)^\top b(i) + \operatorname{tr}\{W(i) C(i)\}.$$

We now readily identify the family of CG densities in the mixed saturated model as forming an exponential model with canonical parameter $\theta = (g, h, K)$ in $(\mathcal{R} \times \mathcal{R}^{|\Gamma|} \times \mathcal{S}_\Gamma^+)^{\mathcal{I}}$ and canonical statistic (t_1, t_2, t_3) where

$$t_1(j; x) = \chi^j(i), \quad t_2(j; x) = \chi^j(i) y, \quad t_3(j; x) = -\chi^j(i) yy^\top/2.$$

If we have $|n|$ independent observations as in the usual sampling situation, we get the basic statistics discussed earlier in the section by adding over cases:

$$\sum_{\nu=1}^{|n|} t_1(j; x^\nu) = \sum_{\nu=1}^{|n|} \chi^j(i^\nu) = n(j)$$

$$\sum_{\nu=1}^{|n|} t_2(j; x^\nu) = \sum_{\nu=1}^{|n|} \chi^j(i^\nu) y^\nu = s(j)$$

$$\sum_{\nu=1}^{|n|} t_3(j; x^\nu) = -\sum_{\nu=1}^{|n|} \chi^j(i^\nu) y^\nu (y^\nu)^\top/2 = -ss(j)/2.$$

The logarithm of the likelihood function in the mixed case can then be written compactly as

$$\log L(g,h,K) = \sum_{i\in\mathcal{I}} g(i)n(i) + \sum_{i\in\mathcal{I}} h(i)^\top s(i) - \sum_{i\in\mathcal{I}} \mathrm{tr}\{K(i)ss(i)/2\}. \quad (6.13)$$

In Proposition 6.8 we chose to use $\{n(i), \bar{y}(i), ssd(i)\}_{i\in\mathcal{I}}$ to represent the minimal sufficient statistic in the model, because these have distributional properties that seem simpler to describe than those of the canonical statistics themselves.

For the model to be minimally represented, we choose a fixed reference cell i^* and exclude the value of $g(i^*)$ in the canonical parameter as well as $\chi^{i^*}(i)$ from the statistic. The cumulant generating function is then equal to $-g(i^*)$ and determined by normalization from the other parameters. If we then let $\mathcal{I}_* = \mathcal{I} \setminus \{i^*\}$, the canonical parameter space in the minimal representation becomes $\mathcal{R}^{\mathcal{I}_*} \times (\mathcal{R}^{|\Gamma|} \times \mathcal{S}_\Gamma^+)^{\mathcal{I}}$.

The exponential model is full because by (6.3) the integral of the expression (6.1) is exactly finite when $K(i)$ is positive definite. The exponential model is regular because the canonical parameter space is an open subset of the vector space $\mathcal{R}^{\mathcal{I}_*} \times (\mathcal{R}^{|\Gamma|} \times \mathcal{S}_\Gamma)^{\mathcal{I}}$. The order of the exponential model is the dimension of this vector space. We calculate the order to be

$$\begin{aligned}
\dim \mathcal{R}^{\mathcal{I}_*} \times (\mathcal{R}^{|\Gamma|} \times \mathcal{S}_\Gamma)^{\mathcal{I}} &= |\mathcal{I}_*| + \{|\Gamma| + |\Gamma|(|\Gamma|+1)/2\}\,|\mathcal{I}| \\
&= |\mathcal{I}|(|\Gamma|+1)(|\Gamma|+2)/2 - 1. \quad (6.14)
\end{aligned}$$

The homogeneous model is obtained from the general model by restricting the concentration matrix $K(i)$ to be constant in i. This is an affine restriction and leads therefore to another full and regular exponential model. Its canonical parameter space in the minimal representation derived from the one above is the smaller space $\mathcal{R}^{\mathcal{I}_*} \times (\mathcal{R}^{|\Gamma|})^{\mathcal{I}} \times \mathcal{S}_\Gamma$ and the third component of the canonical statistic reduces to

$$\tilde{t}_3(x) = \sum_{j\in\mathcal{I}} t_3(j;x) = -yy^\top/2,$$

as is known from Chapter 5. The logarithm of the likelihood function reduces to

$$\log L(g,h,K) = \sum_{i\in\mathcal{I}} g(i)n(i) + \sum_{i\in\mathcal{I}} h(i)^\top s(i) - \mathrm{tr}\{Kss/2\} \quad (6.15)$$

and the order of the exponential model in the homogeneous case becomes

$$\begin{aligned}
\dim \mathcal{R}^{\mathcal{I}_*} \times (\mathcal{R}^{|\Gamma|})^{\mathcal{I}} \times \mathcal{S}_\Gamma &= |\mathcal{I}_*| + |\Gamma||\mathcal{I}| + |\Gamma|(|\Gamma|+1)/2 \\
&= (2|\mathcal{I}| + |\Gamma|)(|\Gamma|+1)/2 - 1.
\end{aligned}$$

6.2 Graphical interaction models

In this section we study graphical models determined by undirected, marked graphs with the vertex set being the set of variables $V = \Delta \cup \Gamma$ and the discrete variables being marked as in Chapter 2. If $\mathcal{G}$ is an undirected, marked graph we let $M(\mathcal{G})$ denote the set of CG distributions that are pairwise Markov with respect to $\mathcal{G}$. Since the densities f of CG distributions are strictly positive and continuous, the different Markov properties all coincide. Hence distributions in $M(\mathcal{G})$ satisfy all types of Markov properties.

A *graphical mixed interaction model* for a sample $x^1, \ldots, x^{|n|}$ is thus obtained by assuming these to be independent observations of identically CG distributed random variables with distribution belonging to $M(\mathcal{G})$.

We also consider models involving only homogeneous CG distributions and use then the notation $M_H(\mathcal{G})$ for the set of such distributions that also satisfy any of the equivalent Markov properties with respect to $\mathcal{G}$. A *homogeneous mixed graphical interaction model* is what we obtain when we replace $M(\mathcal{G})$ with $M_H(\mathcal{G})$ in the above assumptions.

6.2.1 CG interactions

As in the discrete case, it can be illuminating to expand the logarithm of the density into interactions. One alternative is to decompose the appropriate spaces as in Section B.2. Alternatively we might choose a reference cell i^* and define interactions relative to this cell. The latter approach is what we use here.

The situation is somewhat more complex than in previous chapters, due to the presence of both continuous and discrete variables. This gives rise to several types of interaction, some of which are quite different from those previously described. More precisely we define λ, η, and Ψ from g, h, and K as

$$\lambda_d(i) = \sum_{a:a\subseteq d} (-1)^{|d\setminus a|} g(i_a, i^*{}_{\Delta\setminus a})$$

$$\eta_d(i) = \sum_{a:a\subseteq d} (-1)^{|d\setminus a|} h(i_a, i^*{}_{\Delta\setminus a})$$

$$\Psi_d(i) = \sum_{a:a\subseteq d} (-1)^{|d\setminus a|} K(i_a, i^*{}_{\Delta\setminus a}).$$

This representation satisfies

$$\lambda_d(i) = 0 \quad \text{if } i_\delta = i^*{}_\delta \text{ for some } \delta \in d$$
$$\eta_d(i) = 0 \quad \text{if } i_\delta = i^*{}_\delta \text{ for some } \delta \in d$$
$$\Psi_d(i) = 0 \quad \text{if } i_\delta = i^*{}_\delta \text{ for some } \delta \in d.$$

Möbius inversion gives that g, h, and K are determined from λ, η, and Ψ as

$$g(i) = \sum_{d:d\subseteq \Delta} \lambda_d(i), \quad h(i) = \sum_{d:d\subseteq \Delta} \eta_d(i), \quad K(i) = \sum_{d:d\subseteq \Delta} \Psi_d(i).$$

If we expand the expression for the logarithm of the density into single terms, we get

$$\log f(i,y) = \sum_{d\subseteq \Delta} \lambda_d(i) + \sum_{d\subseteq \Delta} \sum_{\gamma \in \Gamma} \eta_d(i)_\gamma y_\gamma - \frac{1}{2} \sum_{d\subseteq \Delta} \sum_{\gamma,\mu \in \Gamma} \psi_d(i)_{\gamma\mu} y_\gamma y_\mu.$$

We refer to the terms $\lambda_d(i)$ as *discrete interactions* among the variables in d. If $|d| = 1$ we also use the term *main effect* of the variable in d. The terms $\eta_d(i)_\gamma$ are *linear interactions* between γ and the variables in d. If $d = \emptyset$ we also use the term *linear main effect* of the variable γ. Finally the terms $\psi_d(i)_{\gamma\mu}$ are *quadratic interactions* between γ, μ and the variables in d. Again we speak of *quadratic main effects* if $\gamma = \mu$ and $d = \emptyset$.

The discrete interactions are exactly as investigated earlier in the discrete case. The term $\lambda_\emptyset$ is constant and determined by normalization. The terms in $\eta_\emptyset$ and $\Psi_\emptyset$ correspond to the interaction terms for covariance selection models in (5.13), whereas for $d \neq \emptyset$, η_d and Ψ_d have no analogue in the pure cases. These terms are *mixed interactions* and describe the interaction between the continuous and discrete variables. A homogeneous CG distribution has no mixed quadratic interactions, i.e. $\Psi_d = 0$ for $d \neq \emptyset$.

The interactions are related to the Markov property in the usual way, stated precisely in the following version of the Hammersley–Clifford theorem.

Theorem 6.11 *A CG distribution is Markov on the graph $\mathcal{G}$ if and only if the density has an expansion into interaction terms which satisfy*

$$\begin{aligned} \lambda_d(i) &= 0 \quad \text{unless } d \text{ is complete in } \mathcal{G} \\ \eta_d(i)_\gamma &= 0 \quad \text{unless } d \cup \{\gamma\} \text{ is complete in } \mathcal{G} \\ \psi_d(i)_{\gamma\mu} &= 0 \quad \text{unless } d \cup \{\gamma,\mu\} \text{ is complete in } \mathcal{G}. \end{aligned}$$

Proof: The proof is based on the general version of the Hammersley–Clifford theorem (Theorem 3.9). It is obvious that if the mentioned interaction terms vanish, then the density factorizes on the graph.

To see the converse we choose, as in the proof of Theorem 3.9, a reference configuration $x^* = (i^*, 0)$ and define interaction terms for $d \subseteq \Delta$ and $c \subseteq \Gamma$ as

$$\begin{aligned} \phi_{d\cup \dot{c}}(x) &= \sum_{a: a \subseteq d\cup c} (-1)^{|(d\cup c)\setminus a|} \log f(x_a, x^*_{V\setminus a}) \\ &= \sum_{a: a \subseteq d} \sum_{b: b \subseteq c} (-1)^{|d\setminus a| + |c\setminus b|} \log f(i_a, i^*_{\Delta\setminus a}, y_b, 0_{\Gamma\setminus b}). \end{aligned}$$

Exploiting the expressions for the CG interactions, we find for $d \subseteq \Delta$, $c \subseteq \Gamma$, and $\gamma, \mu \in \Gamma$ with $\gamma \neq \mu$ that

$$\begin{aligned}
\phi_d(x) &= \lambda_d(i) \\
\phi_{d\cup\{\gamma\}}(x) &= \eta_d(i)_\gamma y_\gamma - \psi_d(i)_{\gamma\gamma} y_\gamma^2/2 \\
\phi_{d\cup\{\gamma,\mu\}}(x) &= -\psi_d(i)_{\gamma\mu} y_\gamma y_\mu \\
\phi_{d\cup c}(x) &= 0 \text{ for } |c| > 2.
\end{aligned}$$

As in the proof of Theorem 3.9, we find that the left-hand sides of these equations must be identically equal to zero unless the corresponding subsets are complete. The conclusions of the theorem follow. □

6.2.2 Maximum likelihood estimation

Canonical statistics

Theorem 6.11 identifies mixed graphical interaction models as full and regular exponential models, since the restrictions imposed are determined by demanding that certain linear functions of the canonical parameters in the saturated models be equal to zero.

To identify the canonical statistics in a graphical interaction model, we introduce the sets

$$\begin{aligned}
\mathcal{C}_\Delta &= \text{the set of cliques in } \mathcal{G}_\Delta \\
\mathcal{C}_\Delta(\gamma) &= \text{the sets } d \subseteq \Delta \text{ with } d \cup \{\gamma\} \text{ a clique in } \mathcal{G}_{\Delta \cup \{\gamma\}} \\
\mathcal{C}_\Delta(\gamma, \mu) &= \text{the sets } d \subseteq \Delta \text{ with } d \cup \{\gamma, \mu\} \text{ a clique in } \mathcal{G}_{\Delta \cup \{\gamma,\mu\}}.
\end{aligned}$$

Next we collect terms appropriately to obtain

$$g(i) = \sum_{d \subseteq \Delta : d \text{ complete}} \lambda_d(i) = \sum_{d \in \mathcal{C}_\Delta} \tilde{\lambda}_d(i_d)$$

$$h(i)_\gamma = \sum_{d \subseteq \Delta : d \cup \{\gamma\} \text{ complete}} \eta_d(i)_\gamma = \sum_{d \in \mathcal{C}_\Delta(\gamma)} \tilde{\eta}_d(i_d)_\gamma$$

$$\psi(i)_{\gamma\mu} = \sum_{d \subseteq \Delta : d \cup \{\gamma,\mu\} \text{ complete}} \psi_d(i)_{\gamma\mu} = \sum_{d \in \mathcal{C}_\Delta(\gamma,\mu)} \tilde{\psi}_d(i_d)_{\gamma\mu}.$$

Using next

$$\sum_{i \in \mathcal{I}} \sum_{d \in \mathcal{C}_\Delta} \tilde{\lambda}_d(i_d) n(i) = \sum_{d \in \mathcal{C}_\Delta} \sum_{i_d \in \mathcal{I}_d} \tilde{\lambda}_d(i_d) n(i_d)$$

and a similar calculation for the other quantities yields the following expression for the likelihood function:

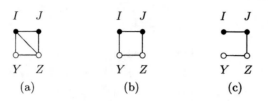

Fig. 6.1. Three mixed graphical interaction models.

$$\log L = \sum_{d \in \mathcal{C}_\Delta} \sum_{i_d \in \mathcal{I}_d} \tilde{\lambda}_d(i_d) n(i_d)$$
$$+ \sum_{\gamma \in \Gamma} \sum_{d \in \mathcal{C}_\Delta(\gamma)} \sum_{i_d \in \mathcal{I}_d} \left\{ \tilde{\eta}_d(i_d)_\gamma s(i_d)_\gamma - \frac{1}{2} \tilde{\psi}_d(i_d)_{\gamma\gamma} ss(i_d)_{\gamma\gamma} \right\}$$
$$- \sum_{\{\gamma,\mu\} \in E} \sum_{d \in \mathcal{C}_\Delta(\gamma,\mu)} \sum_{i_d \in \mathcal{I}_d} \tilde{\psi}_d(i_d)_{\gamma\mu} ss(i_d)_{\gamma\mu}. \quad (6.16)$$

In the homogeneous case the likelihood function further reduces to

$$\log L = \sum_{d \in \mathcal{C}_\Delta} \sum_{i_d \in \mathcal{I}_d} \tilde{\lambda}_d(i_d) n(i_d) + \sum_{\gamma \in \Gamma} \sum_{d \in \mathcal{C}_\Delta(\gamma)} \sum_{i_d \in \mathcal{I}_d} \tilde{\eta}_d(i_d)_\gamma s(i_d)_\gamma$$
$$- \frac{1}{2} \sum_{\gamma \in \Gamma} \psi_{\gamma\gamma} ss_{\gamma\gamma} - \sum_{\{\gamma,\mu\} \in E} \psi_{\gamma\mu} ss_{\gamma\mu}. \quad (6.17)$$

These expressions identify the canonical statistics in the two cases. The statistics can be arranged in a hierarchy as follows, each level possibly involving further marginalization of tables:

1. a set of marginal tables of counts $n(i_d), d \in \mathcal{C}_\Delta$;

2. for each continuous variable, a set of marginal tables of sums and sums of squares $\{s(i_d), ss(i_d)\}, d \in \mathcal{C}_\Delta(\gamma)$;

3. for each edge $\{\gamma, \mu\}$ between continuous variables, a set of marginal tables of sums of products $ss(i_d)_{\gamma\mu}, d \in \mathcal{C}_\Delta(\gamma,\mu)$.

In the homogeneous case only one table of sums of squares and products is needed. As in Chapter 4 we will also here use a double notation system. In specific examples we name the variables I, J, K, Y, Z instead of i_1, i_2, i_3, Y_1, Y_2, write g_{ij} instead of $g(i_1, i_2)$, h^Y_{ij} instead of $h(i_1, i_2)_1$, and so on.

Example 6.12 In this example we investigate the model(s) with graph (a) in Fig. 6.1. The only interactions which are not allowed are those that involve variables Y and J. This does not affect the first component g of the canonical characteristics, but only h and K. The full interaction expansion of the logarithm of the density becomes

$$\log f_{ij}(y,z) = \lambda^\emptyset + \lambda_i^I + \lambda_j^J + \lambda_{ij}^{IJ}$$
$$+ \left(\eta^Y + \eta_i^{IY}\right) y + \left(\eta^Z + \eta_i^{IZ} + \eta_j^{JZ} + \eta_{ij}^{IJZ}\right) z$$
$$- \frac{1}{2} \left\{ \left(\psi^Y + \psi_i^{IY}\right) y^2 + \left(\psi^Z + \psi_i^{IZ} + \psi_j^{JZ} + \psi_{ij}^{IJZ}\right) z^2 \right\}$$
$$- \left(\psi^{YZ} + \psi_i^{IYZ}\right) yz.$$

If we collect terms and reduce as above, the logarithm of the density can be more compactly written as

$$\log f_{ij}(y,z) = \tilde{\lambda}_{ij}^{IJ} + \tilde{\eta}_i^{IY} y + \tilde{\eta}_{ij}^{IJZ} z - (\tilde{\psi}_i^{IY} y^2 + 2\tilde{\psi}_i^{IYZ} yz + \tilde{\psi}_{ij}^{IJZ} z^2)/2.$$

Here we have for the sets

$$\mathcal{C}_\Delta = \mathcal{C}_\Delta(Z) = \{\{I,J\}\}, \quad \mathcal{C}_\Delta(Y) = \mathcal{C}_\Delta(Y,Z) = \{\{I\}\},$$

which is reflected in the canonical statistics

$$n_{ij}, s_{ij}^Z, ss_{ij}^Z, s_{i+}^Y, ss_{i+}^Y, ss_{i+}^{YZ},$$

where

$$s_{i+}^Y = \sum_j s_{ij}^Y$$

and so on. The homogeneous model is similar apart from the quadratic terms. The compact form of the expression for the logarithm of the density reduces in the homogeneous case to

$$\log f_{ij}(y,z) = \tilde{\lambda}_{ij}^{IJ} + \tilde{\eta}_i^{IY} y + \tilde{\eta}_{ij}^{IJZ} z - (\tilde{\psi}^Y y^2 + 2\tilde{\psi}^{YZ} yz + \tilde{\psi}^Z z^2)/2$$

with the corresponding set of canonical statistics being

$$n_{ij}, s_{ij}^Z, s_{i+}^Y, ss_{++}^Y, ss_{++}^{YZ}, ss_{++}^Z.$$

The graph (a) is decomposable and, as we shall see later, the estimates of the parameters in this model can be obtained in closed form. □

Likelihood equations

Since the models are full and regular exponential models, the likelihood equations are obtained by equating the canonical statistics to their expectation. We first calculate the expectation of the canonical statistics for the saturated model and then further reduce to the general graphical case.

If as in Chapter 4 we denote the expected cell counts by $m(i) = |n|p(i)$, we find
$$\mathbf{E}S(i) = |n|\mathbf{E}\left\{\chi^i(I)Y\right\} = |n|\left\{0 + p(i)\mathbf{E}(Y \mid I = i)\right\} = m(i)\xi(i)$$
and similarly
$$\begin{aligned}\mathbf{E}SS(i) &= |n|\mathbf{E}\left\{\chi^i(I)YY^\top\right\} \\ &= m(i)\mathbf{E}(YY^\top \mid I = i) = m(i)\left\{\Sigma(i) + \xi(i)\xi(i)^\top\right\}.\end{aligned}$$

Hence by further summation we find the likelihood equations
$$\begin{aligned}n(i_d) &= m(i_d), \quad d \in \mathcal{C}_\Delta \\ s(i_d)_\gamma &= \sum_{j:j_d=i_d} m(j)\xi(j)_\gamma, \quad \gamma \in \Gamma,\ d \in \mathcal{C}_\Delta(\gamma) \\ ss(i_d)_{\gamma\gamma} &= \sum_{j:j_d=i_d} m(j)\left\{\sigma(j)_{\gamma\gamma} + \xi(j)_\gamma^2\right\}, \quad \gamma \in \Gamma,\ d \in \mathcal{C}_\Delta(\gamma) \\ ss(i_d)_{\gamma\mu} &= \sum_{j:j_d=i_d} m(j)\left\{\sigma(j)_{\gamma\mu} + \xi(j)_\gamma \xi(j)_\mu\right\}, \\ & \quad \{\gamma,\mu\} \in E,\ d \in \mathcal{C}_\Delta(\gamma,\mu).\end{aligned}$$

In the homogeneous case they reduce to
$$\begin{aligned}n(i_d) &= m(i_d), \quad d \in \mathcal{C}_\Delta \\ s(i_d)_\gamma &= \sum_{j:j_d=i_d} m(j)\xi(j)_\gamma, \quad \gamma \in \Gamma,\ d \in \mathcal{C}_\Delta(\gamma) \\ ss_{\gamma\gamma} &= |n|\sigma_{\gamma\gamma} + \sum_j m(j)\xi(j)_\gamma^2, \quad \gamma \in \Gamma \\ ss_{\gamma\mu} &= |n|\sigma_{\gamma\mu} + \sum_j m(j)\xi(j)_\gamma \xi(j)_\mu, \quad \{\gamma,\mu\} \in E.\end{aligned}$$

The first and last two lines of these equations are analogous to similar equations known for contingency tables and covariance selection models whereas the second lines are truly different.

If we introduce the matrix $U = \sum_i m(i)\xi(i)\xi(i)^\top$, the last two lines are equivalent to the system of equations
$$ss_c = |n|\Sigma_c + U_c, \quad c \in \mathcal{C}_\Gamma,$$
where $\mathcal{C}_\Gamma$ are the cliques of the subgraph induced by the set of continuous vertices.

It is important to remember that the equations in all cases should be supplemented with model restrictions. This makes the equations much more complex than they appear at first sight. We have used moment characteristics. Expressing these in terms of permitted interactions makes the equations quite involved.

Example 6.13 We illustrate the derived likelihood equations in the homogeneous model with graph (b) of Fig. 6.1. Apart from the interactions that were forbidden in model (a), no interaction involving both I and Z is allowed. Hence the logarithm of the density can be expressed in terms of maximal permissible interactions as

$$\log f_{ij}(y,z) = \tilde{\lambda}_{ij}^{IJ} + \tilde{\eta}_i^{IY} y + \tilde{\eta}_j^{JZ} z - (\tilde{\psi}^Y y^2 + 2\tilde{\psi}^{YZ} yz + \tilde{\psi}^Z z^2)/2.$$

The corresponding set of likelihood equations becomes

$$n_{ij} = m_{ij}, \quad s_{i+}^Y = \sum_j m_{ij}\xi_{ij}^Y, \quad s_{+j}^Z = \sum_i m_{ij}\xi_{ij}^Z, \quad ss = |n|\Sigma + U.$$

But if we take into account the model restrictions and try to express the equations in maximal permissible interactions, we get, for example, that

$$ss = |n|\{\Psi_\emptyset\}^{-1} + \sum_{ij} m_{ij}\{\Psi_\emptyset\}^{-1} \begin{pmatrix} \tilde{\eta}_i^{IY} \\ \tilde{\eta}_j^{JZ} \end{pmatrix} (\tilde{\eta}_i^{IY}, \tilde{\eta}_j^{JZ}) \{\Psi_\emptyset\}^{-1}$$

and we still need to express m_{ij} in terms of interaction parameters. □

Without further specialization it is not possible to say much more about solving the maximum likelihood equations in a general or homogeneous mixed interaction model than what follows from the fact that it is a full and regular exponential model. That is, the maximum likelihood estimate exists if the sufficient statistic is in the interior of its convex support and the solution to the likelihood equations is then unique.

A sufficient condition for the existence of a solution to the maximum likelihood equations is that the solution exists in the corresponding saturated model. Conditions for this to be true are given in Proposition 6.9 and Proposition 6.10. But as with covariance selection models, this condition is in most cases not necessary. The issue is briefly touched upon in Section 6.3 in connection with Example 6.23.

The likelihood equations have typically no explicit solution unless the graph $\mathcal{G}$ happens to be decomposable; iterative algorithms must therefore be used. Algorithms for solution of likelihood equations, calculation of asymptotic distributions of estimates, and the asymptotic forms of deviance tests are essentially the same whether we consider graphical models or the more general hierarchical models to be investigated in Section 6.4. Hence we postpone the discussion of these issues until then.

Decomposition of mixed graphical models

At this point we first introduce a little more notation. First, recall that if f is a CG density and A and B are disjoint subsets of V, we denote the marginal density of X_A by f_A and the conditional density of X_B given $X_A = x_A$ by $f^{B|A}$. The full sets of densities so obtained are denoted by

$$M(\mathcal{G})_A, \ M(\mathcal{G})^{B|A}, \quad M_H(\mathcal{G})_A, \ M_H(\mathcal{G})^{B|A}$$

in the general and homogeneous cases respectively. Here one should remember that the elements of $M(\mathcal{G})^{B|A}$ and $M_H(\mathcal{G})^{B|A}$ are CG regressions and not just densities. They carry the full information about dependence of the conditional densities on the set of conditioning variables x_B.

In the pure cases we have seen that graph decompositions lead to decompositions of the estimation problems in graphical and hierarchical models. The mixed case is not quite so simple, because the asymmetry between discrete and continuous variables shows up very clearly. Nevertheless, many the important aspects carry over. Our first result is a lemma.

Lemma 6.14 *If (A, B, C) is a decomposition of the graph $\mathcal{G}$ then we have that*

$$M(\mathcal{G})_{A \cup C} = M(\mathcal{G}_{A \cup C}), \quad M_H(\mathcal{G})_{A \cup C} = M_H(\mathcal{G}_{A \cup C})$$

as well as

$$M(\mathcal{G})^{B|(A \cup C)} = M(\mathcal{G}_{B \cup C})^{B|C}, \quad M_H(\mathcal{G})^{B|(A \cup C)} = M_H(\mathcal{G}_{B \cup C})^{B|C}.$$

Proof: We consider only the case of a general graphical interaction model, as the homogeneous case is completely analogous.

We first show that $M(\mathcal{G})_{A \cup C} = M(\mathcal{G}_{A \cup C})$. To show the inclusion $\subseteq$ we must ensure that $f_{A \cup C}$ is a CG density and also that it satisfies the correct Markov properties. The last part of the statement is a direct consequence of Proposition 3.16. To see that it is CG, we first marginalize over the continuous variables in B, which preserves the CG property by Proposition 6.1, next over the discrete variables in B if $B \cap \Delta \neq \emptyset$. Since the triple (A, B, C) forms a decomposition, we must then have $C \subseteq \Delta$. Also, as C separates A from B, it does so from $B \cap \Delta$, and thus $(B \cap \Delta) \perp\!\!\!\perp A \mid C$. By properties (C3) and (C2) of conditional independence (see Section 3.1) it follows that $(B \cap \Delta) \perp\!\!\!\perp (A \cap \Gamma) \mid (\Delta \setminus B)$. Hence Proposition 6.2 ensures that the marginal $f_{A \cup C}$ is of CG type.

To see the reverse inclusion, consider an arbitrary $f' \in M(\mathcal{G}_{A \cup C})$ and construct $f''(x) = f'(x_{A \cup C}) f_B^0(x_B)$, where f^0 is the CG density that has all discrete variables independent and uniformly distributed and the continuous variables independent of the discrete variables, mutually independent and distributed as standard $\mathcal{N}(0, 1)$ variables. Clearly $f'' \in M(\mathcal{G})$ and $f''_{A \cup C} = f'$.

We next realize that $M(\mathcal{G})^{B|(A \cup C)} \subseteq M(\mathcal{G}_{B \cup C})^{B|C}$, which is seen as follows. From Proposition 6.6 at least it is clear that $f^{B|(A \cup C)}$ is a CG distribution. To see that it is in $M(\mathcal{G}_{B \cup C})^{B|C}$ we construct the density f^* by

$$f^*(x) = f^{B|(A \cup C)}(x_B \mid x_{A \cup C}) f_{A \cup C}^0(x_{A \cup C}),$$

where $f^0_{A\cup C}$ is the marginal of the CG density f^0 defined above. That this is a CG density is most easily seen by writing

$$f^*(x) = \frac{f(x)}{f_{A\cup C}(x_{A\cup C})} f^0_{A\cup C}(x_{A\cup C}).$$

As $f_{A\cup C}$ is a CG density by Lemma 6.14, f^* must define a CG density, i.e. it is the exponential function of an expression which is linear and quadratic in the continuous variables. As C separates A from B, the density f factorizes by Proposition 3.16 as

$$f(x) = \frac{f_{A\cup C}(x_{A\cup C}) f_{B\cup C}(x_{B\cup C})}{f_C(x_C)}.$$

Here it should be remembered that neither $f_{B\cup C}$ nor f_C are CG densities in general. However, if we combine the factorization with that of f^*, we get

$$f^* = \frac{f_{B\cup C}}{f_C} f^0_{A\cup C} = f^{B\mid C} f^0_A f^0_C = f^*_{B\cup C} f^0_A.$$

According to f^*, A is independent of $B\cup C$ and therefore $f^*_{B\cup C} \in M(\mathcal{G}_{B\cup C})$, using again Proposition 6.1, Proposition 6.2 and Proposition 3.16. By the same independence and the definition of f^* we have

$$f^{B\mid (A\cup C)} = f^{*B\mid (A\cup C)} = f^{*B\mid C}$$

and hence

$$f^{B\mid (A\cup C)} \in M(\mathcal{G}_{B\cup C})^{B\mid C}.$$

The reverse inclusion $M(\mathcal{G})^{B\mid (A\cup C)} \supseteq M(\mathcal{G}_{B\cup C})^{B\mid C}$ follows by taking an arbitrary $f_1 \in M(\mathcal{G}_{B\cup C})$ and defining $f_2 = f_1 f^0_A$. Then $f_2^{B\mid (A\cup C)} = f_1^{B\mid C}$. □

The conclusions of the lemma are false if the roles of A and B are reversed, unless (B, A, C) is also a decomposition (Frydenberg 1990b). Exploiting the lemma, we get an important connection betwen a decomposition of the graph and a cut as defined in Section A.5.

Proposition 6.15 *If (A, B, C) is a decomposition of the graph $\mathcal{G}$ then the statistic $t_{A\cup C}(x) = x_{A\cup C}$ is a cut in the models $M(\mathcal{G})$ and $M_H(\mathcal{G})$.*

Proof: We first write

$$f(x) = f_{A\cup C}(x_{A\cup C}) f^{B\mid (A\cup C)}(x_B \mid x_{A\cup C}).$$

Trivially, the marginal distribution of $X_{A\cup C}$ depends only on $f_{A\cup C}$ and the conditional distribution of X_B given $X_{A\cup C}$ depends only on $f^{B\mid (A\cup C)}$.

It remains to be shown that these are variation-independent, i.e. vary in a product space. Hence if f_1 is an arbitrary element of $M(\mathcal{G})_{A\cup C}$ and f_2 an arbitrary element of $M(\mathcal{G})^{B\,|\,(A\cup C)}$, we must show that the function $\tilde{f}$ defined as
$$\tilde{f}(x) = f_1(x_{A\cup C})f_2(x_B\,|\,x_{A\cup C})$$
is a valid CG density in $M(\mathcal{G})$ and also that $\tilde{f}_{A\cup C} = f_1$ and $\tilde{f}^{B\,|\,(A\cup C)} = f_2$.

From Lemma 6.14 we know that we may equivalently assume that $f_1 \in M(\mathcal{G}_{A\cup C})$ and $f_2 \in M(\mathcal{G}_{B\cup C})^{B\,|\,C}$. So we take two such elements and write
$$\tilde{f}(x) = f_1(x_{A\cup C})\frac{h(x_{B\cup C})}{h_C(x_C)},$$
where $h \in M(\mathcal{G}_{B\cup C})$. Since (A, B, C) was a decomposition, $B \subseteq \Gamma$ or $C \subseteq \Delta$, whereby h_C is a CG density, which again implies that $\tilde{f}$ is a CG density. That $\tilde{f}$ also has the Markov properties follows from Theorem 6.11, because all interaction terms in f_1 and h are permitted also in $\tilde{f}$ and, since C is a complete subset, this is also true for h_C. Integrating over x_B gives $\tilde{f}_{A\cup C} = f_1$, and division gives $\tilde{f}^{B\,|\,(A\cup C)} = f_2$.

Again the homogeneous case is completely analogous. $\square$

Before we finally proceed to prove the extension of Proposition 4.14 and Proposition 5.6, we need another lemma and some notation. For $A \subseteq V$ let $\hat{f}_{[A]}$ be the maximum likelihood estimate of the CG density under the model $M(\mathcal{G}_A)$, based on the marginal data $x_A^1, \ldots, x_A^{|n|}$ and the marginal likelihood function $L_{[A]}$. Note that $\hat{f}_{[A]}$ will not in general be equal to the weak A-marginal $(\hat{f})_{[A]}$ of $\hat{f}$, so some care should be taken with the notation.

Lemma 6.16 *Consider a graphical interaction model $M(\mathcal{G})$ or $M_H(\mathcal{G})$ and the usual sampling situations. Let C be a complete subset of $\mathcal{G}$. Then, if $\hat{f}$ exists in any of these models, so does $\hat{f}_{[C]}$.*

Proof: Assume that $\hat{f}$ exists. Then, as C is a complete subset of $\mathcal{G}$, we get from the likelihood equations as derived in Section 6.2.2 that $\hat{p}(i_C) = n(i_C)/|n|$ and we must therefore have $n(i_C) > 0$. Assume then that the marginal estimate $\hat{f}$ does not exist. Then, since C is complete, we must by Proposition 6.9 have that $ssd(i_C)$ is singular for some $i_C \in \mathcal{I}_C$, which implies that the likelihood function $L_{[C]}$ for this marginal model is unbounded. But, for any $g \in M(\mathcal{G}_C)$ we must have
$$L(\hat{f}) \geq L(gf^0_{V\setminus C}) = L_{[C]}(g)L_{[V\setminus C]}(f^0_{V\setminus C}),$$
contradicting the unboundedness of $L_{[C]}$. $\square$

Similarly as above, we let $\hat{f}^{[A\mid B]}$ denote the estimate of the conditional density $f^{A\mid B}$ based on the marginal data $x_{A\cup B}$ and the conditional likelihood function $L^{[A\mid B]}$. We are now ready to extend the decomposition results from the pure discrete and continuous cases to the mixed case.

Proposition 6.17 *Consider a graphical interaction model $M(\mathcal{G})$ or its homogeneous version $M_H(\mathcal{G})$. If (A, B, C) is a decomposition of $\mathcal{G}$, then it holds in both cases that the maximum likelihood estimate $\hat{f}$ exists if and only if $\hat{f}_{[A\cup C]}$, $\hat{f}_{[B\cup C]}$ and $\hat{f}^{[B\mid C]}$ all exist and we then have*

$$\hat{f} = \hat{f}_{[A\cup C]}\hat{f}^{[B\mid C]} = \frac{\hat{f}_{[A\cup C]}\hat{f}_{[B\cup C]}}{\hat{f}_{[C]}}. \qquad (6.18)$$

Further, we have for the marginals that

$$\hat{f}_{A\cup C} = (\hat{f})_{[A\cup C]} = \hat{f}_{[A\cup C]}$$

and

$$(\hat{f})_{[B\cup C]} = \hat{f}_{[B\cup C]}, \quad (\hat{f})_{[C]} = \hat{f}_{[C]} = (\hat{f}_{[B\cup C]})_C.$$

Proof: The joint likelihood function can be factorized as

$$\begin{aligned} L(f) &= \prod_{\nu=1}^{|n|} f_{A\cup C}(x^\nu_{A\cup C}) \prod_{\nu=1}^{|n|} f^{B\mid C}(x^\nu_B \mid x^\nu_C) \\ &= L_{[A\cup C]}(f_{A\cup C}) L^{[B\mid C]}\left(f^{B\mid C}\right) \end{aligned}$$

into likelihood functions for the marginal and conditional models. By Lemma 6.15 $f_{A\cup C}$ and $f^{B\mid C}$ are variation-independent and hence the product can be maximized by maximizing the factors separately. We conclude that $\hat{f}$ exists if and only if $\hat{f}_{[A\cup C]}$ and $\hat{f}^{[B\mid C]}$ exist and we have that

$$\hat{f} = \hat{f}_{[A\cup C]}\hat{f}^{[B\mid C]}. \qquad (6.19)$$

But now we reuse this result in the special case $A = \emptyset$ to obtain

$$\hat{f}_{[B\cup C]} = \hat{f}_{[C]}\hat{f}^{[B\mid C]} \qquad (6.20)$$

and that the existence of $\hat{f}_{[B\cup C]}$ is equivalent to the existence of both $\hat{f}_{[C]}$ and $\hat{f}^{[B\mid C]}$. Combining the results concerning the existence, we get that if $\hat{f}$ exists then $\hat{f}_{[A\cup C]}$ and $\hat{f}^{[B\mid C]}$ exist and, using Lemma 6.16, therefore also $\hat{f}_{[B\cup C]}$. Conversely, if $\hat{f}_{[A\cup C]}$ and $\hat{f}_{[B\cup C]}$ both exist, then so does $\hat{f}^{[B\mid C]}$ and hence also $\hat{f}$. Finally, solving (6.20) for $\hat{f}^{[B\mid C]}$ and inserting into (6.19) gives (6.18).

Using that weak and strong marginals to $A \cup C$ coincide by Proposition 6.3, we find by integration over x_B in (6.19) that
$$(\hat{f})_{[A \cup C]} = \hat{f}_{A \cup C} = \hat{f}_{[A \cup C]},$$
and by division in the same equation we get
$$\hat{f}^{B \mid C} = \hat{f}^{[B \mid C]}.$$
As C is assumed complete, $\hat{f}_{[C]}$ is the estimate obtained in a saturated model and it has therefore moment characteristics equal to the corresponding empirical quantities by Proposition 6.9. It follows from the form of the likelihood equations that this also holds for $(\hat{f})_{[C]}$ and hence
$$(\hat{f})_{[C]} = \hat{f}_{[C]}.$$
Finally we get from Proposition 6.7, (6.20), and the relations just derived that
$$(\hat{f})_{[B \cup C]} = (\hat{f})_{[C]} \hat{f}^{B \mid C} = \hat{f}_{[C]} \hat{f}^{[B \mid C]} = \hat{f}_{[B \cup C]}.$$
By Propositions 6.3 and 6.5 we have that
$$(\hat{f})_{[C]} = \left\{ (\hat{f})_{[B \cup C]} \right\}_{[C]} = \left\{ (\hat{f})_{[B \cup C]} \right\}_{C},$$
which completes the proof. □

Theorem 6.17 gives a factorization of the density estimates. Below we translate this factorization into combination formulae for the standard mixed characteristics of the estimated densities.

Let for any $A \subseteq V$ the expressions $(\hat{p}_{[A]}, \hat{h}_{[A]}, \hat{K}_{[A]})$ denote the standard mixed characteristics for $\hat{f}_{[A]}$. As in Chapter 5, we let $[D]^{|\Gamma|}$ be the matrix or vector obtained from D by filling up with zero entries such as to give it full dimension $|\Gamma| \times |\Gamma|$ or $|\Gamma|$. Then we have

Proposition 6.18 *If (A, B, C) is a decomposition of $\mathcal{G}$ and the maximum likelihood estimate exists in the model $M(\mathcal{G})$ or $M_H(\mathcal{G})$, it satisfies*

$$\hat{p}(i) = \frac{\hat{p}_{[A \cup C]}(i_{A \cup C}) \hat{p}_{[B \cup C]}(i_{B \cup C})}{\hat{p}_{[C]}(i_C)} \quad (6.21)$$

$$\hat{f}(y \mid i) = \frac{\hat{f}_{[A \cup C]}(y_{A \cup C} \mid i_{A \cup C}) \hat{f}_{[B \cup C]}(y_{B \cup C} \mid i_{B \cup C})}{\hat{f}_{[C]}(y_C \mid i_C)} \quad (6.22)$$

and

$$\hat{h}(i) = \left[\hat{h}_{[A \cup C]}(i_{A \cup C}) \right]^{|\Gamma|} + \left[\hat{h}_{[B \cup C]}(i_{B \cup C}) \right]^{|\Gamma|} - \left[\hat{h}_{[C]}(i_C) \right]^{|\Gamma|} \quad (6.23)$$

$$\hat{K}(i) = \left[\hat{K}_{[A \cup C]}(i_{A \cup C}) \right]^{|\Gamma|} + \left[\hat{K}_{[B \cup C]}(i_{B \cup C}) \right]^{|\Gamma|} - \left[\hat{K}_{[C]}(i_C) \right]^{|\Gamma|}. \quad (6.24)$$

Proof: Using the usual abbreviated notation introduced, we have that

$$\hat{f}_{[A]}(i_A, y_A) = \hat{p}_{[A]}(i_A)\hat{f}_{[A]}(y_A \mid i_A),$$

where $\hat{f}_{[A]}(\cdot \mid i_A)$ is an $|A \cap \Gamma|$-dimensional Gaussian density with mean $\hat{\xi}_{[A]}(i_A)$ and variance equal to $\hat{\Sigma}_{[A]}(i_A)$. If (A, B, C) is a decomposition of $\mathcal{G}$, we get

$$\hat{p}(i)\hat{f}(y \mid i) = \frac{\hat{p}_{[A\cup C]}(i_{A\cup C})\hat{p}_{[B\cup C]}(i_{B\cup C})}{\hat{p}_{[C]}(i_C)}$$
$$\times \frac{\hat{f}_{[A\cup C]}(y_{A\cup C} \mid i_{A\cup C})\hat{f}_{[B\cup C]}(y_{B\cup C} \mid i_{B\cup C})}{\hat{f}_{[C]}(y_C \mid i_C)}. \quad (6.25)$$

Consider first the case $C \subseteq \Delta$. Then the last factor becomes

$$\hat{f}_{[A\cup C]}(y_A \mid i_{A\cup C})\hat{f}_{[B\cup C]}(y_B \mid i_{B\cup C}).$$

Integration over y_A and y_B then yields

$$\hat{p}(i) = \frac{\hat{p}_{[A\cup C]}(i_{A\cup C})\hat{p}_{[B\cup C]}(i_{B\cup C})}{\hat{p}_{[C]}(i_C)},$$

whereby

$$\hat{f}(y \mid i) = \hat{f}_{[A\cup C]}(y_A \mid i_{A\cup C})\hat{f}_{[B\cup C]}(y_B \mid i_{B\cup C}).$$

Thus if we partition y into (y_A, y_B) we get

$$\hat{\xi}(i) = \begin{pmatrix} \hat{\xi}_{[A\cup C]}(i_{A\cup C}) \\ \hat{\xi}_{[B\cup C]}(i_{B\cup C}) \end{pmatrix}$$

and

$$\hat{\Sigma}(i) = \left\{ \begin{matrix} \hat{\Sigma}_{[A\cup C]}(i_{A\cup C}) & 0 \\ 0 & \hat{\Sigma}_{[B\cup C]}(i_{B\cup C}) \end{matrix} \right\}.$$

In the case $B \subseteq \Gamma$, i.e. $B \cap \Delta = \emptyset$, we use the fact that $\hat{f}_{[C]}$ is the C-marginal of $\hat{f}_{[B\cup C]}$ to obtain

$$\hat{p}_{[C]}(i_C)\hat{f}_{[C]}(y_C \mid i_C) = \hat{p}_{[B\cup C]}(i_C)\int_{y_B} \hat{f}_{[B\cup C]}(y_{B\cup C} \mid i_C)\, dy_B. \quad (6.26)$$

Integrating (6.25) first over y_B and then over y_A, taking into account (6.26), yields

$$\hat{p}(i) = \frac{\hat{p}_{[A\cup C]}(i_{A\cup C})\hat{p}_{[B\cup C]}(i_{B\cup C})}{\hat{p}_{[C]}(i_C)} = \hat{p}_{[A\cup C]}(i_{A\cup C}),$$

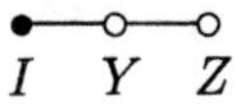

```
●───○───○
I   Y   Z
```

Fig. 6.2. A decomposable mixed graphical model. The triple (I, Z, Y) is a decomposition. The marginal distribution of (Y, Z) is not Gaussian in general.

and we also get

$$\hat{f}(y \mid i) = \frac{\hat{f}_{[A \cup C]}(y_{A \cup C} \mid i_{A \cup C}) \hat{f}_{[B \cup C]}(y_{B \cup C} \mid i_{B \cup C})}{\hat{f}_{[C]}(y_C \mid i_C)}.$$

From these expressions we obtain the desired formulae. □

Combination formulae for g, ξ and Σ can be derived analogously or by translation of those in Proposition 6.18. Note that it follows from (6.22) that

$$\det\left\{\hat{K}(i)\right\} = \frac{\det\left\{\hat{K}_{[A \cup C]}(i_{A \cup C})\right\} \det\left\{\hat{K}_{[B \cup C]}(i_{B \cup C})\right\}}{\det\left\{\hat{K}_{[C]}(i_C)\right\}}. \qquad (6.27)$$

These results are really somewhat surprising, as illustrated below.

Example 6.19 Consider the model given by Fig. 6.2. Even in this simple example the marginal distribution of (Y, Z) is not a CG distribution. Nevertheless the estimates in the corresponding graphical model can be found in the following simple way. First, in the marginal distribution of (I, Y), we get

$$\hat{p}_i = n_i/|n| \quad \hat{\xi}_i^Y = \bar{y}_i \quad \hat{\sigma}_i^{YY} = ssd_i^{YY}/n_i,$$

where ssd_i^{YY} is the sum of squares of deviations of the Y-values in cell i. The estimates in the marginal model of (Y, Z) are given as

$$\hat{\xi}^Y = \bar{y}, \quad \hat{\xi}^Z = \bar{z}, \quad |n|\hat{\Sigma} = \begin{pmatrix} ssd_\emptyset^{YY} & ssd_\emptyset^{YZ} \\ ssd_\emptyset^{ZY} & ssd_\emptyset^{ZZ} \end{pmatrix}, \qquad (6.28)$$

where $ssd_\emptyset$ denotes the total sum of squares of deviations. These estimates are not estimates of parameters in the marginal distribution of (Y, Z), since this is not even Gaussian. Still, it is true that the estimate of the parameters in the conditional distribution of Z given Y is obtained by conditioning in the normal distribution with parameters determined by (6.28), and this is equal to what one would get by performing a standard linear regression analysis of Z on Y. The final estimates of the covariance matrix are found from Proposition 6.18 to be

$$\hat{\Sigma}_i^{-1} = \begin{pmatrix} 1/\hat{\sigma}_i^{YY} + \hat{k}^{YY} - 1/ssd_\emptyset^{YY} & \hat{k}^{YZ} \\ \hat{k}^{ZY} & \hat{k}^{ZZ} \end{pmatrix},$$

where $\hat{K}$ is the estimated concentration matrix in the marginal model for (Y, Z):

$$\hat{K} = \begin{pmatrix} \hat{k}^{YY} & \hat{k}^{YZ} \\ \hat{k}^{ZY} & \hat{k}^{ZZ} \end{pmatrix} = |n| \begin{pmatrix} ssd_\emptyset^{YY} & ssd_\emptyset^{YZ} \\ ssd_\emptyset^{ZY} & ssd_\emptyset^{ZZ} \end{pmatrix}^{-1}.$$

The estimates of the means are

$$\hat{\xi}_i = \hat{\Sigma}_i \begin{pmatrix} \bar{y}_i/\hat{\sigma}_i^{YY} + \bar{y}\hat{k}^{YY} + \bar{z}\hat{k}^{YZ} - |n|\bar{y}/ssd_\emptyset^{YY} \\ \bar{y}\hat{k}^{ZY} + \bar{z}\hat{k}^{ZZ} \end{pmatrix}.$$

In the homogeneous case, $\hat{\sigma}_i^{YY}$ should everywhere be replaced by

$$\hat{\sigma}^{YY} = \frac{1}{|n|} \sum_i ssd_i^{YY},$$

which will make $\hat{\Sigma}(i)$ independent of i. □

Note that for the mixed case, in contrast to the discrete case, we have chosen to treat decomposition in connection with graphical models rather than the more general hierarchical models which are treated later. This is because it is not so clear in the mixed case what a decomposition of a hierarchical model should be.

6.3 Decomposable models

A decomposable graphical model is a graphical interaction model whose graph $\mathcal{G}$ is decomposable. Decomposable models have interesting properties both in the general case where the model is of the form $M(\mathcal{G})$ and in the homogeneous case, where they are of the the form $M_H(\mathcal{G})$.

6.3.1 Basic factorizations

When the graph $\mathcal{G}$ is decomposable, Proposition 2.17 ensures that there is a perfect numbering $C_1, \cdots, C_k$ of the cliques of $\mathcal{G}$, and by Lemma 2.11 this satisfies that for all j,

$$(H_{j-1} \setminus C_j, R_j, S_j) \text{ decomposes } \mathcal{G}_{H_j}, \tag{6.29}$$

where

$$H_j = C_1 \cup \cdots \cup C_j, \quad S_j = (C_1 \cup \cdots \cup C_{j-1}) \cap C_j, \quad R_j = C_j \setminus S_j.$$

Recall that H_j are the histories, S_j the separators, and R_j the residuals of the sequence. Repeated use of Proposition 6.7 yields the following factorization of the joint density into weak marginals to cliques and separators:

$$f(x) = \prod_{j=1}^{k} \frac{f_{[C_j]}(x_{C_j})}{f_{[S_j]}(x_{S_j})}.$$

Within each of these complete sets we can further partition the variables into discrete and continuous to obtain

$$f(x) = f(i, y) = \prod_{j=1}^{k} \frac{p(i_{C_j})}{p(i_{S_j})} \prod_{j=1}^{k} \frac{f_{[C_j]}(y_{C_j} \mid i_{C_j})}{f_{[S_j]}(y_{S_j} \mid i_{S_j})},$$

where we have let $S_1 = \emptyset$ and $f_{[\emptyset]} \equiv 1$.

6.3.2 Maximum likelihood estimation

Decomposable models admit explicit likelihood estimates in the general as well as in the homogeneous case. We use Proposition 6.17 repeatedly to obtain the factorization

$$\hat{f}(x) = \prod_{j=1}^{k} \frac{\hat{f}_{[C_j]}(x_{C_j})}{\hat{f}_{[S_j]}(x_{S_j})}, \tag{6.30}$$

where as before we have let $S_1 = \emptyset$ and $\hat{f}_{[\emptyset]} \equiv 1$.

Because all the sets C_j, S_j are complete, each of the terms on the right-hand side is of the kind treated in Section 6.1.2, and translation of Propositions 6.9 and 6.10 yields:

Proposition 6.20 *In a decomposable model, the maximum likelihood estimate exists if and only if for all cliques C of $\mathcal{G}$ with marginal cells $i_C \in \mathcal{I}_C$, we have that $n(i_C) > 0$ and $ssd_C(i_C)$ is positive definite. This event is almost surely equal to the event that $n(i_C) \geq |C \cap \Gamma|$ for all C and i_C.*

And we get for the homogeneous case

Proposition 6.21 *In a decomposable homogeneous model the maximum likelihood estimate exists if and only if for all cliques C of $\mathcal{G}$ with marginal cells $i_C \in \mathcal{I}_C$, we have that $n(i_C) > 0$ and $ssd_C(C)$ is positive definite. This event is almost surely equal to the event that $n(i_C) > 0$ for all C and i_C, provided $|n| > |C \cap \Gamma| + |\mathcal{I}_C|$.*

Further, we get from Proposition 6.9, Proposition 6.18 and (6.30) that

$$\hat{p}(i) = \prod_{j=1}^{k} \frac{\hat{p}_{[C_j]}(i_{C_j})}{\hat{p}_{[S_j]}(i_{S_j})} = \prod_{j=1}^{k} \frac{n(i_{C_j})}{n(i_{S_j})} \tag{6.31}$$

and
$$\hat{f}(y|i) = \prod_{j=1}^{k} \frac{\hat{f}_{[C_j]}(y_{C_j}|i_{C_j})}{\hat{f}_{[S_j]}(y_{S_j}|i_{S_j})}.$$

The expression (6.31) is the explicit expression for the estimate of p and from Proposition 6.18 we derive those for h and K to be

$$\hat{h}(i) = \sum_{j=1}^{k} \left\{ \left[ssd_{C_j}(i_{C_j})^{-1} s_{C_j}(i_{C_j}) \right]^{|\Gamma|} - \left[ssd_{S_j}(i_{S_j})^{-1} s_{S_j}(i_{S_j}) \right]^{|\Gamma|} \right\}$$
(6.32)

and

$$\hat{K}(i) = \sum_{j=1}^{k} \left\{ n(i_{C_j}) \left[ssd_{C_j}(i_{C_j})^{-1} \right]^{|\Gamma|} - n(i_{S_j}) \left[ssd_{S_j}(i_{S_j})^{-1} \right]^{|\Gamma|} \right\}. \quad (6.33)$$

In the homogeneous case we get

$$\hat{h}(i) = |n| \left\{ \sum_{j=1}^{k} \left[ssd_{C_j}(C_j)^{-1} \overline{y}_{C_j}(i_{C_j}) \right]^{|\Gamma|} - \left[ssd_{S_j}(S_j)^{-1} \overline{y}_{S_j}(i_{S_j}) \right]^{|\Gamma|} \right\}$$

and

$$\hat{K} = |n| \left\{ \sum_{j=1}^{k} \left[ssd_{C_j}(C_j)^{-1} \right]^{|\Gamma|} - \left[ssd_{S_j}(S_j)^{-1} \right]^{|\Gamma|} \right\}.$$

Finally we derive from (6.27) that

$$\det \hat{\Sigma}(i) = \prod_{j=1}^{k} \frac{\det ssd_{C_j}(i_{C_j})}{\det ssd_{S_j}(i_{S_j})}$$

and in the homogeneous case

$$\det \hat{\Sigma} = \prod_{j=1}^{k} \frac{\det ssd_{C_j}(C_j)}{\det ssd_{S_j}(S_j)}.$$

We shall illustrate the issues by an example.

Example 6.22 Consider the model given by the graph (a) in Fig. 6.1. The model is decomposable and the estimates can therefore be obtained in closed form. There are two cliques $\{I, J, Z\}$ and $\{I, Y, Z\}$, and one separator $\{I, Z\}$. Recall that in the formulae C_j is short either for $C_j \cap \Gamma$ or for $C_j \cap \Delta$, whichever is appropriate. Let $\check{K}_i$ denote the estimate of the joint conditional concentration matrix of Y and Z based on the marginal model involving $\{I, Y, Z\}$ only. This is equal to

$$\check{K}_i = \begin{pmatrix} \check{k}_i^{YY} & \check{k}_i^{YZ} \\ \check{k}_i^{ZY} & \check{k}_i^{ZZ} \end{pmatrix} = n_i \begin{pmatrix} ssd_i^{YY} & ssd_i^{YZ} \\ ssd_i^{ZY} & ssd_i^{ZZ} \end{pmatrix}^{-1}.$$

We then get for the final estimates

$$\begin{aligned}
\hat{p}_{ij} &= (n_{ij} n_{i+})/(n_{i+} |n|) = n_{ij}/|n| \\
\hat{h}_i^Y &= \check{k}_i^{YY} s_i^Y + \check{k}_i^{YZ} s_i^Z \\
\hat{h}_{ij}^Z &= \check{k}_i^{ZY} s_i^Y + \check{k}_i^{ZZ} s_i^Z - s_i^Z/ssd_i^Z + s_{ij}^Z/ssd_{ij}^Z \\
\hat{k}_i^{YY} &= \check{k}_i^{YY} \\
\hat{k}_i^{YZ} &= \check{k}_i^{YZ} \\
\hat{k}_{ij}^{ZZ} &= \check{k}_i^{ZZ} + n_{ij}/ssd_{ij}^{ZZ} - n_{i+}/ssd_i^{ZZ}.
\end{aligned}$$

We abstain from deriving the similar formulae in the homogeneous case. □

Propositions 6.20 and 6.21 can be used to throw light on the problem of existence of the maximum likelihood estimates in a graphical mixed interaction model with graph $\mathcal{G}$ which is not necessarily decomposable itself. By adding edges to $\mathcal{G}$ we can obtain a decomposable cover of $\mathcal{G}$, i.e. a decomposable graph $\mathcal{G}^*$ with $E \subseteq E^*$. If the appropriate conditions for existence in the quoted propositions are fulfilled for the graph $\mathcal{G}^*$, the maximum likelihood estimates will exist in $M(\mathcal{G}^*)$ and $M_H(\mathcal{G}^*)$ and therefore also in the smaller models $M(\mathcal{G})$ and $M_H(\mathcal{G})$. But in general a model has many different decomposable covers and the condition is therefore not necessary either. Necessary and sufficient conditions for existence of the maximum likelihood estimates in general graphical or hierarchical models which are reasonably operational have not been found except in special cases.

Example 6.23 Consider the model given by the graph (b) in Fig. 6.1. The model is not decomposable, but the model with graph (a) can be used as a decomposable cover. The conditions for existence of the estimates in model (a) are obtained from Proposition 6.20 to be

$$n_{ij} > 0, \quad ssd_{ij}^Z > 0, \quad ssd_i \text{ positive definite,}$$

or equivalently

$$n_{ij} \geq 2, \quad n_{i+} \geq 3,$$

and these conditions are therefore sufficient for existence of the maximum likelihood estimates in the model with graph (b).

But an alternative decomposable cover could have been made by connecting J and Y instead of I and Z. It follows that an alternative set of sufficient conditions for existence of maximum likelihood estimates is

$$n_{ij} \geq 2, \quad n_{+j} \geq 3$$

which is clearly not equivalent to the one found above. □

From (6.30) it follows that the maximized likelihood function for a decomposable model can be obtained as a suitable product of maximized likelihood functions for saturated models. Using (6.11) and the relation

$$\sum_{j=1}^{k} |\Gamma \cap C_j| - |\Gamma \cap S_j| = |\Gamma|,$$

we find that

$$\begin{aligned} L(\hat{f}) &= (2\pi)^{-|n||\Gamma|/2} |n|^{-|n|} e^{-|n||\Gamma|/2} \\ &\times \prod_{j=1}^{k} \frac{\prod_{i_{S_j} \in \mathcal{I}_{S_j}} \{\det \mathrm{ssd}_{S_j}(i_{S_j})\}^{n(i_{S_j})/2}}{\prod_{i_{C_j} \in \mathcal{I}_{C_j}} \{\det \mathrm{ssd}_{C_j}(i_{C_j})\}^{n(i_{C_j})/2}} \\ &\times \prod_{j=1}^{k} \frac{\prod_{i_{C_j} \in \mathcal{I}_{C_j}} n(i_{C_j})^{n(i_{C_j})(|\Gamma \cap C_j|/2+1)}}{\prod_{i_{S_j} \in \mathcal{I}_{S_j}} n(i_{S_j})^{n(i_{S_j})(|\Gamma \cap S_j|/2+1)}}. \end{aligned} \quad (6.34)$$

In the homogeneous case we similarly use (6.12) and find

$$\begin{aligned} L(\hat{\hat{f}}) &= (2\pi)^{-|n||\Gamma|/2} |n|^{|n|(|\Gamma|/2-1)} e^{-|n||\Gamma|/2} \\ &\times \prod_{j=1}^{k} \frac{\{\det \mathrm{ssd}_{S_j}\}^{|n|/2} \prod_{i_{C_j} \in \mathcal{I}_{C_j}} n(i_{C_j})^{n(i_{C_j})}}{\{\det \mathrm{ssd}_{C_j}\}^{|n|/2} \prod_{i_{S_j} \in \mathcal{I}_{S_j}} n(i_{S_j})^{n(i_{S_j})}}. \end{aligned} \quad (6.35)$$

6.3.3 Exact tests in decomposable models

As in the pure cases, we deal with the problem of testing conditional independence of two variables under the saturated model before we discuss the general testing problem. In the mixed interaction model there is the extra variation that a test of homogeneity could be of interest. As we shall see, it is true in the mixed case that the general testing problem in nested decomposable models can be reduced to a combination of testing problems concerned with pairwise conditional independence in saturated models. As opposed to the pure discrete and continuous cases, there are several different types of test for conditional independence that can be handled by exact methods: independence of two discrete variables, of two continuous variables or of one discrete and one continuous variable. In the two latter cases we must distinguish beween the homogeneous and non-homogeneous cases, so totally there are five types of test for conditional independence in a saturated model.

Conditional independence of discrete variables

The case of two discrete variables can only be handled exactly when the saturated model has no continuous variables, as the constrained model would otherwise not be decomposable by Lemma 2.19. In this case there is therefore no distinction between the homogeneous and non-homogeneous cases and the testing problem can be handled with Monte Carlo methods as already described in Section 4.4.3.

Conditional independence of continuous variables

Next we consider testing conditional independence of two continuous variables γ and μ given the remaining discrete and continuous variables under the saturated model. The maximized likelihood function for the constrained model factorizes into maximized likelihood functions for the three marginal saturated models involving $V \setminus \{\gamma\}$, $V \setminus \{\mu\}$, and $V \setminus \{\gamma, \mu\}$ as in (6.34) and (6.35). The deviance for the hypothesis can then be found from these expressions.

In the non-homogeneous case we find that the deviance is

$$d = -\sum_i n(i) \log \frac{\det ssd(i)_{\Gamma \setminus \{\gamma,\mu\}} \det ssd(i)}{\det ssd(i)_{\Gamma \setminus \{\gamma\}} \det ssd(i)_{\Gamma \setminus \{\mu\}}} = -\sum_i n(i) \log b(i).$$

In the conditional distribution given the statistics which are sufficient under the hypothesis, the quantities $B(i)$ are independent and beta-distributed $\mathcal{B}\{(n(i) - |\Gamma|)/2, 1/2\}$. Hence the conditional distribution of the deviance is equal to the unconditional distribution and it is of Box type. It is generally advisable to examine the terms $-n(i) \log b(i)$ individually but, if desired, the tail probability for the joint deviance test can be computed accurately using the methods in Section C.3 or, alternatively, by Monte Carlo methods. A cruder approximation can be obtained through a Bartlett correction as in Section 5.2.2.

In the homogeneous case the deviance simplifies and the problem becomes identical to the pure continuous case. Hence

$$d = -|n| \log \frac{\det ssd_{\Gamma \setminus \{\gamma,\mu\}} \det ssd}{\det ssd_{\Gamma \setminus \{\gamma\}} \det ssd_{\Gamma \setminus \{\mu\}}} = -|n| \log b$$

and the distribution of B is, conditionally as well as unconditionally, the beta distribution $\mathcal{B}\{(|n| - |\Gamma| - |\mathcal{I}| + 1)/2, 1/2\}$. If desired, the deviance

can be transformed to a t-distributed or an F-distributed statistic. More precisely,
$$|t| = \sqrt{f} \text{ with } f = (|n| - |\Gamma| - |\mathcal{I}| + 1)(b^{-1} - 1).$$

Conditional independence of mixed variables

We now consider testing conditional independence between a discrete variable δ and a continuous variable γ, given the remaining variables. Let $\Delta^* = \Delta \setminus \{\delta\}$ and $\Gamma^* = \Gamma \setminus \{\gamma\}$. The deviance is then found directly from (6.34) using $C_1 = V \setminus \{\gamma\}$ and $C_2 = V \setminus \{\delta\}$ for the non-saturated model. We find

$$\begin{aligned} d &= \sum_{i \in \mathcal{I}} n(i) \log n(i) - \sum_{i \in \mathcal{I}} n(i) \log \frac{\det ssd_\Gamma(i)}{\det ssd_{\Gamma^*}(i)} \\ &+ \sum_{i_{\Delta^*} \in \mathcal{I}_{\Delta^*}} n(i_{\Delta^*}) \log \frac{\det ssd_\Gamma(i_{\Delta^*})}{\det ssd_{\Gamma^*}(i_{\Delta^*})} - \sum_{i_{\Delta^*} \in \mathcal{I}_{\Delta^*}} n(i_{\Delta^*}) \log n(i_{\Delta^*}). \end{aligned}$$

The total deviance D can be decomposed into components $D(i_{\Delta^*})$ that are independent in the conditional distribution given the counts

$$d = \sum_{i_{\Delta^*} \in \mathcal{I}_{\Delta^*}} d(i_{\Delta^*}),$$

where

$$\begin{aligned} d(i_{\Delta^*}) &= \sum_{j: j_{\Delta^*} = i_{\Delta^*}} n(j) \log n(j) - \sum_{j: j_{\Delta^*} = i_{\Delta^*}} n(j) \log \frac{\det ssd_\Gamma(j)}{\det ssd_{\Gamma^*}(j)} \\ &+ n(i_{\Delta^*}) \log \frac{\det ssd_\Gamma(i_{\Delta^*})}{\det ssd_{\Gamma^*}(i_{\Delta^*})} - n(i_{\Delta^*}) \log n(i_{\Delta^*}). \end{aligned}$$

The hypothesis of a missing edge between γ and δ specifies that the conditional expectation and variance of Y_γ given all remaining variables do not depend on the discrete variables j for $j_{\Delta^*} = i_{\Delta^*}$. Therefore the deviance $d(i_{\Delta^*})$ can be further partitioned into $d_1(i_{\Delta^*})$ and $d_2(i_{\Delta^*})$, where d_1 is the deviance for testing partial homogeneity of the residual variances and d_2 is the deviance for testing that slopes and intercepts of the regressions are identical, assuming partial homogeneity of variances. The estimate for the residual variances after regression on Y_{Γ^*} without assuming homogeneity is

$$\hat{\sigma}^2(j) = \frac{1}{n(j)} \frac{\det ssd_\Gamma(j)}{\det ssd_{\Gamma^*}(j)},$$

and under partial variance homogeneity it is

$$\hat{\sigma}^2 = \frac{1}{n(i_{\Delta^*})} \sum_{j: j_{\Delta^*} = i_{\Delta^*}} \frac{\det ssd_\Gamma(j)}{\det ssd_{\Gamma^*}(j)}.$$

Hence the part of the deviance related to variance homogeneity becomes

$$d_1(i_{\Delta^*}) = \sum_{j:j_{\Delta^*}=i_{\Delta^*}} n(j)\log n(j) - \sum_{j:j_{\Delta^*}=i_{\Delta^*}} n(j)\log \frac{\det ssd_\Gamma(j)}{\det ssd_{\Gamma^*}(j)}$$

$$+ n(i_{\Delta^*}) \log \left\{ \sum_{j:j_{\Delta^*}=i_{\Delta^*}} \frac{\det ssd_\Gamma(j)}{\det ssd_{\Gamma^*}(j)} \right\} - n(i_{\Delta^*}) \log n(i_{\Delta^*}),$$

whereby

$$d_2(i_{\Delta^*}) = n(i_{\Delta^*}) \log \frac{\det ssd_\Gamma(i_{\Delta^*})}{\det ssd_{\Gamma^*}(i_{\Delta^*})} - n(i_{\Delta^*}) \log \left\{ \sum_{j:j_{\Delta^*}=i_{\Delta^*}} \frac{\det ssd_\Gamma(j)}{\det ssd_{\Gamma^*}(j)} \right\}.$$

The conditional distribution of $D_1(i_{\Delta^*})$ is one of the Box-type distributions studied in Section C.3. It does not depend on the continuous part of the conditioning statistic, which implies that $D_1(i_{\Delta^*})$ and $D_2(i_{\Delta^*})$ are independent given the counts. The conditional distribution of $D_2(i_{\Delta^*})$ given the counts is well known from standard linear model theory to be the same as that of $-|n|\log B_2$, where B_2 follows the beta distribution with parameters $\mathcal{B}\{(|n|-|\Gamma||\mathcal{I}_\delta|)/2, |\Gamma|(|\mathcal{I}_\delta|-1)/2\}$. Hence the distribution of $D_2(i_{\Delta^*})$ is also of Box type. Note that this distribution does not depend on the counts and therefore $D_2(i_{\Delta^*})$ are independent of these. As all components of the deviance are mutually independent given the counts, the conditional distribution of the total deviance is also of Box type.

In the homogeneous case there is no component of the deviance related to variance homogeneity, and the problem reduces to testing partial homogeneity of the intercepts and slopes in the linear regression of Y_γ on the remaining continuous variables. We find the deviance from (6.35) as

$$d = -|n|\log b = |n|\log \frac{\det ssd_\Gamma \det ssd_{\Gamma^*}(\Delta^*)}{\det ssd_{\Gamma^*} \det ssd_\Gamma(\Delta^*)}.$$

Standard linear model theory gives that B follows the beta distribution $\mathcal{B}\{(|n|-|\Gamma||\mathcal{I}|)/2, |\Gamma||\mathcal{I}_{\Delta^*}|(|\mathcal{I}_\delta|-1)/2\}$ independently of the counts.

Comparing nested decomposable models

The five types of test just considered are the main components of any general test of a model $M(\mathcal{G}')$ assuming $M(\mathcal{G})$ to hold, where $\mathcal{G}'$ is a decomposable subgraph of a decomposable graph $\mathcal{G}$.

If we first consider the situation where $\mathcal{G}'$ differs from $\mathcal{G}$ by one edge only, it follows from Lemma 2.19 that this edge is a member of one clique only, C^*, say. If we order the cliques of $\mathcal{G}$ in a perfect sequence as in Lemma 2.20, this clique has number q and splits into two complete subsets

C_{q1} and C_{q2} when the edge in question is removed. The lemma implies that in the appropriate ordering, the sequence with C_{q1} and C_{q2} inserted as replacement for C^* is perfect. Hence, by Lemma 2.11 the sequence induces successive decompositions of the corresponding graphs.

Now use Proposition 6.17 to obtain expressions similar to (6.34) and (6.35) for the maximum likelihood estimate of the densities under the model determined by $\mathcal{G}'$. Most terms cancel in the likelihood ratio, so the deviance between the two models is equal to the deviance between the submodels generated by the variables in C^* only. An argument analogous to cases already considered gives that the conditional distributions given the sufficient statistics under the hypothesis also coincide. We summarize.

Proposition 6.24 *When comparing a decomposable submodel $M(\mathcal{G}')$ to a decomposable model $M(\mathcal{G})$ with $\mathcal{G}'$ having exactly one edge, e, fewer than $\mathcal{G}$, the likelihood ratio is equal to the likelihood ratio for conditional independence, obtained when comparing $M(\mathcal{G}'_{C^*})$ with the saturated model $M(\mathcal{G}_{C^*})$, where C^* is the unique clique of $\mathcal{G}$ containing e. The ratio depends on the marginal data $(x^1_{C^*}, \ldots, x^{|n|}_{C^*})$ only.*

Example 6.25 Supplementing Example 6.19, we consider the likelihood ratio for the hypothesis that the edge between Y and Z is missing in Fig. 6.2. By Proposition 6.24 this is equal to the likelihood ratio for the similar hypothesis when only the analogous model for (Y, Z) is considered. Thus we get

$$d = -|n|\log \frac{\det ssd_\emptyset}{ssd_\emptyset^{YY} ssd_\emptyset^{ZZ}} = \frac{ssd_\emptyset^{YY} ssd_\emptyset^{ZZ} - (ssd_\emptyset^{YZ})^2}{ssd_\emptyset^{YY} ssd_\emptyset^{ZZ}} = -|n|\log b,$$

where B is beta-distributed $\mathcal{B}\{(|n|-2)/2, 1/2\}$. This is true even though the joint distribution of (Y, Z) is not bivariate Gaussian and the conditional variance of Y depends on the discrete variable. □

Example 6.26 As a more complex example, consider the model (a) in Fig. 6.1 and assume that we want to investigate whether we can also assume that the edge between I and Y is missing, corresponding to the further conditional independence $I \perp\!\!\!\perp Y \mid (J, Z)$. The missing edge belongs only to the clique $\{I, Y, Z\}$ and the deviance can therefore be calculated by just considering this marginal model and the corresponding data as well as the independence hypothesis $I \perp\!\!\!\perp Y \mid Z$. The deviance reduces to

$$d = \sum_i n_i \log n_i - \sum_i n_i \log \frac{\det ssd_i}{ssd_i^{ZZ}} + |n| \log \frac{\det ssd_\emptyset}{ssd_\emptyset^{ZZ}} - |n| \log |n|.$$

This is equal to the usual deviance for testing that all the linear regressions of Y on Z do not depend on i. □

Next we turn to the question of likelihood ratios for comparing decomposable models that differ by more than one edge.

Proposition 6.27 *The deviance statistic for testing a decomposable mixed interaction submodel $M(\mathcal{G}')$, assuming a decomposable model $M(\mathcal{G})$, can be partitioned as*

$$d = -2\log\frac{L(\mathcal{G}')}{L(\mathcal{G})} = \sum_{j=1}^{k} d_j$$

into deviances for conditional independence in suitable marginal saturated models. The partitioning can be made such that the first k_1 terms are deviances for edges missing between discrete variables and the remaining k_2 terms involve at least one continuous variable. Conditionally on the counts, the last k_2 deviances are independent and each of these has a Box-type distribution, whence this holds also for their sum.

Proof: From Lemma 2.22 we can find a sequence $\mathcal{G}' = \mathcal{G}_0 \subset \cdots \subset \mathcal{G}_k = \mathcal{G}$ of decomposable graphs that differ by exactly one edge with discrete edges added first. We then get

$$Q = \frac{L(\mathcal{G}')}{L(\mathcal{G})} = \frac{L(\mathcal{G}_0)}{L(\mathcal{G}_1)}\frac{L(\mathcal{G}_1)}{L(\mathcal{G}_2)} \cdots \frac{L(\mathcal{G}_{k-1})}{L(\mathcal{G}_k)} = \prod_{j=1}^{k} Q_j.$$

All of the factors are by Proposition 6.24 likelihood ratios for conditional independence in the marginal saturated model given by the single clique of which the missing edge is a member. The independence of the last terms in the deviance follows by appealing to Basu's theorem. This time we refrain from giving the details. □

The statement obviously also holds in the homogeneous case when $M_H(\mathcal{G}')$ is compared with $M_H(\mathcal{G})$.

The result points to ways of evaluating the tail probability of a joint test between decomposable models. The most obvious is a pure Monte Carlo procedure: fill out necessary marginal tables by combination of Patefield's algorithm as in the discrete case and calculate the corresponding k_1 terms of the deviance; then, conditionally on the counts so obtained, simulate independent random variables corresponding to the remaining deviances.

Alternatively, combine the Monte Carlo method with the approximations for Box-type distributions by first simulating counts as above and then subsequently calculating the conditional tail probability that

$$D > d_{\text{obs}} - \sum_{j=1}^{k_1} d_j^*,$$

where d_j^* are the simulated values. Averaging these tail probabilities over Monte Carlo samples gives approximately the correct tail probability.

Finally there is a possibility of obtaining a direct saddle-point approximation of the tail probability. It is not clear which of the methods should be preferred.

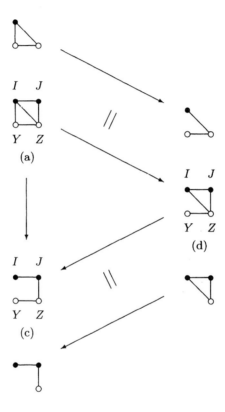

Fig. 6.3. Illustration of the decomposition of deviances. The deviance for the total reduction is the sum of deviances for the stepwise reductions. Each of these is equal to the corresponding deviance for removing one edge in suitable marginal saturated models, as indicated.

Example 6.28 Finally let us consider the model of graph (c) in Fig. 6.1. Here no interactions among I and Y are allowed, and the log-density can therefore be expressed as

$$\log f_{ij}(y, z) = g_{ij} + h^Y y + h_j^{JZ} z - (k^{YY} y^2 + 2k^{YZ} yz + k_j^{ZZ} z^2)/2.$$

This model is again decomposable. The cliques, ordered to form a perfect sequence, are $\{I, J\}, \{J, Z\}, \{Y, Z\}$ with separators $\{J\}, \{Z\}$. Combining the notation

$$\begin{pmatrix} \tilde{k}^{YY} & \tilde{k}^{YZ} \\ \tilde{k}^{ZY} & \tilde{k}^{ZZ} \end{pmatrix} = |n| \begin{pmatrix} ssd_\emptyset^{YY} & ssd_\emptyset^{YZ} \\ ssd_\emptyset^{ZY} & ssd_\emptyset^{ZZ} \end{pmatrix}^{-1}$$

with the formulae (6.31), (6.32), and (6.33) we get

$$\begin{aligned} \hat{p}_{ij} &= n_{ij}/|n| \\ \hat{h}^Y &= \tilde{k}^{YY}\bar{y} + \tilde{k}^{YZ}\bar{z} \\ \hat{h}_j^Z &= \tilde{k}^{ZY}\bar{y} + (\tilde{k}^{ZZ} - |n|/ssd_\emptyset^{ZZ})\bar{z} + s_j^Z/ssd_j^{ZZ} \\ \hat{k}^{YY} &= \tilde{k}^{YY} \\ \hat{k}^{YZ} &= \tilde{k}^{YZ} \\ \hat{k}_j^{ZZ} &= \tilde{k}^{ZZ} + n_{+j}/ssd_j^{ZZ} - |n|/ssd_\emptyset^{ZZ}. \end{aligned}$$

The likelihood ratio for comparing models (a) and (b) as well as that for comparing (b) and (c) cannot be obtained in closed form, essentially because model (b) is not decomposable. However, the deviance for comparing model (c) with model (a) can be obtained in closed form and it can also be partitioned into the sum of those for comparing model (c) with model (d) and for comparing model (d) with model (a), where model (d) is obtained from model (c) by adding an edge between I and Z. Each of these deviances can be calculated from appropriate marginals, and the distribution of the total deviance conditionally on the counts is of Box type. The partitioning is illustrated in Fig. 6.3. □

Test for homogeneity

Finally we briefly consider the deviance test for the hypothesis that an unknown CG density f belongs to the class of homogeneous densities $M_H(\mathcal{G})$, assuming that it is a Markov density in $M(\mathcal{G})$, where $\mathcal{G}$ is a decomposable, undirected marked graph. From (6.11) and (6.12) we immediately find the likelihood ratio as

$$Q = \frac{L(\hat{\hat{f}})}{L(\hat{f})} = |n|^{|n||\Gamma|/2} \prod_{j=1}^{k} \frac{\prod_{i_{C_j} \in \mathcal{I}_{C_j}} n(i_{C_j})^{n(i_{C_j})|\Gamma \cap C_j|/2}}{\prod_{i_{S_j} \in \mathcal{I}_{S_j}} n(i_{S_j})^{n(i_{S_j})|\Gamma \cap S_j|/2}}$$
$$\times \prod_{j=1}^{k} \frac{\{\det ssd_{S_j}\}^{|n|/2} \prod_{i_{C_j} \in \mathcal{I}_{C_j}} \{\det ssd_{C_j}(i_{C_j})\}^{n(i_{C_j})/2}}{\{\det ssd_{C_j}\}^{|n|/2} \prod_{i_{S_j} \in \mathcal{I}_{S_j}} \{\det ssd_{S_j}(i_{S_j})\}^{n(i_{S_j})/2}}.$$

Conditionally on the counts, the deviance can be shown to follow a Box-type distribution as usual.

6.4 Hierarchical interaction models

In this section we shall extend the homogeneous and graphical interaction models to more general types of restriction on the interaction structure. The class of models we define extends the hierarchical models from the discrete case and thus represent a unification of topics discussed earlier in this book.

6.4.1 General properties

As in the case of hierarchical log–linear models for contingency tables, our point of departure is an interaction expansion of the joint density of the random variables as

$$\log f(i,y) = \sum_{d\subseteq\Delta} \lambda_d(i) + \sum_{d\subseteq\Delta}\sum_{\gamma\in\Gamma} \eta_d(i)_\gamma y_\gamma - \frac{1}{2}\sum_{d\subseteq\Delta}\sum_{\gamma,\mu\in\Gamma} \psi_d(i)_{\gamma\mu} y_\gamma y_\mu, \quad (6.36)$$

where all functions with d as subscript only depend on i through i_d.

In the next stage we choose to restrict the set of densities by demanding certain interactions to be equal to zero. However, this time the restrictions may not initially be determined by complete subsets of a graph. On the other hand, we would not want to allow completely arbitrary patterns of restriction. This is because the expansion (6.36) is not uniquely determined from the density alone but has some arbitrariness in the choice of interaction representations. Whether or not an interaction is present may depend on issues that in many cases would appear to be irrelevant, such as a change of scale or location for the continuous variables.

Suppose that we would demand the discrete interaction term λ_d to be identically equal to zero, but for some c with $d \subset c$ allow λ_c to vary freely. Then this would not restrict the class of densities at all, because if the density has been expanded as above in (6.36), we can rewrite the density in a way that fulfils the restriction by redefining

$$\tilde{\lambda}_c(i) = \lambda_c(i) + \lambda_d(i), \quad \tilde{\lambda}_d(i) = 0.$$

To avoid this happening, we demand of our model restriction that if λ_d is restricted to zero and $d \subseteq c$, then λ_c is also restricted to be zero. A similar property must hold for any of the other terms $\eta_d()_\gamma$ and $\psi_d()_{\gamma\mu}$ for the same reason.

Next let us investigate the effect of a change in location and scale on the interaction terms. Let $Y_\gamma = a_\gamma \tilde{Y}_\gamma - b_\gamma$ for $a, b \in \mathcal{R}^{|\Gamma|}$. Then the density of $\tilde{X} = (I, \tilde{Y})$ can be found to be

$$\log \tilde{f}(i, \tilde{y}) = \tilde{g}(i) + \tilde{h}(i)^\top \tilde{y} - \frac{1}{2}\tilde{y}^\top \tilde{K}(i)\tilde{y},$$

where

$$\tilde{g}(i) = g(i) - h(i)^\top b - b^\top K(i)b/2 + \sum_\gamma \log|a_\gamma|$$

$$\tilde{h}(i)_\gamma = a_\gamma h(i)_\gamma + a_\gamma \sum_{\mu\in\Gamma} k(i)_{\gamma\mu} b_\mu$$

$$\tilde{k}(i)_{\gamma\mu} = a_\gamma a_\mu k(i)_{\gamma\mu}.$$

Applying Möbius inversion to these equations yields similar expressions for the interactions:

$$\tilde{\lambda}_d(i) = \lambda_d(i) - \eta_d^\top b - b^\top \Psi_d b/2 + \sum_\gamma \log|a_\gamma| \qquad (6.37)$$

$$\tilde{\eta}_d(i)_\gamma = a_\gamma \eta_d(i)_\gamma + a_\gamma \sum_{\mu\in\Gamma} \psi_d(i)_{\gamma\mu} b_\mu \qquad (6.38)$$

$$\tilde{\psi}_d(i)_{\gamma\mu} = a_\gamma a_\mu \psi_d(i)_{\gamma\mu}. \qquad (6.39)$$

Inspection of (6.37) shows that unless zero restrictions on the interaction terms are to be affected by a change of location, we must demand that whenever λ_d is restricted to be zero, then so are η_d and ψ_d. Further, (6.38) shows that if $\eta_d()_\gamma$ is restricted to zero, then so must $\psi_d()_{\gamma\mu}$ be for all $\mu \in \Gamma$. Finally (6.39) gives no restrictions on the interaction terms. On the other hand, the identities also show that a change of scale has no further effect. A change of scale affects the linear and quadratic interaction terms, but it does not change absence to presence or conversely, as long as we have not restricted any interactions to have fixed values (other than zero). The demands are summarized below:

1. if λ_d is restricted to zero and $d \subseteq c$, then λ_c, η_c and ψ_c are all restricted to zero;

2. if $\eta_d()_\gamma$ is restricted to zero and $d \subseteq c$, then $\eta_c()_\gamma$ and $\psi_c()_{\gamma\mu}$ are both restricted to zero for all $\mu \in \Gamma$;

3. if $\psi_d()_{\gamma\mu}$ is restricted to zero and $d \subseteq c$, then $\psi_c()_{\gamma\mu}$ is restricted to zero.

Models that restrict interactions in such a way that these demands are fulfilled are termed *hierarchical* mixed interaction models.

An alternative route to restricting interactions is to look directly at the expansions used in the proof of Theorem 6.11:

$$\phi_d(x) = \lambda_d(i)$$
$$\phi_{d\cup\{\gamma\}}(x) = \eta_d(i)_\gamma y_\gamma - \psi_d(i)_{\gamma\gamma} y_\gamma^2/2$$
$$\phi_{d\cup\{\gamma,\mu\}}(x) = -\psi_d(i)_{\gamma\mu} y_\gamma y_\mu$$
$$\phi_{d\cup c}(x) = 0 \text{ for } |c| > 2.$$

Then we could restrict these interactions to be zero in certain patterns. A zero restriction on $\phi_{d\cup\gamma}$ simultaneously restricts $\eta_d()_\gamma$ and $\psi_d()_{\gamma\gamma}$. In analogy with the discrete models we could demand that if ϕ_a is restricted to zero and $a \subseteq b$, then ϕ_b is restricted to zero as well. A model of this type would also be a hierarchical model in our sense. But we have further allowed the possibility of restricting the terms $\psi_d()_{\gamma\gamma}$ without insisting on this having consequences for $\eta_d()_\gamma$. This enables us to make parsimonious models by imposing variance homogeneity and similar restrictions. But it is worth noting that such restrictions are of a different nature from what we have otherwise mostly considered.

Note that we have chosen to define a hierarchical interaction model in a way which is different from that of Edwards (1990). In his work it was also demanded that if $\psi_d()_{\gamma\gamma}$ was restricted to zero, so should $\psi_d()_{\gamma\mu}$ be for all μ. Such models will be referred to as *MIM models*, because the program MIM (Edwards 1995) is based upon such models. However, there seems to be no immediate justification for this demand and, as we shall see, it appears that estimation issues are not seriously affected either. Time will show whether this extension of the class of hierarchical models turns out to be useful; see also Example 6.31 in the next subsection.

6.4.2 Generators and canonical statistics

Specification of a hierarchical model is slightly more complex in the mixed case than in the case with all variables being discrete. This is partly because we have interactions of different types, but also because we want to include the possibility of imposing additional restrictions on terms associated with variances, such as is indicated at the end of the previous subsection.

To capture this aspect we introduce additional variable names γ^2 to distinguish between quadratic and linear interaction involving a single continuous variable. A *generator* $a = (d, c)$ has then a discrete component and a continuous component. The discrete component is a subset $d \subset \Delta$ of the discrete variables. The continuous component c can be empty, and we then say that the generator is *discrete*. If this is not the case it is either a continuous variable $\gamma \in \Gamma$, in which case we say that the generator is *linear*, a quadratic variable γ^2, or a pair $\{\gamma, \mu\}$ with $\gamma \neq \mu$. In the last two cases we say that the generator is *quadratic*, either *single* or *pair*. These types of generators suffice to generate all the hierarchical models. However, in examples with many generators the list of quadratic generators can be very long and it is therefore practical also to operate with *quadratic set generators* of the type (d, c^2) with $c \subseteq \Gamma$. The meaning of these is further explained below. If the discrete or continuous component is empty, we write for simplicity just c, c^2, or d instead of $(\emptyset, c)$, $(\emptyset, c^2)$, or $(d, \emptyset)$.

We next introduce a partial ordering on the generators. First we say that $\gamma \prec \gamma^2$ and $\gamma \prec \{\gamma, \mu\}$. For $\gamma, \mu \in c \subseteq \Gamma$ we also let $\gamma^2 \prec c^2$, $\mu^2 \prec c^2$, and $\{\gamma,\mu\} \prec c^2$. If $c_1 \subset c_2 \subseteq \Gamma$ we let $c_1^2 \prec c_2^2$. Transitive extension of this relation defines a partial order on the continuous generators.

The reflexive form of the ordering appears by writing $c_1 \preceq c_2$ if either $c_1 \prec c_2$ or $c_1 = c_2$ for a pair of continuous generators. Finally we extend the ordering to all generators as

$$(d_1, c_1) \preceq (d_2, c_2) \iff (d_1 \subseteq d_2 \text{ and } c_1 \preceq c_2)$$

and

$$(d_1, c_1) \prec (d_2, c_2) \iff \{(d_1, c_1) \preceq (d_2, c_2) \text{ but } (d_1, c_1) \neq (d_2, c_2)\}.$$

We will use the expression that a_2 *contains* a_1 if $a_1 \preceq a_2$, even though the meaning is somewhat different from usual. This should not cause real confusion; the current meaning will be clear from the context.

A *generating class* $\mathcal{A}$ in the mixed case is a set of generators. We will normally assume that it includes all main effects, i.e. all single discrete variables and all generators of the form $\{\gamma^2\}$ are contained in some generator in $\mathcal{A}$. Also it is customary to have all generators pairwise incomparable, although it is not necessarily always practical.

A mixed hierarchical model is determined by restricting an interaction of type (d, c) to zero unless there is an $a \in \mathcal{A}$ with $(d, c) \preceq a$, and we use the notation $M(\mathcal{A})$ for this class of models. Thus if $a_1, a_2 \in \mathcal{A}$ with $a_1 \prec a_2$, a_1 is redundant and may be removed from $\mathcal{A}$ without changing the model. Also, a quadratic set generator (d, c^2) is equivalent to the list of generators

$$(d, \gamma^2), \gamma \in c, \quad (d, \{\gamma, \mu\}), \gamma, \mu \in c.$$

If a quadratic main effect is missing, we will assume the corresponding interaction $\psi_\emptyset()_{\gamma^2}$ to be identically equal to one. This has more formal that practical interest, but it will be exploited later.

Theorem 6.11 readily identifies the generating classes of graphical interaction models corresponding to the canonical statistics in such a model. It follows in particular that a graphical interaction model is always hierarchical. For a general graphical model we use the generators

$$\begin{array}{rl} d & \text{for } d \in \mathcal{C}_\Delta \\ (d, \gamma^2) & \text{for } d \in \mathcal{C}_\Delta(\gamma), \, \gamma \in \Gamma \\ (d, \{\gamma, \mu\}) & \text{for } d \in \mathcal{C}_\Delta(\gamma, \mu), \, \gamma, \mu \in \Gamma, \, \gamma \neq \mu \end{array}$$

and then throw away redundant elements, i.e. those $d \in \mathcal{C}_\Delta$ that for some $\gamma \in \Gamma$ are elements of $\mathcal{C}_\Delta(\gamma)$ as well. If $\mathcal{C}_\Delta(\gamma)$, $\mathcal{C}_\Delta(\mu)$ and $\mathcal{C}_\Delta(\gamma, \mu)$ have a common element d for a range of pairs $\gamma, \mu \in c \subseteq \Gamma$, the corresponding list of generators can be replaced by the single quadratic generator (d, c^2).

A homogeneous model has no mixed quadratic interactions, so here the generators are determined similarly as

$$\begin{aligned} d & \quad \text{for} \quad d \in \mathcal{C}_\Delta \\ (d,\gamma) & \quad \text{for} \quad d \in \mathcal{C}_\Delta(\gamma) \\ c^2 & \quad \text{for} \quad c \in \mathcal{C}_\Gamma, \end{aligned}$$

where $\mathcal{C}_\Gamma$ is the set of cliques of the subgraph $\mathcal{G}_\Gamma$ formed by the continuous variables. We also here reduce by removing redundant generators.

Example 6.29 As illustration we consider the graphical models in Fig. 6.1. As usual we adopt a simplified notation when dealing with any specific example, hoping this to be reasonably transparent.

The graphical model with graph (a) allows discrete interactions IJ, linear and quadratic interactions IJZ and IY, and the quadratic interaction IYZ. Avoiding redundancy in the representation leads to the generating class $\{IJZ^2, IY^2, IYZ\}$. Using quadratic set generators, it could equivalently be represented as $\{IJZ^2, I(YZ)^2\}$. Here we have taken advantage of the fact that $IZ^2 \prec IJZ^2$. The homogeneous version has generating class $\{IJZ, IY, (YZ)^2\}$.

The models in (b) have generating classes $\{IJ, JZ^2, IY^2, YZ\}$ for the general version and $\{IJ, JZ, IY, (YZ)^2\}$ for the homogeneous version.

Finally, the models in (c) have generating classes $\{IJ, JZ^2, Y^2, YZ\}$ or, equivalently, $\{IJ, JZ^2, (YZ)^2\}$ for the general version and $\{IJ, JZ, (YZ)^2\}$ for the homogeneous version. □

For a given generating class $\mathcal{A}$ we define as usual its *interaction graph* by letting α and β be adjacent in $\mathcal{G}(\mathcal{A})$ if there is a generator $a \in \mathcal{A}$, containing both α and β. It follows from Theorem 6.11 that any member of $M(\mathcal{A})$ satisfies the Markov property on $\mathcal{G}(\mathcal{A})$, i.e. $M(\mathcal{A}) \subseteq M\{\mathcal{G}(\mathcal{A})\}$. A hierarchical model is *graphical* if there is equality. As in the discrete case, there is an abundance of hierarchical models that are not graphical.

Example 6.30 The generating class $\{IJ, Y^2, JZ^2, IYZ\}$ has an interaction graph equal to the graphical model in (a) of Fig. 6.1. It is a hierarchical model which is a proper submodel of the graphical model and it is different from the homogeneous graphical model as well. The model is not even a MIM model, as it allows the mixed interaction IYZ without permitting the quadratic terms Y^2 and Z^2 to depend on the variable I. □

The next example illustrates that there might be a place for hierarchical models that are not MIM models.

Example 6.31 Consider one discrete variable I and two continuous variables Y and Z. From Proposition 6.6 we find that the conditional distribution of Z on Y and I has canonical characteristics equal to

$$h_i^{Z|y} = h_i^Z - k_i^{ZY} y, \quad k_i^{Z|y} = k_i^{ZZ}$$

corresponding to the regression of Z on Y for a fixed value $I = i$ being

$$\mathbf{E}(Z\,|\,Y = y, I = i) = h_i^Z/k_i^{ZZ} - k_i^{ZY}/k_i^{ZZ}y, \quad \mathbf{V}(Z\,|\,Y = y, I = i) = 1/k_i^{ZZ}.$$

The hypothesis of variance homogeneity in this regression is given by the restriction $k_i^{ZZ} = k^{ZZ}$ corresponding to the hierarchical model with generator $\{IY^2, IYZ, Z^2\}$. This model is not a MIM model, as it allows the term IYZ without allowing IZ^2. The model has some interest in itself but may also be a useful alternative for testing the fit of the model $\{IY^2, IZ, YZ, Z^2\}$ which corresponds to the further assumption that the regression lines for varying I are parallel because we then also have $k_i^{ZY} = k^{ZY}$. This second model is of MIM type.

Incidentally, both these models admit an explicit likelihood analysis. If we write the joint density of the variables as

$$f(i, y, z) = f(i, y)f(z\,|\,y, i),$$

we note that the first factor only depends on the unknown parameters through $(p_i, \xi_i^Y, \sigma_i^{YY})$ which are allowed to vary in the full saturated model for IY and the conditional distribution depends only on $(h_i^Z, k_i^{ZY}, k^{ZZ})$ which in both models are allowed to vary independently of the former set of parameters. Hence the marginal $t(i, y, z) = (i, y)$ forms a cut in both models, and the joint likelihood factorizes similarly, hence it can be maximized by maximizing each factor separately. The conditional model is in both cases a standard linear normal regression model. We can directly write the estimates for $(p_i, \xi_i^Y, \sigma_i^{YY})$ in both models as

$$\hat{p}_i = n_i/|n|, \quad \hat{\xi}_i^Y = \bar{y}_i, \quad \hat{\sigma}_i^{YY} = ssd_i^{YY}/n_i,$$

provided that $n_i > 0$ for all $i \in \mathcal{I}$.

From standard results for linear normal models we find for the conditional models that the estimates for $(h_i^Z, k_i^{ZY}, k^{ZZ})$ without assuming parallel lines exist if $n_i \geq 2$ for all i and $n > 2|\mathcal{I}|$. For $\hat{k}^{ZZ}$ we have

$$\hat{k}^{ZZ} = \frac{|n|}{\sum_i ssd_i^r},$$

where ssd_i^r is the residual sum of squares in the group i:

$$ssd_i^r = ssd_i^{ZZ} - \frac{(ssd_i^{YZ})^2}{ssd_i^{YY}},$$

and

$$\hat{k}_i^{ZY} = -\hat{k}^{ZZ}\frac{ssd_i^{YZ}}{ssd_i^{YY}}, \quad \hat{h}_i^Z = \hat{k}_i^{ZY}\bar{y}_i + \hat{k}^{ZZ}\bar{z}_i.$$

Assuming parallel lines we get instead

$$\hat{k}^{ZZ} = \frac{|n|}{ssd_0^r},$$

with the total residual sum of squares ssd_0^r being

$$ssd_0^r = \sum_i ssd_i^{ZZ} - \frac{(\sum_i ssd_i^{YZ})^2}{\sum_i ssd_i^{YY}},$$

and further

$$\hat{k}^{ZY} = -\hat{k}^{ZZ} \frac{\sum_i ssd_i^{YZ}}{\sum_i ssd_i^{YY}}, \quad \hat{h}_i^Z = \hat{k}^{ZY} \bar{y}_i + \hat{k}^{ZZ} \bar{z}_i.$$

In the likelihood ratio, the factor corresponding to the marginal distribution of (I, Y) cancels and the likelihood ratio test can be performed by rejecting for large values of the ratio

$$F = \frac{ssd_0^r - \sum_i ssd_i^r}{\sum_i ssd_i^r} \frac{|n| - 2|\mathcal{I}|}{|\mathcal{I}| - 1},$$

which, conditionally on $n_i, i \in \mathcal{I}$, is independent of the remaining sufficient statistics under the smaller model and is F-distributed with degrees of freedom $(|\mathcal{I}| - 1, |n| - 2|\mathcal{I}|)$. □

The generating class of a hierarchical mixed interaction model identifies as in the discrete case the canonical sufficient statistics in an obvious way. A discrete generator d corresponds to a marginal table of counts $n(i_d), i_d \in \mathcal{I}_d$. A linear generator (d, γ) corresponds to the counts, supplemented with sums $s(i_d)_\gamma, i_d \in \mathcal{I}_d$. A quadratic generator of the type (d, γ^2) further supplements with sums of squares $ss(i_d)_\gamma, i_d \in \mathcal{I}_d$, and $(d, \{\gamma, \mu\})$ with sums of products $ss(i_d)_{\gamma\mu}, i_d \in \mathcal{I}_d$. Quadratic set generators (d, c^2) correspond to the matrices of marginal sums of squares and products $ss(i_d)_{cc}, i_d \in \mathcal{I}_d$.

6.4.3 Maximum likelihood estimation

Hierarchical mixed interaction models are clearly full and regular exponential models, being defined through linear restrictions on the canonical parameters of the saturated model. So if the maximum likelihood estimates exist in the saturated models, they also exist in the hierarchical models. The converse need not hold, as demonstrated in Example 6.31. See also the discussion in Example 6.23.

Likelihood equations

The likelihood equations are as usual obtained by equating the set of canonical statistics to their expectations. The canonical statistics are readily identified by the generators in the generating class. We let $m(i) = |n|p(i)$ as earlier and $\mathcal{A}_\Delta$ be the set of maximal permissible discrete interactions, i.e.

$$\mathcal{A}_\Delta = \text{red}\,\{d \mid d \preceq a \text{ for some } a \in \mathcal{A}\}.$$

The equations are

$$n(i_d) = m(i_d) \text{ for } d \in \mathcal{A}_\Delta$$

$$s(i_d)_\gamma = \sum_{j:j_d=i_d} m(j)\xi(j)_\gamma \text{ for } (d,\gamma) \text{ or } (d,\gamma^2) \in \mathcal{A}$$

$$ss(i_d)_{\gamma\gamma} = \sum_{j:j_d=i_d} m(j)\left\{\sigma(j)_{\gamma\gamma} + \xi(j)_\gamma^2\right\} \text{ for } (d,\gamma^2) \in \mathcal{A}$$

$$ss(i_d)_{\gamma\mu} = \sum_{j:j_d=i_d} m(j)\left\{\sigma(j)_{\gamma\mu} + \xi(j)_\gamma\xi(j)_\mu\right\} \text{ for } (d,\{\gamma,\mu\}) \in \mathcal{A}.$$

As usual the equations should be supplemented with equations that ensure the parameters satisfy the model restrictions. And the use of quadratic set generators enables combinations of equations of the last two types to be collected in matrix equations.

The likelihood equations cannot be solved explicitly in general, and iterative methods must be used. Below we describe one such method for calculating the maximum likelihood estimates.

Modified iterative proportional scaling

In the pure discrete and continuous cases we have described how variants of the method of iterative proportional scaling could be applied to calculate the maximum likelihood estimates in graphical models. But in the case of mixed discrete and continuous variables, this method unfortunately cannot be applied. This is essentially because in most mixed models, the generator marginals do not form proper cuts. A version of the more general method of modified iterative proportional scaling, see Section D.1, must be used.

The first element of the method is to describe a value θ_0 of the canonical parameter that is to be used as the fixed value for remaining canonical parameters whenever line sections are to be identified. This parameter value should be chosen so as to make estimation easy when it is fixed. We typically choose θ_0 by setting the pure quadratic interaction parameters $\psi_\emptyset(i)_{\gamma\gamma}$ to be identically equal to 1 for all $\gamma \in \Gamma$ and all other interaction terms equal to zero. This corresponds to the empty generator and it means that under θ_0, all discrete variables are independent and uniformly distributed and the continuous variables are independent of the discrete variables, mutually independent, and distributed as standard $\mathcal{N}(0,1)$ variables.

Next we partition the canonical parameters into possibly overlapping groups. Here we use the generating class $\mathcal{A}$ and simply let a correspond to those canonical parameters that are allowed to vary in the model with just this generator in its generating class. Under some circumstances, advantages can be obtained by grouping some generators together.

Below we show how to calculate the maximum likelihood estimate of the canonical parameters in each of these simple models. We solve the estimation problem and describe the corresponding iterative step of the algorithm in the various cases below.

Discrete generator. A discrete generator $d \subseteq \Delta$ has marginal sufficient statistic $n(i_d), i_d \in \mathcal{I}_d$, and the only interactions present are the pure discrete interactions among variables in d and subsets of these. Estimation in the model is trivial, as

$$\check{p}(i) = \frac{n(i_d)}{|n||\mathcal{I}_d|}$$

and, using (6.4) and the fact that $h(i) = 0$ and $K(i)$ is the the identity, we find

$$\check{g}(i) = -\frac{|\Gamma|}{2}\log(2\pi) + \log\frac{n(i_d)}{|n||\mathcal{I}_d|}.$$

Interactions can be calculated by Möbius inversion if desired.

The iterative step calculates the estimate above as well as the similar quantity just with the current expected value $n^*(i_d)$ of the counts substituted for $n(i_d)$ and then updates the current estimates of the interactions by the difference of interaction parameters in the two cases. In the case of a discrete generator, the marginal counts do form a cut and no line search is needed. Simple calculations show that the updated density $f_d(i,y)$ is related to the current f^* by the relation

$$f_d(i,y) = f^*(i,y)\frac{n(i_d)}{n^*(i_d)},$$

so that for a discrete generator the iterative step is completely analogous to the corresponding step in iterative proportional scaling for the discrete case as in Section 4.3.1.

Linear generator. A linear generator (d,c) identifies on its own a saturated homogeneous model with known variance. The sufficient statistics are the pairs $\{n(i_d), s(i_d)_c\}, i_d \in \mathcal{I}_d$, and all discrete and linear interactions among variables in d and c are permitted. Estimation in this model is trivial as well, and using (6.5) we find

$$\log \tilde{f}(i,y) = -\frac{|\Gamma|}{2}\log(2\pi) + \log\frac{n(i_d)}{|n||\mathcal{I}_d|} - \frac{1}{2}\sum_{\gamma \in c}\left\{\frac{s(i_d)_\gamma}{n(i_d)}\right\}^2$$
$$+ \sum_{\gamma \in c}\frac{s(i_d)_\gamma}{n(i_d)}y_\gamma - \frac{1}{2}\sum_{\mu \in \Gamma}y_\mu^2.$$

Let $n^*(i_d), s^*(i_d)$ denote the expectations under the current parameters. We recall that

$$n^*(i_d) = \sum_{j:j_d=i_d} n^*(j), \quad s^*(i_d) = \sum_{j:j_d=i_d} n^*(j)\xi^*(j),$$

where $\xi^*(j)$ is the current vector of conditional expectations of the continuos variables, given the discrete. Then the estimated density corresponding to these observations is

$$\log \hat{f}(i,y) = -\frac{|\Gamma|}{2}\log(2\pi) + \log\frac{n^*(i_d)}{|n||\mathcal{I}_d|} - \frac{1}{2}\sum_{\gamma \in c}\left\{\frac{s^*(i_d)_\gamma}{n^*(i_d)}\right\}^2$$
$$+ \sum_{\gamma \in c}\frac{s^*(i_d)_\gamma}{n^*(i_d)}y_\gamma - \frac{1}{2}\sum_{\mu \in \Gamma}y_\mu^2.$$

The iterative step updates the canonical parameters by a line search in the direction of the difference between the two set of parameters calculated here. Letting (g^*, h^*, K^*) denote the current parameter values, the discrete component is updated for all $i \in \mathcal{I} \setminus \{i^*\}$ to

$$g_\lambda(i) =$$
$$g^*(i) + \lambda\left(\frac{n(i_d) - n^*(i_d)}{|n||\mathcal{I}_d|} - \frac{1}{2}\sum_{\gamma \in c}\left\{\frac{s(i_d)_\gamma}{n(i_d)}\right\}^2 + \frac{1}{2}\sum_{\gamma \in c}\left\{\frac{s^*(i_d)_\gamma}{n^*(i_d)}\right\}^2\right)$$

and the linear interaction parameters similarly to

$$h_\lambda(i)_\gamma = h^*(i)_\gamma + \lambda\left(\frac{s(i_d)_\gamma}{n(i_d)} - \frac{s^*(i_d)_\gamma}{n^*(i_d)}\right) \quad \text{for } i \in \mathcal{I}, \gamma \in c.$$

The parameter λ is found by line search, i.e. λ is chosen to maximize the likelihood function on the line

$$l(\lambda) = \log L(g_\lambda, h_\lambda, K^*).$$

The updated value of $g(i^*)$ is found by normalization.

In this particular case, the sufficient statistic does not induce a cut, because the distribution of $\{N(i_d), S(i_d)_c\}$ depends not only on the canonical parameters that correspond to the statistics, but, for example, also on the concentration matrix of the distribution.

The possibilities are therefore either to make a full line search or to try the value $\lambda = 1$ and halving to $\lambda = 1/2$, $\lambda = 1/4$ and so on, until a value of λ that increases the likelihood has been found. In this case there are no problems with violating restrictions from the canonical parameter space, but it must be checked that the likelihood function increases after updating.

Linear and quadratic generator. When there explicitly is a linear generator (d, γ) in the generating class $\mathcal{A}$, there must also be a quadratic generator (b, γ^2) with $b \subset d$. There can be advantage in treating the pair of generators $(d, \gamma), (b, \gamma^2)$ simultaneously. This pair identifies a saturated model with some variance homogeneity. The sufficient statistics are the pairs above, combined with $\{ss(i_b)_\gamma\}, i_b \in \mathcal{I}_b$ and all discrete and linear interactions among variables in d and γ are permitted whereas quadratic interactions are allowed only between γ and variables in b. Estimation in this model is as easy as before, and we have that

$$\log \check{f}(i,y) = -\frac{|\Gamma|}{2}\log(2\pi) + \log\frac{n(i_d)}{|n||\mathcal{I}_d|} - \frac{\{s(i_d)_\gamma\}^2 n(i_b)}{2\{n(i_d)\}^2 ssd(i_b)_{\gamma\gamma}}$$
$$+ \frac{s(i_d)_\gamma n(i_b)}{n(i_d) ssd(i_b)_{\gamma\gamma}} y_\gamma - \frac{n(i_b)}{2 ssd(i_b)_{\gamma\gamma}} y_\gamma^2 - \frac{1}{2} \sum_{\mu \in \Gamma \setminus \{\gamma\}} y_\mu^2,$$

where $ssd(i_b)_{\gamma\gamma} = ss(i_b)_{\gamma\gamma} - \{s(i_b)_\gamma\}^2/n(i_b)$ as previously. Let $n^*(i_d)$, $s^*(i_d)$, and $ss^*(i_b)$ denote the expectations of the sufficient statistics under the current parameters. We recall that

$$ss^*(i_b)_{\gamma\gamma} = \sum_{j:j_b=i_b} n^*(j) \left\{\sigma^*(j)_{\gamma\gamma} + \xi^*(j)_\gamma^2\right\}$$

with $\sigma^*(j)_{\gamma\gamma}$ denoting the current estimate of the conditional variance of Y_γ and otherwise we have used the same notation as in the case of a linear generator.

The iterative step calculates the first estimate as well as the estimate with n, s and ss replaced by n^*, s^* and ss^*. If the original hierarchical model determined by $\mathcal{A}$ implies the conditional independence $\gamma \perp\!\!\!\perp \Delta \setminus d \mid d$, the marginal forms a cut. This holds if d separates γ from the remaining discrete variables in the interaction graph of the model. In this case we can update the current canonical parameters (g^*, h^*, K^*) without a line search in the direction of the difference between the parameters calculated here, i.e. the discrete part g of the canonical characteristic is updated for all $i \in \mathcal{I} \setminus \{i^*\}$ as

$$g(i) =$$
$$g^*(i) + \frac{n(i_d) - n^*(i_d)}{|n||\mathcal{I}_d|} - \frac{\{s(i_d)_\gamma\}^2 n(i_b)}{2\{n(i_d)\}^2 ssd(i_b)_{\gamma\gamma}} + \frac{\{s^*(i_d)_\gamma\}^2 n^*(i_b)}{2\{n^*(i_d)\}^2 ssd^*(i_b)_{\gamma\gamma}},$$

the linear characteristics for all $i \in \mathcal{I}$ as

$$h(i)_\gamma = h^*(i)_\gamma + \frac{s(i_d)_\gamma n(i_b)}{n(i_d) ssd(i_b)_{\gamma\gamma}} - \frac{s^*(i_d)_\gamma n^*(i_b)}{n^*(i_d) ssd^*(i_b)_{\gamma\gamma}}$$

and finally the quadratic interactions also for all $i \in \mathcal{I}$ as

$$\psi(i)_{\gamma\gamma} = \psi^*(i)_{\gamma\gamma} + \frac{n(i_b)}{2 ssd(i_b)_{\gamma\gamma}} - \frac{n^*(i_b)}{2 ssd^*(i_b)_{\gamma\gamma}}.$$

Here $ssd^*(i_b)_{\gamma\gamma} = ss^*(i_b)_{\gamma\gamma} - \{s^*(i_b)_\gamma\}^2/n^*(i_b)$. If the conditional independence above does not hold, a line search must be performed as in the case of a linear generator.

Single quadratic generator. A single quadratic generator (d, γ^2) identifies a saturated model without any variance homogeneity. The sufficient statistics are triples $\{n(i_d), s(i_d)_\gamma, ss(i_d)_{\gamma\gamma}\}, i_d \in \mathcal{I}$, and all interactions among variables in d and γ are permitted. Otherwise there are no differences between this case and the case of a combined linear quadratic generator other than what appear when b is everywhere replaced with d.

Quadratic pair generator. The last type of generator involves two continuous variables at a time and is thus of the form $(d, \{\gamma, \mu\})$ with $\gamma \neq \mu$. It identifies a model that permits linear interaction among all variables in d and both of the variables γ and μ, as well as simultaneous quadratic pair interaction, but no quadratic main effects. The sufficient statistics are

$$\{n(i_d), s(i_d)_\gamma, s(i_d)_\mu, ss(i_d)_{\gamma\mu}\}, \quad i_d \in \mathcal{I}_d.$$

The model is somewhat special and estimation is therefore not completely trivial. But, as we shall now show, explicit estimation is in fact possible. To make the algorithm more efficient it seems appropriate to choose θ_0 slightly different here; more precisely we choose it to depend on the observations as

$$k_0(i)_{\kappa\kappa} = \frac{n(i_d)}{ssd(i_d)_{\kappa\kappa}} \quad \text{for } \kappa \in \Gamma \tag{6.40}$$

and all other canonical characteristics equal to zero. We first investigate the case with $d = \emptyset$. Then Y_γ and Y_μ are independent of all other variables and jointly they are bivariate normal with unknown mean $(\xi_\gamma, \xi_\mu)^T$ and concentration matrix

$$\begin{pmatrix} k_{\gamma\gamma} & k_{\gamma\mu} \\ k_{\mu\gamma} & k_{\mu\mu} \end{pmatrix} = \begin{pmatrix} \frac{|n|}{ssd_{\gamma\gamma}} & \frac{|n|\rho}{\sqrt{ssd_{\gamma\gamma}ssd_{\mu\mu}}} \\ \frac{|n|\rho}{\sqrt{ssd_{\gamma\gamma}ssd_{\mu\mu}}} & \frac{|n|}{ssd_{\mu\mu}} \end{pmatrix},$$

where $-1 < \rho < 1$ also is unknown. The covariance matrix is calculated by matrix inversion and is equal to

$$\begin{pmatrix} \sigma_{\gamma\gamma} & \sigma_{\gamma\mu} \\ \sigma_{\mu\gamma} & \sigma_{\mu\mu} \end{pmatrix} = \frac{1}{|n|(1-\rho^2)} \begin{pmatrix} ssd_{\gamma\gamma} & -\rho\sqrt{ssd_{\gamma\gamma}ssd_{\mu\mu}} \\ -\rho\sqrt{ssd_{\gamma\gamma}ssd_{\mu\mu}} & ssd_{\mu\mu} \end{pmatrix}.$$

The model is as usual a full and regular exponential family, and estimation is thus performed by equating the set of canonical statistics to their expectation, which is

$$\mathbf{E}S_\gamma = |n|\xi_\gamma, \quad \mathbf{E}S_\mu = |n|\xi_\mu, \quad \mathbf{E}SS_{\gamma\mu} = \frac{-\rho\sqrt{ssd_{\gamma\gamma}ssd_{\mu\mu}}}{1-\rho^2} + |n|\xi_\gamma\xi_\mu,$$

whereby the equations become

$$s_\gamma = |n|\xi_\gamma, \quad s_\mu = |n|\xi_\mu, \quad ss_{\gamma\mu} = \frac{-\rho\sqrt{ssd_{\gamma\gamma}ssd_{\mu\mu}}}{1-\rho^2} + |n|\xi_\gamma\xi_\mu.$$

If we solve the two first equations for ξ we get

$$\check{\xi}_\gamma = \bar{y}_\gamma, \quad \check{\xi}_\mu = \bar{y}_\mu,$$

and substituting in the third equation reduces this to

$$\frac{-\rho}{1-\rho^2} = \frac{ssd_{\gamma\mu}}{\sqrt{ssd_{\gamma\gamma}ssd_{\mu\mu}}}.$$

If $ssd_{\gamma\mu} = 0$, the last equation has solution $\check{\rho} = 0$, but otherwise it is a second-degree equation in ρ with two real roots. Only one of these satisfies the restriction $-1 < \rho < 1$. If we let

$$r_{\gamma\mu} = \frac{ssd_{\gamma\mu}}{\sqrt{ssd_{\gamma\gamma}ssd_{\mu\mu}}},$$

we find

$$\check{\rho} = \frac{1 - \sqrt{1 + 4(r_{\gamma\mu})^2}}{2r_{\gamma\mu}}.$$

Note that it holds that

$$\frac{1-\sqrt{5}}{2} \leq \check{\rho} \leq \frac{\sqrt{5}-1}{2}.$$

This is due to choosing k_0 depending on the observations as in (6.40).

Next we express everything in the canonical parameters and further get

$$\check{h}_\gamma = \frac{|n|}{ssd_{\gamma\gamma}}\check{\xi}_\gamma + \frac{\check{\rho}|n|}{\sqrt{ssd_{\gamma\gamma}ssd_{\mu\mu}}}\check{\xi}_\mu = \frac{s_\gamma}{ssd_{\gamma\gamma}} + \frac{\check{\rho}s_\mu}{\sqrt{ssd_{\gamma\gamma}ssd_{\mu\mu}}}$$

$$\check{h}_\mu = \frac{\check{\rho}|n|}{\sqrt{ssd_{\gamma\gamma}ssd_{\mu\mu}}}\check{\xi}_\gamma + \frac{|n|}{ssd_{\mu\mu}}\check{\xi}_\mu = \frac{\check{\rho}s_\gamma}{\sqrt{ssd_{\gamma\gamma}ssd_{\mu\mu}}} + \frac{s_\mu}{ssd_{\mu\mu}}.$$

In the general case, when parameters are allowed to vary with the discrete variables we accommodate in the usual way and find for the estimate of the density in this model that

$$\log \check{f}(i,y) = -\frac{|\Gamma|}{2}\log(2\pi) + \log\frac{n(i_d)}{|n||\mathcal{I}_d|}$$
$$-\frac{n(i_d)}{2ssd(i_d)_{\gamma\gamma}}\left\{\frac{s(i_d)_\gamma}{n(i_d)}\right\}^2 - \frac{n(i_d)}{2ssd(i_d)_{\mu\mu}}\left\{\frac{s(i_d)_\mu}{n(i_d)}\right\}^2$$
$$-\frac{n(i_d)\check{\rho}(i_d)}{\sqrt{ssd(i_d)_{\gamma\gamma}ssd(i_d)_{\mu\mu}}}\frac{s(i_d)_\gamma}{n(i_d)}\frac{s(i_d)_\mu}{n(i_d)}$$
$$+\left\{\frac{s(i_d)_\gamma}{ssd(i_d)_{\gamma\gamma}} + \frac{\check{\rho}(i_d)s(i_d)_\mu}{\sqrt{ssd(i_d)_{\gamma\gamma}ssd(i_d)_{\mu\mu}}}\right\}y_\gamma$$
$$+\left\{\frac{\check{\rho}(i_d)s(i_d)_\gamma}{\sqrt{ssd(i_d)_{\gamma\gamma}ssd(i_d)_{\mu\mu}}} + \frac{s(i_d)_\mu}{ssd(i_d)_{\mu\mu}}\right\}y_\mu$$
$$-\frac{\check{\rho}(i_d)n(i_d)y_\gamma y_\mu}{\sqrt{ssd(i_d)_{\gamma\gamma}ssd(i_d)_{\mu\mu}}} - \frac{1}{2}\sum_{\kappa\in\Gamma}\frac{n(i_d)y_\kappa^2}{ssd(i_d)_{\kappa\kappa}}.$$

Here

$$\check{\rho}(i_d) = \frac{1 - \sqrt{1 + 4\{r(i_d)_{\gamma\mu}\}^2}}{2r(i_d)_{\gamma\mu}}$$

with

$$r(i_d)_{\gamma\mu} = \frac{ssd(i_d)_{\gamma\mu}}{\sqrt{ssd(i_d)_{\gamma\gamma}ssd(i_d)_{\mu\mu}}}.$$

As in other cases, we also determine the similar estimates $\tilde{\xi}_\gamma$, $\tilde{\xi}_\mu$, and $\tilde{\rho}$ based on the expected value of the sufficient statistics under the current parameter values and use the differences of the corresponding canonical characteristics for line search.

Here some care should be taken, as the same value of θ_0 should be used as before, and this had a dependence on the observations and not the expected values. So not all observed quantities in the above formulae should be replaced with the corresponding expected values. We find the following expression for $\tilde{f}$:

$$\log \tilde{f}(i,y) = -\frac{|\Gamma|}{2}\log(2\pi) + \log\frac{n^*(i_d)}{|n||\mathcal{I}_d|}$$
$$-\frac{n(i_d)}{2ssd(i_d)_{\gamma\gamma}}\left\{\frac{s^*(i_d)_\gamma}{n^*(i_d)}\right\}^2 - \frac{n(i_d)}{2ssd(i_d)_{\mu\mu}}\left\{\frac{s^*(i_d)_\mu}{n^*(i_d)}\right\}^2$$
$$-\frac{n(i_d)\tilde{\rho}(i_d)}{\sqrt{ssd(i_d)_{\gamma\gamma}ssd(i_d)_{\mu\mu}}}\frac{s^*(i_d)_\gamma}{n^*(i_d)}\frac{s^*(i_d)_\mu}{n^*(i_d)}$$
$$+\left\{\frac{n(i_d)s^*(i_d)_\gamma}{ssd(i_d)_{\gamma\gamma}n^*(i_d)} + \frac{\tilde{\rho}(i_d)n(i_d)s^*(i_d)_\mu}{\sqrt{ssd(i_d)_{\gamma\gamma}ssd(i_d)_{\mu\mu}}n^*(i_d)}\right\}y_\gamma$$
$$+\left\{\frac{\tilde{\rho}(i_d)n(i_d)s^*(i_d)_\gamma}{\sqrt{ssd(i_d)_{\gamma\gamma}ssd(i_d)_{\mu\mu}}n^*(i_d)} + \frac{n(i_d)s^*(i_d)_\mu}{ssd(i_d)_{\mu\mu}n^*(i_d)}\right\}y_\mu$$
$$-\frac{\tilde{\rho}(i_d)n(i_d)y_\gamma y_\mu}{\sqrt{ssd(i_d)_{\gamma\gamma}ssd(i_d)_{\mu\mu}}} - \frac{1}{2}\sum_{\kappa\in\Gamma}\frac{n(i_d)y_\kappa^2}{ssd(i_d)_{\kappa\kappa}}.$$

Here

$$\tilde{\rho}(i_d) = \frac{1 - \sqrt{1 + 4\{\tilde{r}(i_d)_{\gamma\mu}\}^2}}{2\tilde{r}(i_d)_{\gamma\mu}}$$

with

$$\tilde{r}(i_d)_{\gamma\mu} = \frac{ssd^*(i_d)_{\gamma\mu}}{\sqrt{ssd(i_d)_{\gamma\gamma}ssd(i_d)_{\mu\mu}}}.$$

Note that $\tilde{r}(i_d)$ might not be between -1 and 1, but $\tilde{\rho}(i_d)$ will be.

We abstain from giving details for the line search but point out that not all parameter values on the line searched may lead to positive definite covariance matrices, so this must be appropriately checked during the search.

There are other cases where it could be practical to group generators in batches rather than using them one at a time. This is true for example if all quadratic interactions among a subset of continuous variables depend on the same subset of discrete variables. This is typically the case in MIM models. But it would lead too far astray to discuss these issues in full detail here.

6.4.4 Mixed hierarchical model subspaces

As in the discrete case, we formally go through the structure of the subspaces that define mixed hierarchical models. The dimensions of the subspaces determine the order of the models as exponential models and are therefore needed for calculating the degrees of freedom for deviance tests based on asymptotic results. Clearly it is just a question of counting free parameters, but as the models are quite complex it is desirable to adopt a systematic approach.

We first look at the mixed factor subspaces corresponding to single generators of the various types. For a pure discrete generator $(d, \emptyset)$ we define the *mixed factor subspace* as the subspace of $(R \times \mathcal{R}^{|\Gamma|} \times \mathcal{S}_{|\Gamma|})^{\mathcal{I}}$ to be

$$MF_{(d,\emptyset)} = \{(g, h, K) \mid g \in F_d, h() \equiv 0, K() \equiv 0\},$$

where F_d is the discrete factor subspace of $\mathcal{R}^{\mathcal{I}}$ of functions that only depend on i through i_d, i.e.

$$g \in F_d \iff g(i) = g(j) \quad \text{for all } i, j \text{ with } i_d = j_d.$$

The dimension of this space is clearly equal to the dimension of F_d:

$$\dim MF_{(d,\emptyset)} = \dim F_d = |\mathcal{I}_d| = \prod_{\delta \in d} |\mathcal{I}_\delta|.$$

We emphasize that in particular

$$\dim MF_{(\emptyset,\emptyset)} = \dim F_\emptyset = |\mathcal{I}_\emptyset| = 1.$$

Next, for a subspace determined by a linear generator we have

$$MF_{(d,\gamma)} = MF_{(d,\emptyset)} + \{(g, h, K) \mid h()_\gamma \in F_d, g() \equiv 0, K() \equiv 0\}.$$

The sum is direct, so we find

$$\dim MF_{(d,\gamma)} = \dim MF_{(d,\emptyset)} + \dim F_d = 2|\mathcal{I}_d|.$$

For a single quadratic generator we get

$$MF_{(d,\gamma^2)} = MF_{(d,\gamma)} +$$
$$\{(g, h, K) \mid k()_{\gamma\gamma} \in F_d, g() \equiv 0, h() \equiv 0, k()_{\alpha\beta} \equiv 0 \text{ unless } \alpha = \beta = \gamma\},$$

which leads to the dimension

$$\dim MF_{(d,\gamma^2)} = \dim MF_{(d,\gamma)} + \dim F_d = 3|\mathcal{I}_d|.$$

The quadratic pair generator subspaces are

$$MF_{(d,\{\gamma,\mu\})} = MF_{(d,\gamma)} + MF_{(d,\mu)} +$$
$$\{k()_{\gamma\mu} = k()_{\mu\gamma} \in F_d, g() \equiv 0, h() \equiv 0, k()_{\alpha\beta} \equiv 0 \text{ for } \{\alpha,\beta\} \neq \{\gamma,\mu\}\}.$$

where we have abbreviated slightly. The sum is not direct, as we have

$$MF_{(d,\gamma)} \cap MF_{(d,\mu)} = MF_{(d,\emptyset)}$$

and hence

$$\dim MF_{(d,\{\gamma,\mu\})} = \dim MF_{(d,\gamma)} + \dim MF_{(d,\mu)} - \dim MF_{(d,\emptyset)} + \dim F_d$$
$$= 4|\mathcal{I}_d|.$$

Finally we look at the quadratic set generators where

$$MF_{(d,c^2)} = \sum_{\gamma \in c} MF_{(d,\gamma^2)} + \sum_{\gamma,\mu \in c, \gamma \neq \mu} MF_{(d,\{\gamma,\mu\})}.$$

In this case the easiest way to obtain the dimension is to realize that (d, c^2) is the generator of the saturated model with variables $d \cup c$. The dimension is therefore one more than the order of this model. We find from (6.14) that

$$\dim MF_{(d,c^2)} = |\mathcal{I}_d|(|c|+1)(|c|+2)/2,$$

which in the special case of $|c| = 1$ gives $3|\mathcal{I}_d|$ as also derived earlier. We summarize the results in the following system of equations:

$$\dim MF_{(d,\emptyset)} = |\mathcal{I}_d| \qquad (6.41)$$
$$\dim MF_{(d,\gamma)} = 2|\mathcal{I}_d| \qquad (6.42)$$
$$\dim MF_{(d,\gamma^2)} = 3|\mathcal{I}_d| \qquad (6.43)$$
$$\dim MF_{(d,\{\gamma,\mu\})} = 4|\mathcal{I}_d| \qquad (6.44)$$
$$\dim MF_{(d,c^2)} = |\mathcal{I}_d|(|c|+1)(|c|+2)/2, \qquad (6.45)$$

where it should be remembered that $|\mathcal{I}_\emptyset| = 1$.

For a generating class $\mathcal{A}$ we define its *mixed hierarchical model subspace* as the subspace

$$H_\mathcal{A} = \sum_{a \in \mathcal{A}} MF_a.$$

To calculate the dimension of $H_\mathcal{A}$ we first introduce the join and meet of two generating classes $\mathcal{A}$ and $\mathcal{B}$ as

$$\mathcal{A} \vee \mathcal{B} = \text{red}(\mathcal{A} \cup \mathcal{B})$$
$$\mathcal{A} \wedge \mathcal{B} = \text{red}\{a \wedge b \mid a \in \mathcal{A}, b \in \mathcal{B}\},$$

where $a \wedge b$ is the generator that allows those interactions that are allowed by both a and b, and red in front of a list of generators signifies as in the discrete case that all redundant generators are removed from the list. The join $\mathcal{A} \vee \mathcal{B}$ is the smallest generating class that contains all generators in both $\mathcal{A}$ and $\mathcal{B}$, and the meet $\mathcal{A} \wedge \mathcal{B}$ is the largest generating class that is smaller than both $\mathcal{A}$ and $\mathcal{B}$. Here we also have that for two generating classes $\mathcal{A}$ and $\mathcal{B}$ the following relations hold for their hierarchical model subspaces:

$$H_{\mathcal{A} \vee \mathcal{B}} = H_\mathcal{A} + H_\mathcal{B}, \quad H_\mathcal{A} \cap H_\mathcal{B} = H_{\mathcal{A} \wedge \mathcal{B}}$$

which implies that their dimensions correspondingly satisfy

$$\dim H_{A \vee B} = \dim H_A + \dim H_B - \dim H_{A \wedge B}. \qquad (6.46)$$

This latter formula allows recursive computation of dimensions of mixed hierarchical model subspaces by combining with the relations (6.41)–(6.45) just as in the discrete case.

Example 6.32 As illustration we consider some of the generating classes of the graphical models in Example 6.29. The graphical model with graph (a) of Fig. 6.1 had the generating class $\{IJZ^2, IY^2, IYZ\}$ or, equivalently, $\{IJZ^2, I(YZ)^2\}$. The homogeneous version had generating class given as $\{IJZ, IY, (YZ)^2\}$. We find for the first representation that

$$\begin{aligned}
\dim H_{\{IJZ^2,IY^2,IYZ\}} &= \dim H_{\{IJZ^2,IY^2\}} + \dim H_{\{IYZ\}} - \dim H_{\{IZ,IY\}} \\
&= \dim H_{\{IJZ^2\}} + \dim H_{\{IY^2\}} - \dim H_{\{I\}} \\
&\quad + 4|I| - \dim H_{\{IZ,IY\}} \\
&= 3|IJ| + 3|I| - |I| + 4|I| \\
&\quad - \dim H_{\{IZ\}} - \dim H_{\{IY\}} + \dim H_{\{I\}} \\
&= 3|IJ| + 6|I| - 2|I| - 2|I| + |I| \\
&= 3|IJ| + 3|I|.
\end{aligned}$$

The result comes quicker using the second representation:

$$\begin{aligned}
\dim H_{\{IJZ^2,I(YZ)^2\}} &= \dim H_{\{IJZ^2\}} + \dim H_{\{I(YZ)^2\}} - \dim H_{\{IZ^2\}} \\
&= 3|IJ| + |I|(2+1)(2+2)/2 - 3|I| \\
&= 3|IJ| + 3|I|.
\end{aligned}$$

For the homogeneous version we get

$$\begin{aligned}
\dim H_{\{IJZ,IY,(YZ)^2\}} &= \dim H_{\{IJZ,IY\}} + \dim H_{\{(YZ)^2\}} - \dim H_{\{Z,Y\}} \\
&= \dim H_{\{IJZ\}} + \dim H_{\{IY\}} - \dim H_{\{I\}} \\
&\quad + (2+1)(2+2)/2 \\
&\quad - \dim H_{\{Z\}} - \dim H_{\{Y\}} + \dim H_{\{\emptyset\}} \\
&= 2|IJ| + 2|I| - |I| + 6 - 2 - 2 + 1 \\
&= 2|IJ| + |I| + 3.
\end{aligned}$$

Dimensions for other subspaces are obtained in a similar way. □

6.5 Chain graph models

We abstain from giving a full treatment of models for mixed data that have response structure in the variables, corresponding to the Markov properties associated with a chain graph. However, it seems appropriate to discuss a few aspects, as the results are not completely analogous to the pure cases.

6.5.1 CG regressions

The basic distributional element that is used in the construction of chain graph models and hierarchical block-recursive models in the mixed case is that of a CG regression, such as mentioned in connection with Proposition 6.6.

A CG regression describes the dependence of a CG distribution over response variables $x = (i, y)$ on explanatory variables (j, z), where $j \in \mathcal{J}$ is a set of discrete explanatory variables and $z \in \mathcal{Z} = \mathcal{R}^{\Gamma_E}$ a set of continuous explanatory variables. A CG regression can for example be characterized by a sextuple (u, v, W, c, C, D) where for all (i, j)

$u(i \mid j)$ is a real number

$v(i \mid j) = \{v(i \mid j)_\gamma\}_{\gamma \in \Gamma}$ is a $|\Gamma|$-vector

$W(i \mid j) = \{W(i \mid j)_{\gamma\mu}\}_{\gamma,\mu \in \Gamma}$ is a symmetric $|\Gamma_E| \times |\Gamma_E|$-matrix

$c(i \mid j) = \{c(i \mid j)_\gamma\}_{\gamma \in \Gamma}$ is a $|\Gamma|$-vector

$C(i \mid j) = \{C(i \mid j)_{\gamma\mu}\}_{\gamma,\mu \in \Gamma}$ is a $|\Gamma| \times |\Gamma_E|$-matrix

$D(i \mid j) = \{D(i \mid j)_{\gamma\mu}\}_{\gamma,\mu \in \Gamma}$ is a symmetric and positive definite $|\Gamma| \times |\Gamma|$-matrix.

The moment characteristics of the CG distribution of the response variables given the explanatory variables then have the form

$$\log p(i \mid j, z) = u(i \mid j) + v(i \mid j)^\top z - z^\top W(i \mid j) z - \log \kappa(j, z)$$
$$\xi(i \mid z) = c(i \mid j) + C(i \mid j) z$$
$$\Sigma(i \mid j, z) = D(i \mid j).$$

This form of the moment characteristic of an arbitrary conditional CG distribution was derived in Section 6.1.1. What is less straightforward is that the converse holds: for any sextuple as above there is a joint CG distribution on $(\mathcal{I} \times \mathcal{J}) \times (\mathcal{R}^{\Gamma \cup \Gamma_E})$ such that conditioning on (j, z) in this distribution yields the CG regression specified. The apparent gap is that it could appear that the matrices $W(i \mid j)$ should be positive definite. However, if we let

$$\tilde{W}(i \mid j) = W(i \mid j) + \lambda I$$

and choose λ sufficiently large, then $\tilde{W}$ is certainly positive definite and it still holds that

$$\log p(i \mid j, z) = u(i \mid j) + v(i \mid j)^\top z - z^\top \tilde{W}(i \mid j) z - \log \tilde{\kappa}(j, z),$$

where $\tilde{\kappa}$ is a modified normalization constant. We refer to Lauritzen and Wermuth (1989) for further details.

6.5.2 Estimation in chain graph models

A general *graphical CG regression model* is determined by the Markov assumption relative to a chain graph $\mathcal{G}$ and the distributional assumption that any conditional distribution of the variables in a chain component τ given the parents $\mathrm{pa}(\tau)$ is a CG regression as described above.

General graphical chain models determined in this way lead to likelihoods that factorize into likelihoods of the variables in each chain component given the parents. Each of these likelihood functions are exponential family likelihoods and standard theory therefore applies, but currently no suitable software is available for maximizing the likelihood functions in general.

However, in special cases we can exploit the notion of a decomposition as follows. Let $\hat{f}^*_{[\tau_*]}$ denote the maximum likelihood estimate of the density in the model $M(\tau_*)$ based on the marginal data $x_{\tau \cup \mathrm{pa}(\tau)}$. As before, τ_* denotes the undirected graph that has $\tau \cup \mathrm{pa}(\tau)$ as vertices and undirected edges between a pair (α, β) if either both of these are in $\mathrm{pa}(\tau)$ or there is an edge, directed or undirected, between them in the chain graph $\mathcal{G}$. Further, let $\hat{f}^*_{[\mathrm{pa}(\tau)]}$ denote the estimate of the density in the saturated model for the variables in $\mathrm{pa}(\tau)$. Then we have

Proposition 6.33 *Let $M(\mathcal{G})$ be a graphical chain model such that all chain components τ satisfy*

$$\tau \subseteq \Gamma \vee \mathrm{pa}(\tau) \subseteq \Delta.$$

Then the maximum likelihood estimate of the joint density f exists if and only if $\hat{f}^*_{[\tau_*]}$ exist for all $\tau \in \mathcal{T}$, in which case we have

$$\hat{f} = \prod_{\tau \in \mathcal{T}} \frac{\hat{f}^*_{[\tau_*]}}{\hat{f}^*_{[\mathrm{pa}(\tau)]}}. \tag{6.47}$$

Proof: If the condition is satisfied, $(\emptyset, \tau, \mathrm{pa}(\tau))$ is a decomposition of $M(\tau_*)$. From Proposition 6.17 we then have that the maximum likelihood estimate of the conditional density $f_{\tau|\mathrm{pa}(\tau)}$ in the conditional model obtained from $M(\tau_*)$ by conditioning on the variables in $\mathrm{pa}(\tau)$ can be found by conditioning on $\mathrm{pa}(\tau)$ in the estimate of $f_{\tau \cup \mathrm{pa}(\tau)}$ in the model $M(\tau_*)$. It further holds that this estimate is equal to the ratio between $\hat{f}^*_{[\tau_*]}$ and $\hat{f}^*_{[\mathrm{pa}(\tau)]}$.

Combining with the factorization (3.25) for a general chain Markov probability gives the result. □

The proposition holds true also if only the homogeneous case is considered.

The formula (6.47) is the analogue of the similar formulae (4.68) and (5.52) in the pure cases. As the condition in Proposition 6.33 is automatically fulfilled in the pure cases, the results just given contain the similar results in the pure cases.

The simplest example where the formula (6.47) does not hold is the saturated chain graph model with just two variables: a continuous explanatory variable with a discrete response. This model can be described in familiar terms as a logistic regression with normally distributed explanatory variables. In this particular case the maximum likelihood estimate can be calculated by using standard software for logistic regression. The simplest model for which no standard software will give the maximum likelihood estimate is the model with graph as below.

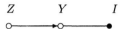

It is a challenge to design and implement a good algorithm for maximizing the likelihood in a general chain graph model. The right-hand side of formula (6.47) can be used to give excellent starting values for an iterative algorithm of Newton type.

6.6 Notes

6.6.1 Collapsibility

Also in the general case there are notions of collapsibility which are completely analogous to the pure cases. Wermuth (1989b) and Geng and Asano (1993) discuss notions related to parametric collapsibility and Frydenberg (1990b) studies stability of models under marginalization in detail.

The condition for graphical model collapsibility is slightly more complex in the mixed case. Frydenberg (1990b) shows that a graphical model is collapsible onto a subset A if and only if every connected component B of A^c is strongly simplicial in $\mathcal{G}$. Recalling that a subset B is strongly simplicial if and only if $\mathrm{bd}(B)$ is complete and

$$B \subseteq \Gamma \vee \mathrm{bd}(B) \subseteq \Delta.$$

This also is a clear unification of the pure discrete and continuous cases. In fact, Frydenberg (1990b) contains a complete extension of the graphical part of Asmussen and Edwards (1983).

Collapsibility for mixed hierarchical models is somewhat more subtle. A straightforward modification of the condition for discrete models was given in Edwards (1990). This condition is necessary but may not capture the phenomenon completely.

6.6.2 Bibliographical notes

The homogeneous CG distribution and its statistical properties in the case of a saturated model were studied in detail by Olkin and Tate (1961). Sometimes the term location model has been used for models for discrete and continuous variables based upon this distribution. They have been studied previously in connection with discrimination problems (Krzanowski 1975, 1976). Dempster (1973) studied distributions that are essentially equivalent to homogeneous CG regressions with unrestricted covariance and gave computational procedures for determining asymptotic variances and covariances for estimated parameters in CG regression models.

Mixed graphical models corresponding to undirected graphs, directed acyclic graphs and chain graphs were introduced by Lauritzen and Wermuth (1984, 1989) and their potential application was described in Wermuth and Lauritzen (1990). The extension to hierarchical mixed interaction models is due to Edwards (1990) although the version given here follows up on suggestions by Whittaker (1990a) and Dawid (1990).

It was shown in Lauritzen and Wermuth (1984, 1989) that if the graph is decomposable and undirected the maximum likelihood estimate could be calculated explicitly, and a corresponding result for the problem of hypothesis testing was derived in Lauritzen (1985). These results were extended in Frydenberg and Lauritzen (1989). Lauritzen (1989b) surveys some of the fundamental properties also described in this book and also contains such results as Proposition 6.33.

7
Further topics

In this chapter we give a brief discussion of a number of topics that have not been covered in the central part of the book but nevertheless illuminate essential aspects of the theory and application of graphical models.

7.1 Probabilistic expert systems

Graphical models have become the basic formal structure for constructing and representing expert systems within domains characterized by inherent uncertainty.

Expert systems are computer programs meant to assist or replace humans performing complicated tasks in areas where human expertise is fallible or scarce. Some such programs work in domains where the task involves performing a sequence of steps according to specified logical rules. However, other programs work in domains that are characterized by inherent uncertainty. This uncertainty can be due to imperfect understanding of the domain, incomplete knowledge of the state of the domain at the time where the task is to be performed, randomness in the mechanisms governing the behaviour of the domain, or a combination of these.

Probabilistic methods were for some time discarded in this context as requiring too complex specification and computation, and alternative formalisms for handling uncertainty in expert systems were in focus, the most prominent being those of fuzzy sets (Zadeh 1965, 1978) and belief functions (Shafer 1976).

However, Pearl (1986*b*) and Lauritzen and Spiegelhalter (1988) showed that these difficulties could be overcome by exploiting the natural modular structure of graphical models.

Probabilistic expert systems are based on a graphical model for the variables associated with the domain. The reasoning is then performed by updating the various probabilities in the light of specific knowledge according to the laws of conditional probability.

For a simple example, consider a fictitious expert system meant to be used for supporting the diagnosis of lung diseases at a chest clinic. The

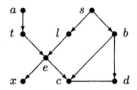

Fig. 7.1. Graph representing structural aspects of medical knowledge concerning lung diseases.

example is adapted from one used by Lauritzen and Spiegelhalter (1988). The diagnostic task to be performed is the following:

> A patient, who has recently been to Asia, presents at the clinic with cough and shortness-of-breath (dyspnoea). The patient is a non-smoker. What are the likely diagnoses?

The graph in Fig. 7.1 represents relevant medical knowledge, verbally expressed as:

> Cough (c) can be caused by tuberculosis (t), lung cancer (l), bronchitis (b), or a combination of them, or none of them. Dyspnoea (d) is associated with cough and can be caused by bronchitis. An X-ray investigation (x) does not distinguish between tuberculosis or lung cancer and these two diseases also have similar risks for developing cough. A recent visit to Asia (a) could be a risk factor for tuberculosis, and smoking (s) increases the risk both for lung cancer and bronchitis.

The node e in the graph represents 't or l' and the state of e is logically determined by the states of t and l. It expresses the fact that neither cough nor X-ray distinguishes between t and l.

In this example each node of the graph represents a binary variable with possible states 'yes' or 'no'. To complete the graphical model, a joint distribution of the variables must be specified. This joint distribution should match the graph by satisfying the appropriate Markov property.

Based on a full specification of the joint distribution, the marginal node probabilities can be calculated and displayed as in Fig. 7.2. Next, the diagnostic task described is solved by calculating the conditional probabilities of the various diseases, given that s is 'no', a is 'yes', c is 'yes', and d is 'yes'. This leads to the probabilities in Fig. 7.3. The most likely diagnosis is bronchitis, but the other lung diseases could be present; in particular, tuberculosis seems a possible contender.

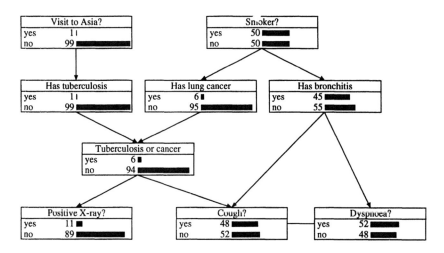

Fig. 7.2. Marginal node probabilities in the chest-clinic example.

If an X-ray is taken subsequently, the probabilities can again be updated in the light of this observation. Figures 7.4 and 7.5 display the revised probabilities in the case of a positive and negative result. It is clearly seen that the chest X-ray investigation has a high diagnostic value. A positive X-ray, in the light of other evidence, will point the suspicion strongly towards tuberculosis, whereas a negative X-ray would rule out the serious diseases as likely diagnoses.

The graph plays several roles in connection with the expert system:

- it gives a visual picture of the domain information;
- it gives a concise representation of domain information in terms of conditional independence restrictions on the joint distribution of the variables;
- it enables simple specification of the joint distribution;
- it enables rapid computation and revision of interesting probabilities.

We shall further detail some of these aspects below.

7.1.1 Specification of the joint distribution

The specification of the joint distribution of the variables in the domain exploits the factorizations (3.23) and (3.24). Only the first of these is needed for chain components that are singletons. Table 7.1 gives conditional odds for each of the non-deterministic variables being in the state 'yes", given the states of the parent nodes. In addition, the conditional probabilities of

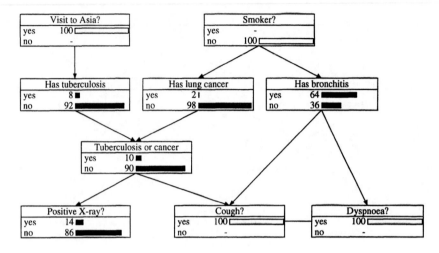

Fig. 7.3. Conditional node probabilities in the chest-clinic example, given that the patient does not smoke, has been to Asia, coughs, and suffers from dyspnoea.

Table 7.1. Odds for non-deterministic singleton chain components in the chest-clinic example.

Parent			a		s		s		e	
State			y	n	y	n	y	n	y	n
Variable	a	s	t		l		b		x	
Odds	1:99	1:1	5:95	1:99	1:9	1:99	3:2	3:7	98:2	5:95

e given t and l are determined by the truth table for logical disjunction.

The specification of the last chain component is more complex, as the factorization (3.24) also must be taken into account. This is done by specifying interaction terms of the conditional probabilities.

The main effects are represented as odds for presence of the features, given that neighbours in the same chain components are in a chosen state, here 'no'. Hence the main effect for c is the odds for cough, given that there is no dyspnoea. The main effect for c is allowed to depend on the states of both parents b and e, whereas the main effect for d is allowed to depend on b only, since there is no edge in the graph between d and e.

The interaction is represented as the ratio between odds for presence of dyspnoea given presence of cough and odds given absence of cough. This interaction is also only allowed to depend on b, as $\{b, e, c, d\}$ is not a complete subset. Table 7.2 gives the numbers used in the example.

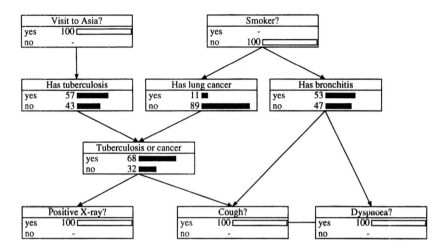

Fig. 7.4. Conditional node probabilities in the chest-clinic example updated after observation of a positive chest X-ray.

Table 7.2. Odds and odds ratio for second chain component in the chest-clinic example.

	Main effects			Odds ratio
	c		d	
State of e	'yes'	'no'		
b='yes'	5:1	2:1	4:1	1
b='no'	3:1	1:3	1:3	2

From the odds and odds ratios the probabilities can be computed by suitable normalization. For example, if both b and e are 'yes', the conditional probabilities for cough and dyspnoea are proportional to the numbers in Table 7.3. Thus by exploiting the conditional independence relations in the graph, the full joint distribution of the 9 variables has been specified by a total of 22 numbers, including the deterministic probabilities. Without these relations $2^9 - 1 = 511$ numbers were needed. In larger networks the saving can be much more dramatic.

The probabilities involved in the specification of an expert system are usually based on a combination of subjective judgement, empirical findings, and other domain knowledge.

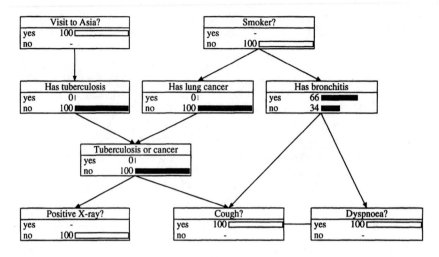

Fig. 7.5. Conditional node probabilities in the chest-clinic example updated after observation of a negative chest X-ray.

Table 7.3. Joint conditional odds for cough and dyspnoea when bronchitis and one of the more severe diseases are present, obtained from appropriate multiplication of numbers in Table 7.2.

	Cough	
Dyspnoea	'yes'	'no'
'yes'	20	4
'no'	5	1

7.1.2 Local computation algorithm

The next aspect where the graph structure gives a dramatic reduction of complexity is in the actual computation of the conditional probabilities. A direct calculation of each of the marginal node probabilities involves the summation of 512 terms and this number grows exponentially with the number of variables in the network, so at first sight this seems prohibitive.

The key to making probability a feasible formalism for uncertainty in expert systems is the exploitation of the graph structure to perform the computation in steps, each of which is local, i.e. involves only a few variables at a time. This computation involves several stages. The first of these is to form a moralized, triangulated cover for the chain graph in question. This is done by first forming the moral graph and then triangulating the graph by adding edges until all cycles of length four or more possess chords. The

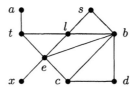

Fig. 7.6. Moralized and triangulated graph for the chest-clinic example.

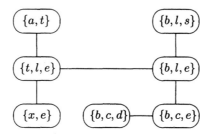

Fig. 7.7. Junction tree for the chest-clinic example.

last step can be done in several ways. The resulting graph in the example is shown in Fig. 7.6.

The set $\mathcal{C}$ of cliques of the triangulated graph is arranged in a junction tree having the property that the intersection $C \cap D$ of any two cliques is a subset of all cliques on the path between C and D in the tree. A junction tree for the example is shown in Fig. 7.7.

There are now various related algorithms available for performing the computations exploiting the junction tree. We describe the algorithm which is implemented in the expert system shell HUGIN (S.K. Andersen et al. 1989), as described by F.V. Jensen et al. (1990). For a slightly modified version of the algorithm, see Dawid (1992). The computation so performed is referred to as *probability propagation*.

From the factorization of p, a factorization of the type

$$p(x) = \frac{\prod_{C \in \mathcal{C}} \phi_C(x)}{\prod_{S \in \mathcal{S}} \phi_S(x)} \qquad (7.1)$$

is obtained, where $\mathcal{S}$ is the list of separators. The 'same' separator set may appear several times in $\mathcal{S}$. The functions ϕ_U, $U \in \mathcal{C} \cup \mathcal{S}$, are stored as *potential tables* associated with each clique or separator.

The calculations are performed via a message passing scheme. The basic operation is that a clique A *absorbs* information from a neighbour B as follows:

$$\phi_A(x) \leftarrow \phi_A(x) \frac{\sum_{y_B : y_{B \cap A} = x_{B \cap A}} \phi_B(y)}{\phi_{A \cap B}(x)}$$

$$\phi_{A \cap B}(x) \leftarrow \sum_{y_B : y_{B \cap A} = x_{B \cap A}} \phi_B(y).$$

Thus the potential functions associated with the absorbing clique and the separator change. The expression (7.1) remains invariant under the operation.

The message passing scheme involves two passes. First a clique R is chosen as the root of the tree and R *collects evidence* by asking each of its neighbours for a message. Before these neighbours send, they in turn ask their neighbours and so on until the requests reach the leaves of the tree. Messages are then sent towards the root R.

When the root clique has received its messages it *distributes evidence* by sending to all its neighbours, who again send to their neighbours and so on. When the distributed evidence reaches the leaves of the tree, the process terminates and all cliques hold potentials ϕ_U that satisfy

$$\phi_C(x) = p(x_C) \text{ for all } C \in \mathcal{C} \cup \mathcal{S},$$

i.e. these are the marginal probabilities. Marginals to single nodes can then easily be calculated. The computation of conditional probabilities is essentially the same as the one described. Modified potentials are first found by setting potentials equal to zero for states that are known to be impossible. The same calculation then gives modified clique potentials that are proportional to the desired conditional probabilities.

7.1.3 Extensions

In the example given, we have pretended that all probabilities were known exactly and the same was true for the graph itself. In general both the model and the probabilities involved must be partly inferred from suitable data, exploiting various procedures for estimation and model selection. In the expert system literature these activities are known as *learning* quantitative and structural aspects.

The algorithms can be modified and generalized in a variety of ways, to include continuous variables based on the CG distributions described in this book (Lauritzen 1992), optimizing decisions in influence diagrams

(F. Jensen et al. 1994) and many other things. There is also a completely abstract version of the local computation theory (Shenoy and Shafer 1990).

For further reading, the collections Oliver and Smith (1990) and Shafer and Pearl (1990), the book by Neapolitan (1990), and the review paper by Spiegelhalter et al. (1993) are recommended. See also recent books by F.V. Jensen (1996) and Almond (1995).

7.2 Model selection

When performing statistical analyses using graphical models, it becomes apparent that formal methods and strategies for investigating or selecting relevant graphical models are crucial.

In short, a model selection method is composed of a *criterion* that judges the feasibility of any model and a *search procedure* that leads the investigator (or the machine) through a list of possible models. Strictly speaking, the specification of the class of models is also a part of the model selection method.

Some methods are purely automatic, while others involve an element of human judgement. Some methods are based upon a fixed *training set* of data, some are considered in a sequential setting, where data come in gradually and learning then takes place little by little. The more automatic and sequential the methods are, the more it seems appropriate to describe them as machine learning procedures.

Traditionally, the term model selection is not used when the structural elements of the model are fixed and only unknown parameters are estimated. However, from a general point of view there is no real reason for the distinction. The formal difference is that model selection will typically refer to a parameter space which is discrete, whereas estimation traditionally is used for activities concerned with identification of continuous parameters.

There is a general theory for model selection and model choice, not necessarily centred around graphical models. Some of this literature is concentrating on the discussion of criteria only. An important point is that pure maximum likelihood is not feasible, as it does not take account of the dimension of the parameter space in an appropriate way.

There are a number of criteria that simply trade off goodness-of-fit, measured by the deviance, against the complexity of the model, measured by the number of free parameters or, equivalently, the dimension of the associated parameter space. These have the form

$$IC_\kappa(M) = -2\log L_M(\hat{\theta}) + \kappa \dim(M),$$

where M denotes the model in question, $\hat{\theta}$ is the maximum likelihood estimate of the unknown parameters under the model, $\dim M$ is the number of parameters, and κ is a constant that gives the exchange rate for trading fit and parameters.

The criterion AIC given by Akaike (1974) is derived from asymptotic decision theoretic principles and leads to $\kappa = 2$ independently of the number of observations. This criterion was applied in the contingency table context by Sakamoto and Akaike (1978).

An asymptotic Bayesian argument leads to BIC, which has $\kappa = \log n$, where n is the number of observations (Schwarz 1978).

An alternative criterion is MDL (minimum description length), which is based on information theoretic considerations (Rissanen 1987). This criterion was used successfully in an expert system context by Wedelin (1993). It is asymptotically equivalent to BIC as described above. The same holds for a criterion which is based on the predictive power of the selected model (Dawid 1984) or even cross-validatory methods for choice (Stone 1974b). For an application of the latter in a contingency table context, see Stone (1974a).

A full Bayesian approach will use suitable conjugate prior distributions. Such were described by Spiegelhalter and Lauritzen (1990) for the directed and discrete case and Dawid and Lauritzen (1993) in the general case. These distributions match the Markov property at a higher level and are therefore termed hyper Markov distributions.

In the expert system literature the Bayesian methods have become very popular as methods for structural learning, and they have for example been studied in the case of directed a graph by Cooper and Herskovits (1992) and Heckerman et al. (1994). Strategies for search in undirected graphical models using hyper Markov distributions have been investigated by Madigan and Raftery (1994); see also Madigan and York (1995).

The papers Wermuth (1976b), Edwards and Havránek (1985, 1987), Whittaker (1984), Kreiner (1987), and P.W.F. Smith (1992) all focus on the specification of search procedures but use significance testing as the fundamental criterion for judging the fit of a model.

The programs CoCo made by Badsberg (1991, 1995) and MIM by Edwards (1995) have a number of the various search procedures built in; see also Badsberg (1992) for a discussion of many different strategies. Lauritzen et al. (1994) compare the predictive performance of some of the methods in an expert system context where the model selection program BIFROST was used (Højsgaard and Thiesson 1995).

7.3 Modelling complexity

An important property of graphical models is their ability to describe quite complex structures by stringing together local relationships in a modular way — as parent–child relations or graph adjacencies. This modularity can then be exploited for efficient model specification and computation.

Probabilistic expert systems, described in Section 7.1, are examples of such complex systems where computations in many cases can be done explicitly using exact probability propagation.

However, the modularity of the graphical models can be exploited in much greater generality based on essentially two simple ideas. Firstly, represent the unknown parameters directly in the graph as separate vertices and secondly, use Markov chain Monte Carlo methods as the basic computational tool.

These ideas appear to play most elegantly within a Bayesian approach, as the parameters then have status of random variables, but also likelihood-based methods can exploit the ideas (Thompson and Guo 1991; Geyer and Thompson 1992). Below we give a brief description of some elements in this technique as well as some basic references.

7.3.1 Markov chain Monte Carlo methods

Modern Markov chain Monte Carlo methods (MCMC) have their origin in statistical physics (Metropolis *et al.* 1953), where they have been used to simulate the behaviour of large systems of particles and complex molecules. For a general description and survey, see for example the papers by Gelfand and Smith (1990), A.F.M. Smith and Roberts (1993) and Besag and Green (1993). Also the paper by Hastings (1970) is strongly recommended reading.

There is an abundance of related but different MCMC algorithms and here we shall only explain one of them, the *Gibbs sampler*, introduced in this form by Geman and Geman (1984).

The basis of the algorithm is a collection $(X_v)_{v \in V}$ of random variables having a joint distribution P with positive probability density. The computations then proceed as follows. First, pick an arbitrary starting configuration $x^0 = (x_v^0)_{v \in V}$. Then number the elements of V as $V = \{1, \ldots, |V|\}$ and make successive random drawings from the full conditional distributions $\mathcal{L}(X_v \mid X_{V \setminus \{v\}})$ as:

pick x_1^1 from $\mathcal{L}(X_1 \mid x_{V \setminus \{1\}}^0)$;

pick x_2^1 from $\mathcal{L}(X_2 \mid x_{V \setminus \{1,2\}}^0, x_1^1)$;

pick x_3^1 from $\mathcal{L}(X_3 \mid x_{V \setminus \{1,2,3\}}^0, x_{1,2}^1)$;

continue in this manner until you pick $x_{|V|}^1$ from $\mathcal{L}(X_{|V|} \mid x_{V \setminus \{|V|\}}^1)$.

Each of the steps above is referred to as a *site visit*. This reflects the origin of the ideas in physics where the variables correspond to states of particles placed at different sites. When all sites have been visited, a transition from $x^0 = (x_v^0)_{v \in V}$ to $x^1 = (x_v^1)_{v \in V}$ has taken place. Iteration of the procedure creates successive values $x^0, x^1, \ldots, x^n, \ldots$. The point of the method is that under very general conditions these successive values form a realization from a Markov chain which has the original distribution P of X as equilibrium distribution. Then integrals with respect to P can be approximated by averages of the Gibbs sample

$$\mathbf{E}_P\{f(X)\} \approx \frac{1}{n} \sum_{\nu=1}^{n} f(x^\nu).$$

The key aspects that make the method practically applicable are:

- When P is assumed to satisfy the local Markov property with respect to a graph $\mathcal{G}$, the full conditional distributions simplify as

$$\mathcal{L}(X_v \mid X_{V \setminus \{v\}}) = \mathcal{L}(X_v \mid X_{\text{bd}(v)}).$$

  Hence if the graph $\mathcal{G}$ is sparse, these distributions might easily be represented and amenable to simulation, even though the joint distribution P is very complex.

- If the object of interest is the conditional distribution of X given an observed set of variables $X_A = x_A^*$ where $A \subseteq V$, the only modifications needed are that the starting configuration must satisfy $x_v = x_v^*$ for all $v \in A$ and that sites in A are not visited. This is clearly of interest for computing Bayesian posterior distributions.

Problems with the methods are associated with the fact that convergence to equilibrium can be very slow and it can be quite difficult to assess such convergence in a practical situation. Also, if the joint density is not everywhere positive as assumed above, the method is not generally applicable. Then updating in blocks instead of single sites can be helpful (C.S. Jensen *et al.* 1995).

7.3.2 Applications

There is an abundance of applications of MCMC methods and it is outside the scope of this book to give a complete survey. Apart from the original applications in statistical physics, there are a number of rather direct and successful applications in image analysis and spatial statistics (Besag *et al.* 1991).

But a more recent class of applications deserves special mention. Here the idea is that for modelling many standard but complex problems, distributions satisfying a Markov property with respect to a directed acyclic graph $\mathcal{G}$ seem natural and immediately appealing. Then the model is specified through the conditional densities $f(x_v \,|\, x_{\text{pa}(v)})$ and the Gibbs sampler is then applied on the moral graph $\mathcal{G}^m$. In this case the full conditional distributions simplify to

$$f(x_v \,|\, x_{V \setminus \{v\}}) \propto f(x_v \,|\, x_{\text{pa}(v)}) \prod_{u \in \text{ch}(v)} f(x_u \,|\, x_{\text{pa}(u)}). \qquad (7.2)$$

It is here important that in many methods for generating random numbers one only needs to know the density function up to a constant of proportionality (Ripley 1987).

A major application of this approach is in the area of human and animal genetics, where care must be taken, as the joint density typically is not positive (Sheehan and Thomas 1993). But the scope of application is much more general than that; see for example Gilks *et al.* (1993) for a wealth of interesting examples. The applicability of the method has been drastically enhanced by the availability of the program BUGS (Gilks *et al.* 1994), which essentially allows the user to specify conditional distributions quite generally and then automatically proceeds to construct a Gibbs sampler based on (7.2).

7.4 Missing-data problems

When dealing with missing-data problems in graphical models there are essentially two possibilities. One is to use a full Bayesian approach and exploit methods such as those described in the previous section.

Alternatively the basic tool for dealing with missing-data problems within a likelihood-based approach is the EM algorithm (Dempster *et al.* 1977). In connection with log–linear models for contingency tables, the algorithm was studied by Fuchs (1982), extending work of Chen and Fienberg (1974, 1976) and Hocking and Oxspring (1974).

7.4.1 The EM algorithm

The EM algorithm in the form that is needed here is based on forming the conditional expectation of the log-likelihood function for complete data, given the observed data

$$Q(\theta' \,|\, \theta) = E_\theta \{\log f(X \,|\, \theta') \,|\, y\}, \qquad (7.3)$$

where X is the random variable corresponding to the complete (unobserved) data having density f, whereas $y = g(x)$ is the observed data. When θ is fixed, the process of determining this expectation as a function of θ' is referred to as the E-step. The algorithm then alternates between the E-step and the M-step, which maximizes Q in θ'. The algorithm has generalizations called GEM algorithms which appear by not necessarily maximizing Q but only finding a value of θ' that makes $Q(\theta' \mid \theta)$ strictly increase over $Q(\theta \mid \theta)$. It is also possible to add a penalty to the log-likelihood function, calculating instead

$$Q^*(\theta' \mid \theta) = Q(\theta' \mid \theta) - J(\theta')$$

at the E-step, where $J(\theta)$ is a penalty, for example obtained from a prior density proportional to $\exp\{-J(\theta)\}$ (Green 1990). This leads to maximization of the penalized log-likelihood function $\log L(\theta) - J(\theta)$.

7.4.2 Hierarchical log–linear models

In the case of a log–linear model, the log-likelihood function for the complete data is a linear function of the set of sufficient marginals

$$n(i_a), \quad a \in \mathcal{A}, \ i_a \in \mathcal{I}_a,$$

where $\mathcal{A}$ is the generating class of the model. Therefore the E-step is equivalent to calculating the expected marginal counts

$$n^*(i_a) = \mathbf{E}_p\{N(i_a) \mid \text{observed data}\}.$$

Similarly the M-step can be identified with solving

$$|n| p(i_a) = n^*(i_a), \quad a \in \mathcal{A}, \ i_a \in \mathcal{I}_a, \tag{7.4}$$

for p, which maximizes the likelihood function, assuming the expected counts to be the true counts.

The M-step is computationally equivalent to solving the likelihood equations.

The computational effort involved in the E-step can be considerable, and efficient methods are needed, since this computation must be repeated many times. Some gain of efficiency can be obtained by exploiting collapsibility (Geng and Asano 1988).

It was shown in Lauritzen (1995) how to exploit the computational scheme of Lauritzen and Spiegelhalter (1988) to perform the E-step of the algorithm in log–linear models and recursive models for contingency tables. The method is as follows.

Assume that we have $|n|$ independent observations $i_{b^1}^1, i_{b^2}^2, \ldots, i_{b^{|n|}}^{|n|}$, such that for case ν we have only observed the value of variables in the set b^ν. Then we find

$$n^*(i_a) = \sum_{\nu=1}^{|n|} E_p\{\chi^\nu(i_a) \mid i_{b^1}^1, \ldots, i_{b^{|n|}}^{|n|}\}$$

$$= \sum_{\nu=1}^{|n|} E_p\{\chi^\nu(i_a) \mid i_{b^\nu}^\nu\}$$

$$= \sum_{\nu=1}^{|n|} p(i_a \mid i_{b^\nu}^\nu), \tag{7.5}$$

where we have let

$$\chi^\nu(i_a) = \begin{cases} 1 & \text{if } i_a^\nu = i_a \\ 0 & \text{otherwise.} \end{cases}$$

What remains to be observed is that the Lauritzen–Spiegelhalter procedure for probability propagation used in expert systems is an efficient method of calculating the individual terms in (7.5). For computational efficiency one would of course collect cases with identical observations in groups and only calculate $p(i_a \mid i_b)$ once for each of these identical terms.

In the case of a decomposable hierarchical model the M-step is trivial, since it follows from (4.40) that as potentials in the next iteration we can use

$$\psi_u(i_u) = n^*(i_u)/|n| \text{ for } u \in \mathcal{C} \cup \mathcal{S},$$

where $\mathcal{C}$ is the set of cliques and $\mathcal{S}$ the system of separators associated with a perfect ordering of $\mathcal{C}$.

In the general case, the next values of $\psi_u(i_u)$ must be iteratively calculated from $n^*(i_u)$, $u \in \mathcal{A}$, using for example iterative proportional scaling. Note that it is not necessary to complete a full iterative proportional scaling. By using only a single cycle of proportional scaling, the expected log-likelihood increases strictly, leading to a GEM algorithm.

7.4.3 Recursive models

It is also possible to use the EM algorithm for estimating probabilities in recursive graphical models. This is of particular interest, since these are commonly used for constructing probabilistic expert systems and such systems often have a large number of variables.

To identify the E-step in this case, we use the factorization of the likelihood function for complete data in (4.61).

We then take conditional expectation of the log-likelihood function and get, using (7.3),

$$Q(p' \mid p) = \sum_{\delta \in \Delta} \sum_{i_{\text{cl}(\delta)} \in \mathcal{I}_{\text{cl}(\delta)}} n^*\left(i_{\text{cl}(\delta)}\right) \log p'\left(i_\delta \mid i_{\text{pa}(\delta)}\right).$$

Thus the E-step can be performed exactly as before, with the only difference that the junction tree is initialized with potential functions from the conditional probability tables as described for example by F.V. Jensen et al. (1990).

To identify the M-step, note that the expression for Q is identical to the log-likelihood only with observed data replaced by estimated. So as usual, this joint likelihood function can be maximized by maximizing each of the factors. We therefore obtain that the M-step becomes

$$p(i_\delta \mid i_{\text{pa}(\delta)}) = n^*(i_{\text{cl}(\delta)})/n^*(i_{\text{pa}(\delta)}). \qquad (7.6)$$

In this formula $n(i_\emptyset) = |n|$. This will appear in the denominator whenever a variable δ has no parents.

If data are missing in a massive fashion it seems favourable to penalize the likelihood. Assuming the conditional probabilities $p(\cdot \mid i_{\text{pa}(\delta)})$ to be independent and Dirichlet-distributed with parameters $\alpha(i_{\text{cl}(\delta)})$ leads to the penalty

$$-J(p) = \sum_{\delta \in \Delta} \sum_{i_{\text{cl}(\delta)} \in \mathcal{I}_{\text{cl}(\delta)}} \alpha\left(i_{\text{cl}(\delta)}\right) \log p\left(i_\delta \mid i_{\text{pa}(\delta)}\right).$$

Maximization of the penalized likelihood can be made iteratively as before by replacing (7.6) with

$$p(i_\delta \mid i_{\text{pa}(\delta)}) = \frac{n^*(i_{\text{cl}(\delta)}) + \alpha(i_{\text{cl}(\delta)})}{n^*(i_{\text{pa}(\delta)}) + \alpha(i_{\text{pa}(\delta)})}.$$

Here $\alpha(i_{\text{cl}(\delta)})$ can be interpreted as counts from investigations performed prior to the current sample.

The algorithm has been implemented (Thiesson 1991) using the program HUGIN for the propagation calculations and tested in several examples. Experience indicates that with data missing massively and systematically, the likelihood function has a number of local maxima and straight maximum likelihood gives results with unsuitably extreme probabilities. The penalized likelihood seems to perform much better.

The EM algorithm is known to converge relatively slowly when it is getting close. It turns out that the gradient of the likelihood function can be calculated with essentially the same amount of work as is involved in the E-step of the EM algorithm. Speeding up the EM algorithms by exploiting gradient information is therefore preferable.

The procedure can be generalized to the mixed and continuous cases by using the propagation scheme of Lauritzen (1992) to calculate the conditional moments needed in the E-step.

Appendix A
Various prerequisites

A.1 Inequalities

Some elementary inequalities will be of frequent use. First recall that for $x > 0$

$$\log x \leq x - 1 \qquad (A.1)$$

and (A.1) is strict unless $x = 1$. Here, and throughout the text, log denotes the natural logarithm. This follows from Taylor's formula, expanding around $x = 1$:

$$\log x = x - 1 - \frac{(x-1)^2}{2\theta^2}$$

for some θ between 1 and x. It follows that for $x, y \geq 0$ we have

$$x^y e^{-x} \leq y^y e^{-y} \qquad (A.2)$$

and (A.2) is strict unless $x = y$. This is seen by first considering the cases where at least one of x and y is zero, and using (A.1) to see that if x and y are both positive, we have

$$y \log\left(\frac{x}{y}\right) \leq y\left(\frac{x}{y} - 1\right) = x - y.$$

Taking exponentials on both sides yields (A.2). Also we will make much use of the following inequality.

Lemma A.1 *Let* $(a(i))_{i \in \mathcal{I}}$ *and* $(b(i))_{i \in \mathcal{I}}$ *be vectors of non-negative numbers with* $\sum_i a(i) = \sum_i b(i) < \infty$. *Then*

$$\prod_{i \in \mathcal{I}} a(i)^{b(i)} \leq \prod_{i \in \mathcal{I}} b(i)^{b(i)}. \qquad (A.3)$$

The inequality is strict unless $a(i) = b(i)$ *for all* $i \in \mathcal{I}$.

Proof: From (A.2) we get

$$\prod_{i \in \mathcal{I}} a(i)^{b(i)} e^{-a(i)} \leq \prod_{i \in \mathcal{I}} b(i)^{b(i)} e^{-b(i)}.$$

The inequality (A.3) now follows upon dividing both sides with the common positive constant $\exp\{-\sum_i a(i)\}$. □

The inequality (A.3) is a version of the so-called *information inequality*. Taking logarithms on both sides of (A.3) yields the version

$$\sum_{i \in \mathcal{I}} b(i) \log a(i) \leq \sum_{i \in \mathcal{I}} b(i) \log b(i), \qquad (A.4)$$

holding for non-negative a and b with $\sum_i a(i) = \sum_i b(i)$. It should be noticed that $0 \log 0 = 0$ and the inequality is strict unless $a(i) \equiv b(i)$.

A.2 Kullback–Leibler divergence

The Kullback-Leibler *divergence* $D(P, Q)$ of the probability measure P from Q is defined as

$$D(P, Q) = \int p \log \frac{p}{q} \, d\mu,$$

where P and Q have densities p and q with respect to the σ-finite measure μ. If P has density with respect to Q this may also be written as

$$D(P, Q) = \int \frac{dP}{dQ} \log \frac{dP}{dQ} \, dQ.$$

If not, $D(P, Q) = \infty$. If also Q has density with respect to P, the divergence can alternatively be expressed as

$$D(P, Q) = \int \log \frac{dP}{dQ} \, dP.$$

The notion was introduced by Kullback and Leibler (1951) although they use the term divergence for the symmetrized version $D(P, Q) + D(Q, P)$.

The divergence is always non-negative and $D(P, Q) = 0$ if and only if $P = Q$. In the discrete case this is a direct consequence of the inequality (A.4). Even though neither D nor its symmetrized version is a true metric as they do not satisfy the triangle inequality, the divergence D has a clear geometric interpretation as a non-symmetric distance measure, and a version of Pythagoras's theorem applies to the divergence (Chentsov 1972; Csiszár 1975).

When in the discrete case P is taken to be the empirical distribution and Q is varying in a family specified by a statistical model, then

$$D(P, Q) = -\frac{1}{n} \log L(Q),$$

where L denotes the likelihood function. Hence minimizing $D(P, Q)$ over Q corresponds to maximum likelihood estimation. Reversing the role of

P and Q leads to minimum discriminant information estimation (Kullback 1959; Ireland *et al.* 1969). See also Brown (1986) and E.S. Christensen (1989).

A.3 Möbius inversion

An important combinatorial trick is contained in the following

Lemma A.2 (Möbius inversion) *Let Ψ and Φ be functions defined on the set of all subsets of a finite set V, taking values in an Abelian group. Then the following two statements are equivalent:*

(1) *for all $a \subseteq V$:* $\Psi(a) = \sum_{b:b\subseteq a} \Phi(b)$;

(2) *for all $a \subseteq V$:* $\Phi(a) = \sum_{b:b\subseteq a} (-1)^{|a\setminus b|} \Psi(b)$.

Proof: We show (2) $\implies$ (1):

$$\sum_{b:b\subseteq a} \Phi(b) = \sum_{b:b\subseteq a} \sum_{c:c\subseteq b} (-1)^{|b\setminus c|} \Psi(c)$$

$$= \sum_{c:c\subseteq a} \Psi(c) \left\{ \sum_{b:c\subseteq b\subseteq a} (-1)^{|b\setminus c|} \right\}$$

$$= \sum_{c:c\subseteq a} \Psi(c) \left\{ \sum_{h:h\subseteq a\setminus c} (-1)^{|h|} \right\}.$$

The latter sum is equal to zero unless $a \setminus c = \emptyset$, i.e. if $c = a$, because any finite, non-empty set has the same number of subsets of even as of odd cardinality. The proof of (1) $\implies$ (2) is performed analogously. $\square$

The Abelian group referred to in the lemma can be the real numbers, but often also just the additive group of a real vector space V, the vector space of linear maps on V or the vector space of symmetric $k \times k$ matrices etc. More general versions of the lemma exist that relate to general lattices rather than the lattice of subsets of a set; see for example Aigner (1979).

A.4 Iterative partial maximization

For computation of maximum likelihood estimates we shall rely on procedures involving iterative partial maximization in the sense that the likelihood function is maximized over different sections in the parameter space. This is then repeated cyclically.

We consider a continuous real-valued function L on a compact set Θ, and assume that the value $\hat{\theta}$ that maximizes L is uniquely determined.

We assume further that there for all $\theta^* \in \Theta$ are sections $\Theta_i(\theta^*), i = 1, \ldots, k$ in Θ in such a way that L is globally maximized at θ^* if and only if L is maximized over all of the sections.

Finally we assume that the operations of maximizing L over sections is continuous and well defined, i.e. there are continuous transformations T_i of Θ into itself such that if $\theta \in \Theta_i(\theta^*)$ for $i = 1, \ldots, k$,

$$L\{T_i(\theta^*)\} > L(\theta), \quad \text{if } \theta \neq T_i(\theta^*).$$

In other words, $T_i(\theta^*)$ is the uniquely determined point where L is maximized over the section $\Theta_i(\theta^*)$.

Now let θ_0 be arbitrary and define recursively

$$\theta_{n+1} = T_1 \cdots T_k(\theta_n), \quad n \geq 0.$$

Then we can show

Proposition A.3 *Under the assumptions given above the sequence (θ_n) converges to $\hat{\theta}$, the unique point where L attains its maximum.*

Proof: Since Θ is assumed compact, the sequence (θ_n) has a convergent subsequence (θ_{n_k}) with limit θ^*, say. We need to show that $\hat{\theta} = \theta^*$. Let $S = T_1 \cdots T_k$. Since each T-operation is a partial maximization, $L(\theta_n)$ must be non-decreasing in n. Since also each operation is continuous, we have

$$L\{S(\theta^*)\} = \lim_{k \to \infty} L\{S(\theta_{n_k})\} \leq \lim_{k \to \infty} L(\theta_{n_k+1}) = L(\theta^*).$$

But using that T_i partially maximizes L gives

$$L\{S(\theta^*)\} \geq L\{T_2 \cdots T_k(\theta^*)\} \geq \cdots \geq L(\theta^*).$$

Thus there must everywhere be equality. Uniqueness of the partial maxima yields, when the chain of inequalities is read from right to left, that

$$\theta^* = T_k(\theta^*) = \cdots = T_1(\theta^*).$$

Finally, since the global maximum was uniquely determined by maximizing L over all sections, the proof is complete. □

The above result is the basis of a class of algorithms used in this book to maximize likelihood functions. Sections are chosen appropriately such that the partial maximization problems are relatively simple. A starting value θ_0 is found and is iteratively changed by partial maximization over sections. In all cases the existence, uniqueness and necessary continuity properties will be established separately, but convergent algorithms necessarily appear.

A.5 Sufficiency

If the random variable X has distribution P which is unknown apart from belonging to a class of distributions $\mathcal{P}$, a statistic $T = t(X)$ is said to be *sufficient* for $\mathcal{P}$ if the conditional distributions

$$P^{t_0} = \mathcal{L}\{X \mid t(X) = t_0, P\}$$

can all be chosen so that they do not depend on $P \in \mathcal{P}$.

If the measures $P \in \mathcal{P}$ have densities $f(x \mid P)$ with respect to a σ-finite measure μ which factorize as

$$f(x \mid P) = g(x)h\{t(x), P\}, \tag{A.5}$$

i.e. into a term that does not depend on P and a term that depends on x through $t(x)$ only, then T is sufficient. A suitable converse is also true. The criterion (A.5) is known as *Neyman's factorization criterion*.

We say that a statistic T as above is (boundedly) *complete* with respect to $\mathcal{P}$ if it holds for any bounded and measurable function f that

$$[\mathbf{E}\{f(T) \mid P\} = 0 \text{ for all } P \in \mathcal{P}] \implies \left[f(t) = 0 \ (Pt^{-1}\text{-a.s. for all } P \in \mathcal{P})\right].$$

The following important result is Theorem 2 of Basu (1955) and will be used extensively.

Theorem A.4 (Basu) *If T is sufficient and complete with respect to $\mathcal{P}$ and the distribution of $Y = s(X)$ does not depend on $P \in \mathcal{P}$, then T and Y are independent.*

Proof: We have for all $P \in \mathcal{P}$,

$$g(A) = P(Y \in A) = \mathbf{E}\{\chi_A(Y) \mid P\} = \mathbf{E}\left[\mathbf{E}\{\chi_A(Y) \mid T, P\} \mid P\right].$$

Since T is sufficient, the inner conditional expectation does not depend on P and we can define a function f as

$$f(t) = \mathbf{E}\{\chi_A(Y) \mid T = t, P\} - g(A) = P(Y \in A \mid T = t) - P(Y \in A).$$

The first equation then reads

$$\int_{\mathcal{X}} f\{t(x)\} P(dx) = 0.$$

Since this holds for all $P \in \mathcal{P}$, we conclude that

$$P(Y \in A \mid T = t) = P(Y \in A)$$

and hence Y and T are independent. □

Complete sufficient statistics exist typically in exponential models; see Appendix D.

An important notion related to sufficiency is that of Y being a *cut* in (X, Y) (Barndorff-Nielsen 1978), which means that the family $\mathcal{P}$ can be parametrized by a pair (κ, λ) where κ and λ are variation-independent and the joint density of (X, Y) factorizes as

$$f(x, y \,|\, P) = f(x \,|\, y, \kappa) f(y \,|\, \lambda).$$

That Y is a cut implies that for any fixed value of κ, $t(x, y) = y$ is a sufficient statistic for λ, but many other interesting properties follow. In particular the likelihood function factorizes, so that the joint likelihood can be maximized by maximizing each factor separately.

Similarly we say that Z is a *split* in (X, Y, Z) if $\mathcal{P}$ can be parametrized with the triple (κ, λ, μ) with all variation-independent and

$$f(x, y, z \,|\, P) = f(x \,|\, z, \kappa) f(y \,|\, z, \lambda) f(z \,|\, \mu).$$

Hence in particular X and Y are conditionally independent given Z, written as $X \perp\!\!\!\perp Y \,|\, Z$; see Chapter 3.

Appendix B

Linear algebra and random vectors

B.1 Matrix results

Lemma B.1 *A symmetric block diagonal matrix Σ with*

$$\Sigma = \begin{pmatrix} A & B \\ B^\mathsf{T} & D \end{pmatrix},$$

where D is assumed to be regular, is positive (non-negative) definite if and only if $E = A - BD^{-1}B^\mathsf{T}$ and D are both positive (non-negative) definite.

Proof: If we let

$$u = \begin{pmatrix} x \\ y - D^{-1}B^\mathsf{T} x \end{pmatrix},$$

we find

$$\begin{aligned} u^\mathsf{T}\Sigma u &= x^\mathsf{T} A x + 2x^\mathsf{T} B(y - D^{-1}B^\mathsf{T} x) \\ &\quad + (y - D^{-1}B^\mathsf{T} x)^\mathsf{T} D(y - D^{-1}B^\mathsf{T} x) \\ &= x^\mathsf{T}(A - BD^{-1}B^\mathsf{T})x + y^\mathsf{T} Dy. \end{aligned}$$

Hence the result follows. □

The determinant of a partitioned $n \times n$ matrix can be factorized as

$$\det \begin{pmatrix} A & B \\ C & D \end{pmatrix} = (\det E)(\det D), \qquad (B.1)$$

where A and D are supposed to be $p \times p$ and $q \times q$ matrices respectively, and $E = A - BD^{-1}C$ if D is non-singular. The correctness of (B.1) follows from the calculation

$$\det \begin{pmatrix} A & B \\ C & D \end{pmatrix} = \det \begin{pmatrix} A & B \\ C & D \end{pmatrix} \det \begin{pmatrix} I_p & 0 \\ -D^{-1}C & I_q \end{pmatrix} =$$

$$= \det\left\{\begin{pmatrix} A & B \\ C & D \end{pmatrix}\begin{pmatrix} I_p & 0 \\ -D^{-1}C & I_q \end{pmatrix}\right\}$$

$$= \det\begin{pmatrix} E & B \\ 0 & D \end{pmatrix} = (\det E)(\det D).$$

Finally, the inverse of a partitioned matrix is given as

$$\begin{pmatrix} A & B \\ C & D \end{pmatrix}^{-1} = \begin{pmatrix} E^{-1} & -E^{-1}G \\ -FE^{-1} & D^{-1} + FE^{-1}G \end{pmatrix}. \quad (B.2)$$

The inverse on the left-hand side exists if and only if the inverses on the right-hand side exist. Here $E = A - BD^{-1}C$, $F = D^{-1}C$, and $G = BD^{-1}$. Performing the multiplication shows the correctness.

The Kronecker product of matrices

Let A be an $m \times n$ and B a $p \times q$ matrix. We define the Kronecker product $A \otimes B$ as the $mp \times nq$ matrix with entries

$$(A \otimes B)_{ir,js} = a_{ij} b_{rs},$$

where $A = \{a_{ij}\}$ and $B = \{b_{rs}\}$. When displaying $A \otimes B$ it looks slightly different, depending on how we choose to arrange the indices ij and rs. For example, we can arrange it as

$$\begin{pmatrix} a_{11}B & \cdots & a_{1\nu}B & \cdots & a_{1n}B \\ \vdots & & \vdots & & \vdots \\ a_{\mu 1}B & \cdots & a_{\mu\nu}B & \cdots & a_{\mu n}B \\ \vdots & & \vdots & & \vdots \\ a_{m1}B & \cdots & a_{m\nu}B & \cdots & a_{mn}B \end{pmatrix}, \quad (B.3)$$

i.e. as a block matrix with $m \times n$ blocks of size $p \times q$, or as $p \times q$ blocks of size $m \times n$, as

$$\begin{pmatrix} Ab_{11} & \cdots & Ab_{1\kappa} & \cdots & Ab_{1q} \\ \vdots & & \vdots & & \vdots \\ Ab_{\pi 1} & \cdots & Ab_{\pi\kappa} & \cdots & Ab_{\pi q} \\ \vdots & & \vdots & & \vdots \\ Ab_{p1} & \cdots & Ab_{p\kappa} & \cdots & Ab_{pq} \end{pmatrix}. \quad (B.4)$$

The two ways of writing are equivalent and the important point is how to use the Kronecker product in the calculations.

As A and B can be interpreted as matrices of linear maps from $\mathcal{R}^n$ to $\mathcal{R}^m$ and $\mathcal{R}^q$ to $\mathcal{R}^p$, the product $A \otimes B$ can be interpreted as a linear map from $R^{n \times q}$ to $R^{m \times p}$ as follows. An element x in $R^{n \times q}$ is an $n \times q$ matrix and we simply let

$$(A \otimes B)(x) = AxB^\top, \tag{B.5}$$

where the expression on the right-hand side involves usual matrix products. The correctness of this interpretation follows from the calculation

$$\begin{aligned}(AxB^\top)_{ir} &= \sum_\nu \sum_\kappa a_{i\nu} x_{\nu\kappa} b^\top_{\kappa r} \\ &= \sum_\nu \sum_\kappa a_{i\nu} b_{r\kappa} x_{\nu\kappa} = \sum_{\nu\kappa} (A \otimes B)_{ir,\nu\kappa} x_{\nu\kappa}.\end{aligned}$$

Expressions such as $(A \otimes B)(U)$ will sometimes refer to the usual matrix product and sometimes to the map $(A \otimes B)$ evaluated at U. The dimensions of the elements will always identify which.

Proposition B.2 *The Kronecker product satisfies:*

T1: $0 \otimes A = A \otimes 0 = 0$

T2: $A \otimes (B_1 + B_2) = A \otimes B_1 + A \otimes B_2$

T3: $(A_1 + A_2) \otimes B = A_1 \otimes B + A_2 \otimes B$

T4: $(\lambda A) \otimes (\mu B) = (\lambda\mu)(A \otimes B)$ *if* $\lambda, \mu \in \mathcal{R}$

T5: $(A \otimes B)^\top = A^\top \otimes B^\top$

T6: $(A_1 \otimes B_1)(A_2 \otimes B_2) = (A_1 A_2) \otimes (B_1 B_2)$

T7: $(A \otimes B)^- = A^- \otimes B^-$

T8: $\mathrm{tr}(A \otimes B) = (\mathrm{tr}\, A)(\mathrm{tr}\, B)$ *if A and B are symmetric matrices*

T9: $\det(A \otimes B) = (\det A)^p (\det B)^n$ *if A is $n \times n$ and B is $p \times p$*

T10: *If Φ and Σ are both positive (semi)definite matrices, then so is $\Phi \otimes \Sigma$.*

Proof: This is straightforward algebraic manipulation. □

Here A^- denotes any generalized inverse of A, i.e. a matrix satisfying $AA^-A = A$, and **T7** should be read: if A^- and B^- are g-inverses of A and B, then $A^- \otimes B^-$ is a g-inverse of $A \otimes B$. If Φ and Σ are symmetric matrices of dimensions $n \times n$ and $p \times p$ respectively, then $\Phi \otimes \Sigma$ is a symmetric matrix. The corresponding bilinear form is given as

$$\langle x, y \rangle_{\Phi \otimes \Sigma} = \langle x, (\Phi \otimes \Sigma)(y) \rangle = \mathrm{tr}(x^\top \Phi y \Sigma).$$

The outer product

Consider two Euclidean spaces V and W with inner products $\langle \cdot, \cdot \rangle_V$ and $\langle \cdot, \cdot \rangle_W$. For vectors $v \in V$ and $w \in W$ we define the *outer product* $w \odot v$ as the linear map from V to W that has

$$(w \odot v)(u) = \langle u, v \rangle_V w. \tag{B.6}$$

Thus the range of the map $w \mapsto w \odot v$ is the subspace span$\{w\}$.

B.2 Factor subspaces and interactions

We spend some time here considering interaction spaces and generating classes from an algebraic point of view, inspired by Darroch and Speed (1983).

Let Δ be a finite set and assume that for each $\delta \in \Delta$ a finite set $\mathcal{I}_\delta$ of possible levels is given. We often refer to the elements of Δ as *variables*. The elements $i = (i_\delta)_{\delta \in \Delta}$ of the product $\mathcal{I}$ of the level sets

$$i \in \mathcal{I} = \times_{\delta \in \Delta} \mathcal{I}_\delta$$

are denoted *cells* of the *table* $\mathcal{I}$.

The vector space $L = \mathcal{R}^\mathcal{I}$ of real-valued functions defined on $\mathcal{I}$. L is a Euclidean space with inner product

$$\langle x, y \rangle = \sum_{i \in \mathcal{I}} x(i) y(i).$$

An *a-marginal table* is obtained for $a \subset \Delta$ by only considering the variables in a. It has *marginal cells*

$$i_a \in \mathcal{I}_a = \times_{\delta \in a} \mathcal{I}_\delta.$$

For an arbitrary $x \in \mathcal{R}^\mathcal{I}$ and an arbitrary subset a of Δ, the a-marginal of x is defined as

$$x(i_a) = \sum_{j : j_a = i_a} x(j) = \sum_{j_{\Delta \setminus a} \in \mathcal{I}_{\Delta \setminus a}} x(i_a, j_{\Delta \setminus a})$$

and $|x| = \sum_{i \in \mathcal{I}} |x(i)|$. For $a = \emptyset$ we write $x(i_\emptyset) = \sum_{i \in \mathcal{I}} x(i)$, so that if $x(i) \geq 0$ for all i then we also have that $|x| = x(i_\emptyset)$.

If a is a subset of Δ, we consider the *factor subspace* F_a of L of functions that only depend on i through i_a, i.e.

$$x \in F_a \iff x(i) = x(j) \quad \text{for all } i, j \text{ with } i_a = j_a.$$

The space F_Δ is L itself and $F_\emptyset$ is the space of constants. The orthogonal projections onto F_a are given as

$$\Pi_a x(i) = x(i_a)/|\mathcal{I}_{\Delta\setminus a}|. \tag{B.7}$$

From this it directly follows that

$$\Pi_a \Pi_b = \Pi_b \Pi_a = \Pi_{a\cap b}. \tag{B.8}$$

Subspaces of the form

$$H_\mathcal{A} = \sum_{a \in \mathcal{A}} F_a,$$

where $\mathcal{A}$ is a class of subsets of Δ, are called *hierarchical model subspaces*. Denote the orthogonal projection onto $H_\mathcal{A}$ by $\Pi_\mathcal{A}$. We then have the following useful lemma.

Lemma B.3 *For $x, y \in L$ the equation*

$$\Pi_\mathcal{A} x = \Pi_\mathcal{A} y \tag{B.9}$$

is equivalent to the system of equations

$$\Pi_a x = \Pi_a y \quad \text{for all } a \in \mathcal{A}. \tag{B.10}$$

Proof: It is obvious that the equation (B.9) implies the system of equations (B.10), since $\Pi_a \Pi_\mathcal{A} = \Pi_a$. Conversely, assume that the pair (x, y) satisfies (B.10). To establish (B.9) we just have to show that

$$\langle x - \Pi_\mathcal{A} y, u \rangle = 0$$

or, equivalently, that

$$\langle y, u \rangle = \langle x, u \rangle$$

for an arbitrary $u \in H_\mathcal{A}$. If $u = \sum_a u_a$ with $u_a \in F_a$, we have

$$\begin{aligned}
\langle y, u \rangle &= \sum_{a \in \mathcal{A}} \langle y, u_a \rangle = \sum_{a \in \mathcal{A}} \langle \Pi_a y, u_a \rangle \\
&= \sum_{a \in \mathcal{A}} \langle \Pi_a x, u_a \rangle = \sum_{a \in \mathcal{A}} \langle x, u_a \rangle = \langle x, u \rangle,
\end{aligned}$$

and the result is established. □

Introduce the subspaces for $a \subseteq \Delta$:

$$E_a = F_a \cap \left(\bigcap_{b : b \subset a} F_b^\perp \right).$$

The space E_a is called the *space of interactions* among criteria in a and its elements are called $|a|$-*factor interactions* or *interactions of order* $|a| - 1$. For $|a| = 1$ we also speak of *main effects*. Note that (B.7) implies that

$$x \in E_a \iff x \in F_a \text{ and } x(i_b) = 0 \text{ for all } b \subset a, \tag{B.11}$$

which is the familiar relation that 'summing an interaction term over any index gives 0'.

We assume in the following that $F_\delta \subseteq H_{\mathcal{A}}$ for all $\delta \in \Delta$, i.e. all main effects are represented in $\mathcal{A}$.

If we let Q_a denote the orthogonal projection onto the interaction subspace E_a, we get

$$\Pi_d Q_a = \begin{cases} Q_a & \text{if } d \supseteq a \\ 0 & \text{otherwise} \end{cases} \tag{B.12}$$

since the first of these assertions is obvious and the second is a consequence of (B.11) and the calculation

$$\Pi_d Q_a = \Pi_d \Pi_a Q_a = \Pi_{d \cap a} Q_a = 0.$$

Since

$$\bigcap_{b: b \subset a} F_b^\perp = \bigcap_{\delta: \delta \in a} F_{a \setminus \{\delta\}}^\perp,$$

we also have

$$E_a = F_a \cap \left(\bigcap_{\delta: \delta \in a} F_{a \setminus \{\delta\}}^\perp \right).$$

From this we obtain

$$Q_a = \Pi_a \prod_{\delta: \delta \in a} \left(I - \Pi_{a \setminus \{\delta\}} \right), \tag{B.13}$$

which in particular implies that all Q_a and Π_b commute. As a consequence we derive

$$Q_a Q_b = Q_b Q_a = 0 \quad \text{if } a \neq b, \tag{B.14}$$

since $a \neq b$ implies either $b \not\subseteq a$ or the converse, whereby

$$Q_a Q_b = \Pi_a Q_a Q_b = Q_a \Pi_a Q_b = 0.$$

Thus the interaction subspaces are mutually orthogonal. By direct multiplication we obtain

$$\Pi_a \prod_{\delta: \delta \in a} \left(I - \Pi_{a \setminus \{\delta\}} \right) = \sum_{b: b \subseteq a} (-1)^{|b|} \Pi_{a \setminus b} = \sum_{b: b \subseteq a} (-1)^{|a \setminus b|} \Pi_b.$$

A consequence is the formula
$$Q_a = \sum_{b:b\subseteq a} (-1)^{|a\setminus b|} \Pi_b. \tag{B.15}$$

Using Lemma A.2 (Möbius inversion) on (B.15) we obtain

Proposition B.4 *The factor subspaces F_a are direct orthogonal sums of the interaction subspaces $(E_b)_{b\subseteq a}$ and for all $a \subseteq \Delta$ we have*
$$\Pi_a = \sum_{b:b\subseteq a} Q_b.$$

In particular, any vector in $\mathcal{R}^\mathcal{I}$ has a unique expansion into its interactions, in symbols expressed as
$$\mathcal{R}^\mathcal{I} = \oplus_{a:a\subseteq \Delta} E_a. \tag{B.16}$$

Letting $\tilde{\mathcal{A}}$ denote the permissible interactions
$$c \in \tilde{\mathcal{A}} \iff c \subseteq a \text{ for some } a \in \mathcal{A},$$
we further have the expressions
$$H_\mathcal{A} = \oplus_{c\in\tilde{\mathcal{A}}} E_c, \quad \Pi_\mathcal{A} = \sum_{c\in\tilde{\mathcal{A}}} Q_c. \tag{B.17}$$

If $\mathcal{A}$ is a collection of subsets of Δ, we let $\operatorname{red}\mathcal{A}$ denote the elements of $\mathcal{A}$ that are maximal with respect to inclusion, i.e. $\operatorname{red}\mathcal{A}$ is obtained from $\mathcal{A}$ by throwing away all subsets that are contained in other subsets in $\mathcal{A}$. Since
$$H_\mathcal{A} = H_{\operatorname{red}\mathcal{A}},$$
this reduced collection is the most economical way of specifying the model space. If no sets in $\mathcal{A}$ are subsets of other sets, $\mathcal{A}$ is reduced already, and the collection is called a *generating class*.

As shown in Section 2.2, the set of generating classes forms a distributive lattice with join and meet operations given as
$$\begin{aligned} \mathcal{A} \vee \mathcal{B} &= \operatorname{red}(\mathcal{A} \cup \mathcal{B}) \\ \mathcal{A} \wedge \mathcal{B} &= \operatorname{red}\{a \cap b \mid a \in \mathcal{A}, b \in \mathcal{B}\}. \end{aligned}$$

We then have

Proposition B.5 *Let $\mathcal{A}$ and $\mathcal{B}$ be generating classes. Then the following relations hold for their hierarchical model subspaces:*
$$H_{\mathcal{A}\vee\mathcal{B}} = H_\mathcal{A} + H_\mathcal{B}, \quad H_{\mathcal{A}\wedge\mathcal{B}} = H_\mathcal{A} \cap H_\mathcal{B}$$
and correspondingly for the orthogonal projections
$$\Pi_\mathcal{A}\Pi_\mathcal{B} = \Pi_\mathcal{B}\Pi_\mathcal{A} = \Pi_{\mathcal{A}\wedge\mathcal{B}}. \tag{B.18}$$

Proof: It is trivial that $H_{\mathcal{A}\wedge\mathcal{B}} \subseteq H_{\mathcal{A}} \cap H_{\mathcal{B}}$. Let $u \in H_{\mathcal{A}} \cap H_{\mathcal{B}}$. Then, from (B.17) we get

$$u \in H_{\mathcal{A}} \implies u = \Pi_{\mathcal{A}} u = \sum_{c \in \tilde{\mathcal{A}}} Q_c u.$$

Since also $u \in H_{\mathcal{B}}$,

$$u = \Pi_{\mathcal{B}} u = \sum_{d \in \tilde{\mathcal{B}}} Q_d \sum_{c \in \tilde{\mathcal{A}}} Q_c u.$$

From (B.14) it now follows that $u \in H_{\mathcal{A}\wedge\mathcal{B}}$. That the projections commute follows directly from (B.14) and (B.17). □

In the special case where the generating class $\mathcal{C}$ is the direct join of generating classes $\mathcal{A}$ and $\mathcal{B}$ with $\mathcal{A} \wedge \mathcal{B} = \{A \cap B\}$, we find

$$\Pi_{\mathcal{A}} \Pi_{\mathcal{B}} = \Pi_{\mathcal{B}} \Pi_{\mathcal{A}} = \Pi_{A \cap B}. \tag{B.19}$$

B.3 Random vectors

It seems convenient for the purposes in this book to deal with multivariate distributions in some generality, allowing these to be discussed in general Euclidean vector spaces V rather than in the particular case of $V = \mathcal{R}^n$ with the standard inner product. In particular, when discussing the exact and asymptotic distribution of maximum likelihood estimates, it is most natural to work with random variables that, for example, take values in the vector space of symmetric matrices.

We do not intend to dwell on formal details of the theory. Therefore it will create few difficulties and indeed be very close to the more usual matrix formulation. The reader is referred to Eaton (1983) for a comprehensive and detailed exposition along similar lines.

First we have to discuss the notion of mean and covariance of a random vector X taking values in V, where V is a Euclidean space with inner product $\langle \cdot, \cdot \rangle$.

Definition B.6 An element ξ of V is said to be the *mean vector* or *expectation* of X if it holds that

$$\langle v, \xi \rangle = \mathbf{E} \langle v, X \rangle \quad \text{for all } v \in V.$$

We allow ourselves to write $\xi = \mathbf{E} X$ and have therefore that $\langle v, \mathbf{E} X \rangle = \mathbf{E} \langle v, X \rangle$.

Definition B.7 A bilinear form Σ on V is said to be the *covariance* of X if it holds that

$$\text{Cov}\left(\langle u, X \rangle, \langle v, X \rangle\right) = \Sigma(u, v) \quad \text{for all } u, v \in V.$$

We write $\mathbf{V}X = \Sigma$. Note that the covariance, as well as any other bilinear form, is determined from its values on the diagonal $\Sigma(u,u)$, for the bilinearity gives

$$\Sigma(u,v) = \frac{1}{4}\left\{\Sigma(u+v,u+v) - \Sigma(u-v,u-v)\right\}.$$

To any such bilinear form there is a linear operator, which we also denote by Σ, such that
$$\Sigma(u,v) = \langle u, \Sigma v\rangle.$$
This is referred to as the *covariance operator* of X. If $(e_1,\ldots,e_p)$ is an orthonormal basis of V, we let

$$\sigma_{ij} = \Sigma(e_i,e_j) = \langle e_i, \Sigma e_j\rangle = \operatorname{Cov}\left(\langle e_i,X\rangle, \langle e_j,X\rangle\right).$$

The $p \times p$-matrix of these numbers is the *covariance matrix* of X and we also denote this by $\mathbf{V}X = \Sigma$. So the same symbol is used to refer to the covariance, the covariance operator and the covariance matrix, and the context will determine the exact meaning of the symbol.

The covariance Σ is called *regular* if $\Sigma(u,u) > 0$ for all $u \neq 0$. In this case its matrix is positive definite and the covariance determines an inner product on V which we shall denote as $\langle\cdot,\cdot\rangle_\Sigma$, i.e.

$$\langle u,v\rangle_\Sigma = \Sigma(u,v) = \langle u,\Sigma v\rangle.$$

When the covariance is regular, the inverse operator $K = \Sigma^{-1}$ is called the *concentration operator* and its matrix with respect to a chosen basis is called the *concentration matrix*. The concentration operator determines a symmetric bilinear form as usual by

$$K(u,v) = \langle u, Kv\rangle.$$

This bilinear form is called the *concentration* of the distribution.

The concentration operator K is equivalently defined through the relation

$$\langle u,v\rangle = \langle Ku, \Sigma v\rangle. \tag{B.20}$$

If Σ is not regular, any K which satisfies $\Sigma = \Sigma K\Sigma$, i.e. K is a generalized inverse to Σ, can be used as the concentration operator, and the relation (B.20) then holds for all u,v in the range of Σ, since

$$\langle K\Sigma x, \Sigma\Sigma y\rangle = \langle \Sigma K\Sigma x, \Sigma y\rangle = \langle \Sigma x, \Sigma y\rangle.$$

Note that the concentration and the covariance operator depend on the given inner product on V, and the covariance and concentration matrices further depend on a chosen orthonormal basis. A fully invariant approach to random vectors and the normal distribution on vector spaces avoids introducing the first inner product, but we have chosen not to proceed to this level of abstraction.

In most cases the space V will be $\mathcal{R}^n$ with the usual inner product and standard orthonormal basis, but we also frequently deal with the space $\mathcal{R}^{n \times p}$ of $n \times p$-matrices with inner product

$$\langle A, B \rangle = \mathrm{tr}(A^\top B) \tag{B.21}$$

and canonical basis formed by the matrices E_{ij} with ij-th entry equal to one and the remaining entries equal to zero. In the case where $n = p$, an interesting subspace is formed by the set $\mathcal{S}_p$ of symmetric $p \times p$ matrices where the transpose in (B.21) becomes unnecessary. An orthonormal basis for this space consists of the symmetric matrices

$$\tilde{E}_{ii} = E_{ii}, \quad \tilde{E}_{ij} = (E_{ij} + E_{ji})/\sqrt{2} \quad \text{for } i \neq j. \tag{B.22}$$

If $V = \mathcal{R}^n$, the mean vector is of the form $\xi = (\xi_1, \ldots, \xi_n)^\top$ and we have

$$\langle e_i, \xi \rangle = \mathbf{E} \langle e_i, X \rangle = \xi_i = \mathbf{E} X_i,$$

where $X = (X_1, \ldots, X_n)^\top$. Similarly for the covariance we get

$$\sigma_{ij} = \Sigma(e_i, e_j) = \mathrm{Cov}\left(\langle e_i, X \rangle, \langle e_j, X \rangle\right) = \mathrm{Cov}(X_i, X_j).$$

These formulae indicate how the notation conforms with that used in most statistical literature.

Suppose we have two Euclidean spaces V and W where to avoid confusion we denote their inner products by $\langle \cdot, \cdot \rangle_V$ and $\langle \cdot, \cdot \rangle_W$ respectively. Let A be a linear map from V to W and b an element of W. Then $Y = AX + b$ is a random vector in W. Its mean and covariance are given below.

Proposition B.8 *If the random vector X has mean ξ and covariance Σ, then the mean and covariance operator of Y are*

$$\mathbf{E} Y = A\xi + b, \quad \mathbf{V} Y = A \Sigma A^\top.$$

In the special case of $V = \mathcal{R}^n$ and $W = \mathcal{R}^m$, the same expressions hold for matrices.

Proof: Direct calculation gives

$$\begin{aligned}
\mathbf{E} \langle w, Y \rangle_W &= \mathbf{E} \langle w, AX + b \rangle_W = \mathbf{E} \langle A^\top w, X \rangle_V + \langle w, b \rangle_W \\
&= \langle A^\top w, \xi \rangle_V + \langle w, b \rangle_W = \langle w, A\xi + b \rangle_W,
\end{aligned}$$

which gives the result for the mean, and similarly,

$$\begin{aligned}
\mathrm{Cov}\left(\langle w, Y \rangle_W, \langle y, Y \rangle_W\right) &= \mathrm{Cov}\left(\langle w, AX \rangle_W, \langle y, AX \rangle_W\right) \\
&= \mathrm{Cov}\left(\langle A^\top w, X \rangle_V, \langle A^\top y, X \rangle_V\right) \\
&= \langle A^\top w, \Sigma A^\top y \rangle_V = \langle w, A \Sigma A^\top y \rangle_W,
\end{aligned}$$

which gives the covariance operator. We abstain from repeating the calculations in the matrix case. □

Finally we mention that the distribution of a random vector is uniquely determined by the distribution of all linear functions of the vector. More precisely, the following holds.

Proposition B.9 *If X and Y are two random vectors in V and*

$$\langle v, X \rangle \stackrel{\mathcal{D}}{=} \langle v, Y \rangle \quad \text{for all } v \text{ in } V,$$

then $X \stackrel{\mathcal{D}}{=} Y$.

This is essentially equivalent to the fact that the characteristic function of X,

$$\psi(v) = \mathbf{E} e^{i \langle v, X \rangle},$$

determines the distribution. Here i is the complex unit, i.e. $i^2 = -1$. This result can be found in Cramér (1946). That this is equivalent to the statement in Proposition B.9 follows from the uniqueness of the Fourier transform in the case where $V = \mathcal{R}$.

Appendix C

The multivariate normal distribution

The exposition of the multivariate normal distribution and derived distributions is close to that given in Eaton (1983). Proofs not given here can be found there or in Anderson (1984).

C.1 Basic properties

We first formally define what it means for a random vector in V to be normally distributed:

Definition C.1 A random vector X on a Euclidean space V is said to have a *normal distribution on V* if there exists an element $\xi \in V$ and a bilinear form Σ on V such that

$$\langle v, X \rangle \sim \mathcal{N}\{\langle v, \xi \rangle, \Sigma(v, v)\} \quad \text{for all } v \text{ in } V,$$

where $\mathcal{N}(\mu, \sigma^2)$ denotes the univariate normal distribution with mean μ and variance σ^2.

From Proposition B.9 it follows that the definition is unambiguous and the preceding pages show that then ξ is the mean and Σ the covariance of the random vector X. If X is normally distributed on V we write

$$X \sim \mathcal{N}_V(\xi, \Sigma).$$

In the special cases $V = \mathcal{R}^p$ and $V = \mathcal{R}^{n \times p}$ we write

$$X \sim \mathcal{N}_p(\xi, \Sigma) \quad \text{and} \quad X \sim \mathcal{N}_{n \times p}(\xi, \Sigma).$$

The mean ξ and covariance Σ determine a unique normal distribution on V. Conversely, to any pair (ξ, Σ), where ξ is a vector in V and Σ is a bilinear form on V which is non-negative, i.e. $\Sigma(v, v) \geq 0$ for all $v \in V$, there is a normal distribution with these as mean and covariance.

If Σ is regular, the normal distribution has density with respect to the Lebesgue measure on V that gives mass 1 to a unit cube. This density is equal to

$$f_{\xi, \Sigma}(x) = (2\pi)^{-p/2} (\det \Sigma)^{-1/2} e^{-\langle x - \xi, K(x - \xi) \rangle / 2}, \tag{C.1}$$

where $K = \Sigma^{-1}$ is the concentration operator of the normal distribution. If $\langle \cdot, \cdot \rangle$ is standard inner product on $\mathcal{R}^n$, we have in matrix notation that

$$\langle x - \xi, K(x - \xi)\rangle = (x - \xi)^\top K(x - \xi),$$

where K is the inverse of the covariance matrix. Note that we have then assumed an orthonormal basis in V to be chosen and K depends on this choice.

Adding two independent normal random vectors gives a normal random vector. More accurately:

Proposition C.2 *If $X_1 \sim \mathcal{N}_V(\xi_1, \Sigma_1)$ and $X_2 \sim \mathcal{N}_V(\xi_2, \Sigma_2)$ are independent, then*

$$X_1 + X_2 \sim \mathcal{N}_V(\xi_1 + \xi_2, \Sigma_1 + \Sigma_2).$$

Proof: For $v \in V$ it holds that

$$\langle v, X_1 + X_2 \rangle = \langle v, X_1 \rangle + \langle v, X_2 \rangle.$$

The terms on the right-hand side are independent and univariate normal. Hence the sum is univariate normal. Definition C.1 implies that $X_1 + X_2$ is a normal random vector and the expressions for mean and covariance follow by direct calculation. □

Another important fact about the normal distribution is that an affine transformation of a normal random vector is itself a normal random vector. We consider a situation analogous to that in Proposition B.8.

Proposition C.3 *If A is a linear map from V to W, b an element of W, and $X \sim \mathcal{N}_V(\xi, \Sigma)$, then*

$$Y = AX + b \sim \mathcal{N}_W(A\xi + b, A\Sigma A^\top).$$

Proof: The mean and covariance of Y have been given in Proposition B.8. What remains to be established is that Y follows a normal distribution. But for all $w \in W$ we have

$$\langle w, Y \rangle_W = \langle w, AX + b \rangle_W = \langle A^\top w, X \rangle_V + \langle w, b \rangle_W.$$

Since X has a normal distribution, $\langle A^\top w, X \rangle_V$ is univariate normally distributed. This is not changed by adding the constant $\langle w, b \rangle_W$. Definition C.1 then establishes the result. □

A special case of this result is of interest. Suppose $V = \mathcal{R}^n$ and assume the random vector X partitioned into components X_1 and X_2, where $X_1 \in \mathcal{R}^p$

and $X_2 \in \mathcal{R}^q$ with $p+q = n$. The mean vector and covariance matrix can then be partitioned accordingly into blocks as

$$\xi = \begin{pmatrix} \xi_1 \\ \xi_2 \end{pmatrix} \quad \text{and} \quad \Sigma = \begin{pmatrix} \Sigma_{11} & \Sigma_{12} \\ \Sigma_{21} & \Sigma_{22} \end{pmatrix}$$

such that Σ_{11} has dimensions $p \times p$ and so on. If A in Proposition C.3 is the linear map that sends X into X_2, we obtain:

Proposition C.4 *Let X be distributed as $\mathcal{N}_n(\xi, \Sigma)$, where X, ξ and Σ are partitioned as above. Then the marginal distribution of X_2 is $\mathcal{N}_q(\xi_2, \Sigma_{22})$.*

The conditional distribution of X_1 given $X_2 = x_2$ can be found as follows.

Proposition C.5 *Let X be distributed as $\mathcal{N}_n(\xi, \Sigma)$, where X, ξ and Σ are partitioned as above. Then the conditional distribution of X_1 given $X_2 = x_2$ is $\mathcal{N}_p(\xi_{1|2}, \Sigma_{1|2})$, where*

$$\xi_{1|2} = \xi_1 + \Sigma_{12}\Sigma_{22}^{-}(x_2 - \xi_2) \quad \text{and} \quad \Sigma_{1|2} = \Sigma_{11} - \Sigma_{12}\Sigma_{22}^{-}\Sigma_{21}. \quad (C.2)$$

Here Σ_{22}^{-} is an arbitrary generalized inverse to Σ_{22}.

Proof: We abstain from giving the proof of this well-known result in the general case and assume Σ to be regular. Then let $K = \Sigma^{-1}$ denote the concentration matrix and assume this to be partitioned in the same fashion as Σ. The conditional density is proportional to the joint density of X_1 and X_2. Hence, exploiting that x_2 is fixed, we find by direct calculation that

$$f(x_1 \mid x_2) \propto f_{\xi, \Sigma}(x)$$
$$\propto \exp\left\{-(x_1 - \xi_1)^\top K_{11}(x_1 - \xi_1)/2 - (x_1 - \xi_1)^\top K_{12}(x_2 - \xi_2)\right\}.$$

The linear term involving x_1 has coefficient equal to

$$K_{11}\xi_1 - K_{12}(x_2 - \xi_2) = K_{11}\left\{\xi_1 - K_{11}^{-1}K_{12}(x_2 - \xi_2)\right\}.$$

Using (B.2) we find that

$$K_{11}^{-1} = \Sigma_{11} - \Sigma_{12}\Sigma_{22}^{-1}\Sigma_{21} \quad (C.3)$$

and further that

$$K_{11}^{-1}K_{12} = -\Sigma_{12}\Sigma_{22}^{-1}, \quad (C.4)$$

which then gives

$$f(x_1 \mid x_2) \propto \exp\left\{-(x_1 - \xi_{1|2})^\top K_{11}(x_1 - \xi_{1|2})/2\right\}$$

and the result follows. □

Note that the proportionality constant may in principle depend on the parameters as well as on x_2. But as the distribution is normal, this turns out not to be the case. It follows from (B.1) that we have

$$\det \Sigma = \det \Sigma_{1|2} \det \Sigma_{22} = \frac{\det \Sigma_{22}}{\det K_{11}}. \tag{C.5}$$

Note also the identities (C.3) and (C.4), which are quite useful in their own right. The first expresses that the concentration matrix of the conditional distribution is obtained from the concentration matrix of the joint distribution by deleting rows and columns corresponding to the variables conditioned upon. This gives a formulation of Proposition C.5 which is symmetric to that of Proposition C.4.

Corollary C.6 *Let X be distributed as $\mathcal{N}_n(\xi, \Sigma)$, where X, ξ and Σ are partitioned as above. Then X_1 and X_2 are independent if and only if $\Sigma_{12} = 0$. If Σ is regular, this holds if and only if $K_{12} = 0$.*

Proof: The first statement follows directly from the expression for the conditional mean $\xi_{1|2}$ in Proposition C.5. The second statement then follows from (C.4). □

Consider $X^1, \ldots, X^n$ independent random vectors in $\mathcal{R}^p$ and assume X^i to be distributed as $\mathcal{N}_p(\xi^i, \Sigma)$ such that the covariances are identical. If these random vectors are arranged in an $n \times p$ array with $(X^i)^\top$ as rows, the resulting random matrix will be distributed as $\mathcal{N}_{n \times p}(\xi, I_n \otimes \Sigma)$, where ξ has $(\xi^i)^\top$ as its rows. For this reason, it is of special interest to consider normal distributions with covariance structure of the form $\Phi \otimes \Sigma$.

Proposition C.7 *Let A and B be $m \times n$ and $q \times p$ matrices and $X \sim \mathcal{N}_{n \times p}(\xi, \Phi \otimes \Sigma)$. Then*

$$(A \otimes B)(X) = AXB^\top \sim \mathcal{N}_{m \times q}\left\{A\xi B^\top, (A\Phi A^\top) \otimes (B\Sigma B^\top)\right\}.$$

Proof: The result follows directly from Proposition C.3 by noting that

$$(A \otimes B)(\xi) = A\xi B^\top$$

and that

$$(A \otimes B)(\Phi \otimes \Sigma)(A \otimes B)^\top = (A\Phi A^\top) \otimes (B\Sigma B^\top),$$

where we have used T6 and T5 of Proposition B.2. □

Next we partition the matrix X into $n \times r$ and $n \times s$ matrices X_1 and X_2 with $p = r + s$ and partition the mean ξ and covariance matrix Σ accordingly. Further, let $\Sigma_{1|2}$ be given as in (C.2).

Proposition C.8 *The conditional distribution of X_1 given $X_2 = x_2$ is*

$$\mathcal{N}_{n\times r}(\xi_1 + (x_2 - \xi_2)\Sigma_{22}^{-}\Sigma_{21}, \Phi \otimes \Sigma_{1|2}).$$

Proof: This follows directly from translation of Proposition C.5. □

Finally we mention that in the case when Φ and Σ are both regular, the density becomes

$$f_{\xi,\Phi\otimes\Sigma}(x) = (2\pi)^{-np/2}(\det \Phi)^{-p/2}(\det \Sigma)^{-n/2} e^{-\operatorname{tr}\{(x-\xi)^\top \Phi^{-1}(x-\xi)\Sigma^{-1}\}/2}. \quad \text{(C.6)}$$

This follows from the properties of the direct product, the expression (C.1) and the calculation

$$\langle u, (\Phi^{-1} \otimes \Sigma^{-1})(u) \rangle = \operatorname{tr}(u^\top \Phi^{-1} u \Sigma^{-1}).$$

C.2 The Wishart distribution

The Wishart distribution is the sampling distribution of the empirical covariance matrix when the variables themselves follow a multivariate normal distribution.

Definition C.9 *A random $p \times p$ matrix S is said to follow a p-dimensional Wishart distribution with parameter Σ and n degrees of freedom if*

$$S \stackrel{\mathcal{D}}{=} X^\top X, \quad \text{where } X \sim \mathcal{N}_{n\times p}(0, I_n \otimes \Sigma).$$

We then write $S \sim \mathcal{W}_p(n, \Sigma)$. If $X_i^\top$ denote the rows of X, then

$$X^\top X = \sum_{i=1}^{n} X_i X_i^\top$$

such that the empirical covariance matrix is equal to $n^{-1}S$. In the case $p = 1$ the Wishart distribution is identical to the χ^2 distribution:

$$\mathcal{W}_1(n, \sigma^2) = \sigma^2 \chi^2(n).$$

Below we list some well-known properties of the Wishart distribution. Proofs not given here can be found in Eaton (1983).

Proposition C.10 *If $S \sim \mathcal{W}_p(n, \Sigma)$, its mean and covariance operator with respect to trace inner product are equal to*

$$\mathbf{E}(S) = n\Sigma, \quad \operatorname{Cov}(S) = 2n\Sigma \otimes \Sigma.$$

The inverse S^{-1} has mean and covariance operator equal to

$$\mathbf{E}(S^{-1}) = \frac{1}{n-p-1}K \quad \text{for } n > p+1$$

$$\text{Cov}(S^{-1}) = \frac{2\left(K \otimes K + \frac{1}{n-p-1}K \odot K\right)}{(n-p)(n-p-1)(n-p-3)} \quad \text{for } n > p+3,$$

where $K = \Sigma^{-1}$.

It is important that the covariance operators are interpreted with respect to the trace inner product

$$\langle A, B \rangle = \text{tr}(AB)$$

on the space $\mathcal{S}_p$ of symmetric $p \times p$ matrices. In particular, the outer product is determined as in (B.6) such that

$$(K \odot K)(A) = \text{tr}(AK)K.$$

If for example we wish to calculate the variance of S_{ij} for $i \neq j$ we must first write

$$S_{ij} = \text{tr}\left\{(\sqrt{2}/2)\tilde{E}_{ij}S\right\} = \langle \tilde{E}_{ij}, S \rangle/\sqrt{2},$$

where $\tilde{E}_{ij}$ is defined in (B.22), and then its variance is calculated as

$$\mathbf{V}(S_{ij}) = \langle \tilde{E}_{ij}, 2n(\Sigma \otimes \Sigma)(\tilde{E}_{ij}) \rangle/2 = n\langle \tilde{E}_{ij}, \Sigma \tilde{E}_{ij} \Sigma \rangle$$
$$= n\,\text{tr}(\tilde{E}_{ij}\Sigma\tilde{E}_{ij}\Sigma) = n(\sigma_{ij}^2 + \sigma_{ii}\sigma_{jj}).$$

The general expression for the covariance of the entries in S becomes

$$\text{Cov}(S_{ij}, S_{uv}) = n(\sigma_{iu}\sigma_{jv} + \sigma_{iv}\sigma_{ju}). \tag{C.7}$$

Similarly for the inverse we have

$$\mathbf{V}\left\{\text{tr}(AS^{-1})\right\} = \frac{2\left(\text{tr}(AKAK) + \frac{\{\text{tr}(AK)\}^2}{(n-p-1)}\right)}{(n-p)(n-p-1)(n-p-3)}$$

and for the single elements of the inverse matrix we find

$$\text{Cov}(S^{ij}, S^{uv}) = \frac{k_{iu}k_{jv} + k_{iv}k_{ju} + \frac{2k_{ij}k_{uv}}{(n-p-1)}}{(n-p)(n-p-1)(n-p-3)}.$$

A direct consequence of the definition of the Wishart distribution is the convolution property.

Proposition C.11 *If S_1 and S_2 are independent and Wishart-distributed as $S_i \sim \mathcal{W}_p(n_i, \Sigma)$, then*

$$S_1 + S_2 \sim \mathcal{W}_p(n_1 + n_2, \Sigma).$$

Next, a transformation property:

Proposition C.12 *If A is a $q \times p$ matrix and $S \sim \mathcal{W}_p(n, \Sigma)$, then*

$$ASA^\top \sim \mathcal{W}_q(n, A\Sigma A^\top).$$

In particular, for $q = 1$ we get that when $S \sim \mathcal{W}_p(n, \Sigma)$, then for any $v \in R^p$,

$$v^\top S v \sim \sigma_v^2 \chi^2(n),$$

where $\sigma_v^2 = v^\top \Sigma v$. The next fact relates the Wishart distribution to a slightly different normal distribution.

Proposition C.13 *If $X \sim \mathcal{N}_{n \times p}(0, \Phi \otimes \Sigma)$ and Φ^- is a generalized inverse to Φ, then*

$$X^\top \Phi^- X \sim \mathcal{W}_p(f, \Sigma),$$

where $f = \operatorname{rank} \Phi$.

And there is a partitioning theorem analogous to that for the χ^2 distribution:

Proposition C.14 *Assume that $X \sim \mathcal{N}_{n \times p}(\xi, I_n \otimes \Sigma)$ and that $A_1, \ldots, A_k$ are $n \times n$ matrices for orthogonal projections onto subspaces $L_1, \ldots, L_k$ of $\mathcal{R}^n$, that is,*

$$A_i A_j = \delta_{ij} A_i \quad \text{and} \quad A_i^\top = A_i.$$

Then $A_1 X, \ldots, A_k X$ are independent and

$$A_i X \sim \mathcal{N}_{n \times p}(A_i \xi, A_i \otimes \Sigma).$$

Further, if $A_i \xi = 0$, then

$$S_i = X^\top A_i X \sim \mathcal{W}_p(f_i, \Sigma),$$

where $f_i = \dim L_i = \operatorname{rank} A_i = \operatorname{tr} A_i$. Further, $S_1, \ldots, S_k$ are independent.

A special case of interest is when A represents the projection onto the $n-1$-dimensional subspace of R^n of vectors that are orthogonal to the constant. Then under the assumption that $X \sim \mathcal{N}_{n \times p}(\mu, I_n \otimes \Sigma)$ with $\mu_{ij} = \xi_j$ for all i, j, we have that

$$SSD = \sum_{i=1}^n (X^i - \bar{X})(X^i - \bar{X})^\top = X^\top A X \sim \mathcal{W}_p(n-1, \Sigma). \tag{C.8}$$

The above properties are all natural extensions of similar results for the χ^2 distribution. The next result is genuinely for the Wishart distribution. We assume Σ to be partitioned as

$$\Sigma = \begin{pmatrix} \Sigma_{11} & \Sigma_{12} \\ \Sigma_{21} & \Sigma_{22} \end{pmatrix},$$

where Σ_{11} has dimensions $r \times r$, $\Sigma_{12} = \Sigma_{21}^T$ is $r \times s$ and Σ_{22} is an $s \times s$ matrix. Further assume S to be partitioned similarly and let

$$\Sigma_{1|2} = \Sigma_{11} - \Sigma_{12}\Sigma_{22}^{-1}\Sigma_{21}, \quad S_{1|2} = S_{11} - S_{12}S_{22}^{-1}S_{21},$$

provided Σ_{22} and S_{22} are both regular.

Proposition C.15 *Let* $S \sim \mathcal{W}_p(n, \Sigma)$ *with* Σ *regular and* $n > s$. *Then* S_{22} *is regular with probability one and*

(a) $S_{1|2}$ *is independent of* (S_{21}, S_{22});

(b) $S_{1|2} \sim \mathcal{W}_r(n - s, \Sigma_{1|2})$;

(c) $S_{22} \sim \mathcal{W}_s(n, \Sigma_{22})$;

(d) $\mathcal{L}(S_{21} \mid S_{22} = s_{22}) = \mathcal{N}_{s \times r}(s_{22}\Sigma_{22}^{-1}\Sigma_{21}, s_{22} \otimes \Sigma_{1|2})$.

In particular, if we consider the statement for the case where $p = s$ and thus $S = S_{22}$, we deduce

Corollary C.16 *If* $S \sim \mathcal{W}_p(n, \Sigma)$, *where* Σ *is regular, then* S *is regular with probability one if and only if* $n \geq p$.

If the condition of the corollary is satisfied, then the distribution has density equal to

$$w_{p,n,\Sigma}(s) = c(p, n)^{-1}(\det \Sigma)^{-n/2}(\det s)^{(n-p-1)/2} \exp\{-\operatorname{tr}(\Sigma^{-1}s)/2\} \tag{C.9}$$

with respect to Lebesgue measure on the set of positive definite matrices. When the Lebesgue measure giving mass one to the unit cube in the space $\mathcal{S}_p$ of symmetric $p \times p$ matrices with inner product $\langle A, B \rangle = \operatorname{tr}(AB)$, the Wishart constant $c(p, n)$ is

$$c(p, n) = 2^{np/2}(2\pi)^{p(p-1)/4} \prod_{i=1}^{p} \Gamma\{(n + 1 - i)/2\}.$$

Finally the result in Proposition C.15 can be strengthened in a special case:

Corollary C.17 *Suppose that in Proposition C.15 we further have that* $\Sigma_{12} = 0$. *Then* $S_{1|2}$, $S_{12}S_{22}^{-1}S_{21}$ *and* S_{22} *are mutually independent and distributed as*

$$S_{1|2} \sim \mathcal{W}_r(n - s, \Sigma_{11}), \quad S_{12}S_{22}^{-1}S_{21} \sim \mathcal{W}_r(s, \Sigma_{11}), \quad S_{22} \sim \mathcal{W}_s(n, \Sigma_{22}).$$

C.3 Other derived distributions

C.3.1 Box-type distributions

Important distributions associated with the Wishart distribution have a common structure (Box 1949). We say that the distribution of a positive random variable X is of *Box type* if it has Laplace transform (defined for sufficiently small t) equal to

$$\phi(t) = \mathbf{E}\,e^{tX}$$
$$= e^{2tz} \prod_{i=1}^m \frac{\Gamma\{u_i(1-2t)+\xi_i\}}{\Gamma\{u_i+\xi_i\}} \prod_{j=1}^n \frac{\Gamma\{v_j+\eta_j\}}{\Gamma\{v_j(1-2t)+\eta_j\}}, \quad \text{(C.10)}$$

where $\sum_i u_i = \sum_j v_j$, $z = \sum_i u_i \log u_i - \sum_j v_j \log v_j$, and further u_i, $u_i+\xi_i$, v_j, $v_j+\eta_j$ are all positive. If $X_1, \ldots, X_k$ are independent and have Box-type distributions, so does $a_1 X_1 + \cdots + a_k X_k$ for any non-negative numbers $a_1, \ldots, a_k$ which are not all zero. This follows from straightforward direct verification as the Laplace transforms multiply.

If B has a beta distribution with parameters (α, β), then the distribution of $X = -\log B$ is of Box type, because we get for the Laplace transform that

$$\phi(t) = \mathbf{E}B^{-t} = \frac{\int_0^1 s^{\alpha-t-1}(1-s)^{\beta-1}\,ds}{\int_0^1 s^{\alpha-1}(1-s)^{\beta-1}\,ds} = \frac{\Gamma(\alpha-t)\Gamma(\alpha+\beta)}{\Gamma(\alpha)\Gamma(\alpha-t+\beta)}.$$

This is of Box type with $m = n = 1$, $u = v = 1/2$, $\xi = \alpha - 1/2$ and $\eta = \alpha + \beta - 1/2$. The negative logarithm of a beta distribution is probably the simplest distribution of Box type. But most distributions associated with testing problems in the multivariate normal distribution are of Box type.

An important feature of Box-type distributions is that they are a unified technique can be used to approximate them. The approximations given below in (C.13) are simple to use and more accurate than the version of the original approximations given by Box (1949) used in Anderson (1984). The new approximations are due to J.L. Jensen (1991, 1995), to whom we refer for further details.

The simplest approximation uses a gamma distribution with correct mean and variance. The mean and variance of X can be found as the first and second derivatives of the logarithm of the Laplace transform, evaluated at $s = 0$. From (C.10) we find the logarithmic derivatives as

$$\mu(t) = 2z - 2\sum_{i=1}^m u_i \psi\{u_i(1-2t)+\xi_i\} + 2\sum_{j=1}^n v_j \psi\{v_j(1-2t)+\eta_j\} \quad \text{(C.11)}$$

and

$$\sigma^2(t) = 4\sum_{i=1}^{m} u_i \psi'\{u_i(1-2t) + \xi_i\} - 4\sum_{j=1}^{n} v_j \psi'\{v_j(1-2t) + \eta_j\}, \quad \text{(C.12)}$$

where ψ is the digamma function and ψ' the trigamma function; see for example Abramowitz and Stegun (1965). The digamma and trigamma functions are the first and second logarithmic derivatives of the gamma function. Algorithms for their computation are given in Bernardo (1976) and Schneider (1978). Abbreviate $\mu(0) = \mu$ and $\sigma^2(0) = \sigma^2$. Then the distribution of X is approximated by a gamma distribution $\Gamma(\mu^2/\sigma^2, \mu/\sigma^2)$, where $\Gamma(\alpha, \beta)$ is the gamma distribution with density

$$f(x) = \frac{\beta^\alpha}{\Gamma(\alpha)} x^{\alpha-1} e^{-\beta x}.$$

A very accurate approximation to $P(X \geq x)$ can be obtained for a fixed x as follows. First determine the value t^* such that $\mu(t^*) = x$ by the iteration $t_{n+1} = t_n + \{x - \mu(t_n)\}/\sigma^2(t_n)$. Next let $\lambda^* = \mu(t^*)^2/\sigma^2(t^*) = x^2/\sigma^2(t^*)$. Then it holds to a high degree of accuracy that

$$P(X \geq x) \approx \phi(t^*) \left(\frac{\lambda^*}{xt^* + \lambda^*}\right)^{\lambda^*} \gamma(xt^* + \lambda^*, \lambda^*), \quad \text{(C.13)}$$

where $\gamma(s, \lambda)$ is the tail of the gamma distribution

$$\gamma(s, \lambda) = \int_s^\infty \Gamma(\lambda)^{-1} u^{\lambda-1} e^{-u} \, du.$$

This can be evaluated from the gamma function and the incomplete gamma function (Bhattacharjee 1970). Note that the expression (C.13) differs slightly from (2.3) in J.L. Jensen (1991), partly because the latter contains a misprint (J.L. Jensen 1995), partly because we have reduced the expression exploiting that $\mu(t^*) = x$. The approximation is particularly accurate for large values of x, as the relative error tends to zero when x tends to ∞.

C.3.2 Wilks's distribution

In multivariate testing problems concerned with linear hypotheses for the mean and block independence for the covariance, the following distributional result is important (Wilks 1932, 1935).

Proposition C.18 *If $S_1 \sim \mathcal{W}_p(f_1, \Sigma)$ and $S_2 \sim \mathcal{W}_p(f_2, \Sigma)$, where S_1 and S_2 are independent, Σ is regular, and $f_1 \geq p$, then*

$$\Lambda = \frac{\det S_1}{\det(S_1 + S_2)} \stackrel{\mathcal{D}}{=} \prod_{i=1}^{p} B_i, \quad \text{(C.14)}$$

where $B_1, \ldots, B_p$ are independent and $B_i \sim \mathcal{B}\{(f_1 + 1 - i)/2, f_2/2\}$.

The distribution so obtained does obviously not depend on Σ and is called *Wilks's distribution* with parameters p, f_1, f_2, and is denoted by $\Lambda(p, f_1, f_2)$. The distribution of $-\log \Lambda$ is clearly of Box type with $u_i = v_j = 1/2$, $\xi_i = (f_1 - i)/2$, and $\eta_j = (f_1 + f_2 - j)/2$, hence the approximations described above apply. Wilks's distribution satisfies the identity

$$\Lambda(p, f_1, f_2) = \Lambda(f_2, f_1 + f_2 - p, p).$$

C.3.3 Test for identical covariances

If $S_1, \ldots, S_k$ are independent with $S_i \sim \mathcal{W}_p(f_i, \Sigma_i)$, where $p \geq f_i$ and Σ_i are all regular, then the deviance for testing identical covariances is

$$d = p\sum_{i=1}^{k} f_i \log f_i - pf \log f + f \log \det s - \sum_{i=1}^{k} f_i \log \det s_i, \qquad (C.15)$$

where $f = \sum_i f_i$ and $s = \sum_i s_i$. If $\Sigma_1 = \cdots = \Sigma_k = \Sigma$, then the distribution of D does not depend on Σ and it has moment-generating function equal to

$$\phi(t) = \exp(2tz) \prod_{i=1}^{m} \prod_{j=1}^{k} \frac{\Gamma\{f_j(1-2t)/2 + (1-i)/2\}}{\Gamma\{(f_j + 1 - i)/2\}}$$

$$\times \prod_{l=1}^{k} \frac{\Gamma\{(f + 1 - l)/2\}}{\Gamma\{f(1-2t)/2 + (1-l)/2\}}$$

with

$$z = \frac{p}{2}\left(\sum_{j=1}^{k} f_j \log f_j - f \log f\right).$$

Hence the distribution is of Box type with $u_{ij} = f_j/2$, $\xi_{ij} = (1-i)/2$, $v_l = f/2$, and $\eta_l = (1-l)/2$.

If $S_1, \ldots, S_k$ are sums of squares and products of residuals from independent linear models each with n_i observations and dimensions q_i of the linear expectation spaces, then $f_i = n_i - q_i$ and the deviance for testing identical covariances becomes

$$d_w = p\sum_{i=1}^{k} n_i \log n_i - pn \log n + n \log \det s - \sum_{i=1}^{k} n_i \log \det s_i, \qquad (C.16)$$

where $n = \sum_i n_i$. If $\Sigma_1 = \cdots = \Sigma_k = \Sigma$, then the distribution of D_w does not depend on Σ and it has moment-generating function equal to

$$\phi(t) = \exp(2tz_w) \prod_{i=1}^{m} \prod_{j=1}^{k} \frac{\Gamma\{n_j(1-2t)/2 + (1 - q_j - i)/2\}}{\Gamma\{(f_j + 1 - i)/2\}}$$

$$\times \prod_{l=1}^{k} \frac{\Gamma\{(f+1-l)/2\}}{\Gamma\{n(1-2t)/2 + (1-q-l)/2\}}$$

with $q = \sum q_j$ and

$$z_w = \frac{p}{2}\left(\sum_{j=1}^{k} n_j \log n_j - n \log n\right).$$

So this distribution is also of Box type with $u_{ij} = n_j/2$, $\xi_{ij} = (1-q_j-i)/2$, $v_l = n/2$, and $\eta_l = (1-q-l)/2$.

Appendix D
Exponential models

In this appendix we give a brief survey of some important results in the exact theory of regular exponential models and the asymptotic theory of curved exponential models.

The exact theory is described in detail in Barndorff-Nielsen (1978) and important parts of it also in Johansen (1979), Brown (1986) and S.T. Jensen (1989). The asymptotic theory was first described by A.H. Andersen (1969) and Berk (1972). The description given here of both cases is mainly based upon Johansen (1979) and S.T. Jensen (1989).

All the results will be stated without proof. The reader should consult the above references for documentation.

D.1 Regular exponential models

D.1.1 Basic terminology

Let $\mathcal{X}$ be a sample space. In the present text $\mathcal{X}$ is typically either a discrete set, $\mathcal{R}^k$, or a product of these. Also let t be a statistic defined on $\mathcal{X}$ taking values in a real Euclidean vector space V with inner product $\langle \cdot, \cdot \rangle$. Further, let μ be a fixed σ-finite measure on $\mathcal{X}$ and

$$\Theta = \left\{ \theta \in V \mid \int e^{\langle \theta, t(x) \rangle} \mu(dx) < \infty \right\}.$$

For $\theta \in \Theta$ we let

$$\psi(\theta) = \log \int e^{\langle \theta, t(x) \rangle} \mu(dx).$$

The *exponential model* generated by (t, μ) has parameter space Θ and densities for P_θ with respect to μ equal to

$$f(x, \theta) = \exp\left\{ \langle \theta, t(x) \rangle - \psi(\theta) \right\}. \tag{D.1}$$

The statistic t is referred to as a *canonical statistic*, θ is a *canonical parameter*, μ is a *base measure*, and ψ is the *cumulant function* of the model.

In the following we assume that Θ is not contained in an affine proper subspace of V. If the set Θ is open, we say that the exponential model is *regular*. The *convex support* of the statistic t is the convex hull of the support of the measure μt^{-1} and is denoted by C. The representation (D.1) of a regular exponential model is *minimal* if no affine proper subspace of V contains $t(X)$ with P_θ-probability one, i.e. if no affine proper subspace of V contains C. In general we will allow representations to be non-minimal and possible over-parametrization of the model. The case where the representation is not minimal is referred to as the *singular case*.

The *order* of an exponential model is the dimension of V in any minimal representation of the model.

The *relative interior* of C is the interior of C when C is considered as a subset of the smallest affine subspace containing it. Thus if the representation is minimal, the relative interior is just the interior. We denote the relative interior of C by $\mathrm{ri}\, C$.

D.1.2 Analytic properties

Much of the structure in a regular exponential model is contained in the analytic properties of the function ψ. We list some of them here. First let $\tau(\theta)$ and $v(\theta)$ denote the gradient vector and Hessian map of $\psi(\theta)$. That is, for $\eta, \rho \in V$ we have

$$\frac{\partial}{\partial h}\psi(\theta + h\eta)\Big|_{h=0} = \langle \tau(\theta), \eta \rangle$$

and

$$\frac{\partial^2}{\partial h \partial k}\psi(\theta + h\eta + k\rho)\Big|_{h=k=0} = \langle \eta, v(\theta)\rho \rangle.$$

Both these are well defined in regular exponential models and they determine the mean and covariance of the statistic $t(X)$ as

$$\mathbf{E}_\theta \langle \eta, t(X) \rangle = \langle \eta, \tau(\theta) \rangle, \quad \mathbf{V}_\theta\{\langle \eta, t(X) \rangle, \langle \rho, t(X) \rangle\} = \langle \eta, v(\theta)\rho \rangle. \quad (D.2)$$

If a basis is chosen for the vector space V and everything is expressed in coordinates, these will be the usual vector of means and covariance matrix of $t(X)$. We therefore allow ourselves to write

$$\tau(\theta) = \mathbf{E}_\theta\{t(X)\}, \quad v(\theta) = \mathbf{V}_\theta\{t(X)\}$$

instead of (D.2). The covariance $v(\theta)$ will be positive definite if and only if the model is minimally represented. In a minimally represented exponential model the map τ is strictly increasing in the sense that

$$(\theta_1 - \theta_2)^\top \{\tau(\theta_1) - \tau(\theta_2)\} > 0. \quad (D.3)$$

The mean value map τ can be used to parametrize the model, since one can show that

$$\tau(\theta) = \tau(\theta^*) \iff P_\theta = P_{\theta^*}.$$

Such a parametrization is called a *mean value parametrization* of the model.

D.1.3 Maximum likelihood estimation

Suppose an observation $X = x_0$ from a regular exponential model is given. From (A.5), $t(X)$ is sufficient, so we let $t_0 = t(x_0)$. If the model is minimally represented, $t(X)$ is complete with respect to $\mathcal{P} = \{P_\theta, \theta \in \Theta\}$. The main result about maximum likelihood estimation is the following.

Theorem D.1 *The likelihood function attains its maximum in Θ if and only if $t_0 \in \operatorname{ri} C$. The maximum is then attained exactly at points that solve the equation*
$$\tau(\theta) = t_0. \tag{D.4}$$
If $\hat{\theta}$ is any particular solution to this equation, any other solution $\hat{\theta}^*$ satisfies $\hat{\theta}^* = \hat{\theta} + u$ where $\langle u, v \rangle$ is constant for $v \in C$ and thus $P_{\hat{\theta}^*} = P_{\hat{\theta}}$.

In particular we note that in the non-singular case there is a unique solution to the equation (D.4).

A consequence of the above result is that the parameter space $\tau(\Theta)$ for the mean value parameter in a regular exponential model is equal to $\operatorname{ri} C$. Also, in the non-singular case, we have the following useful lemma.

Lemma D.2 *Suppose $t_0 \in \operatorname{ri} C$ in a minimally represented, regular exponential model. Then for any $\theta_0 \in \Theta$ the set*
$$\Theta^* = \{\theta \mid L(\theta) \geq L(\theta_0)\} \tag{D.5}$$
is compact.

D.1.4 Affine hypotheses

Consider a regular exponential model in the situation where we want to test the hypothesis that $\theta \in \Theta_0$, where
$$\Theta_0 = \Theta \cap R_0$$
for some affine subspace $R_0 = r + V_0$ of V. Such a hypothesis is called an *affine hypothesis*. The calculation
$$\begin{aligned}
\langle \theta, t(x) \rangle &= \langle \theta - r, t(x) \rangle + \langle r, t(x) \rangle \\
&= \langle \theta - r, \Pi_0 t(x) \rangle + \langle r, t(x) \rangle \\
&= \langle \theta, \Pi_0 t(x) \rangle + \langle r, t(x) - \Pi_0 t(x) \rangle,
\end{aligned}$$
where Π_0 denotes the orthogonal projection onto V_0, shows that the reduced model $\{P_\theta, \theta \in \Theta_0\}$ is again a regular exponential model with $\Pi_0 t(x)$ as its canonical statistic, $\theta \in \Theta_0$ as its canonical parameter and μ_r as its base measure, where
$$\mu_r(dx) = e^{\langle r, t(x) - \Pi_0 t(x) \rangle} \mu(dx).$$

When the original model is regular and minimally represented, so is the submodel. Further, if the maximum likelihood estimate exists in the original model, it will also exist in the smaller model. The standard test statistic to use is the *deviance* given as

$$d(x) = -2\log \frac{L(\hat{\hat{\theta}})}{L(\hat{\theta})} = 2\left\{\langle \hat{\theta} - \hat{\hat{\theta}}, t(x)\rangle + \psi(\hat{\hat{\theta}}) - \psi(\hat{\theta})\right\},$$

where $\hat{\hat{\theta}}$ denotes the maximum likelihood estimate of θ in the reduced model. Large deviances speak against the hypothesis.

Other test statistics can be of interest for a variety of reasons, in particular to obtain high power against specific alternatives.

In most cases, asymptotic theory is needed to evaluate test probabilities for the deviance and other statistics; cf. Section D.2. Occasionally, however, exact results can be obtained by considering the distribution of the test statistic conditionally on $\Pi_0 t(X) = \Pi_0 t_0$. By considering this conditional distribution, the sufficiency of $\Pi_0 t(X)$ under the hypothesis ensures that the test probabilities are independent of the unknown parameter $\theta \in \Theta_0$, thus leading to similar rejection regions. There are other reasons for choosing to evaluate the test probability in this conditional distribution, but it is outside the scope of the present text to discuss general principles of inference in any detail.

In some cases the unconditional distribution of the deviance D does not depend on θ for $\theta \in \Theta_0$. Then, since the canonical statistic $\Pi_0 t(X)$ is complete with respect to the submodel, Basu's theorem (Theorem A.4) ensures that D and $\Pi_0 t(X)$ are independent. It therefore makes no difference whether the conditional or unconditional distribution is considered.

Tests based on probabilities in the conditional distribution given $\Pi_0 t(X)$ will be referred to as *exact conditional tests* or simply *exact tests*.

D.1.5 Iterative computational methods

None of the results so far give direct guidance as to how to compute the maximum likelihood estimates in a full and regular exponential model if it exists. Here iterative methods are in general necessary. The most common is Newton's method applied to the logarithm of the likelihood function. This begins at a value θ_0 and then lets

$$\theta_{n+1} = \theta_n - v(\theta_n)^-(\tau(\theta_n) - t_0),$$

where $v(\theta)^-$ is a generalized inverse to $v(\theta)$, i.e. it satisfies $vv^- v = v$. The convergence is rapid when the starting value is sufficiently close to $\hat{\theta}$.

It is often convenient and more stable to use Newton's method on the reciprocal likelihood function, which leads to the iteration

$$\theta_{n+1} = \theta_n - \tilde{v}(\theta_n)^-(\tau(\theta_n) - t_0),$$

where
$$\tilde{v}(\theta) = v(\theta) + \{\tau(\theta) - t_0\}^{\otimes 2},$$
i.e. $\tilde{v}(\theta)$ is determined as
$$\langle \eta, \tilde{v}(\theta)\rho \rangle = \langle \eta, v(\theta)\rho \rangle + \langle \eta, \tau(\theta) - t_0 \rangle \langle \rho, \tau(\theta) - t_0 \rangle.$$

In fact it has been shown by S.T. Jensen *et al.* (1991) that the latter iteration is always convergent in regular exponential models of order one.

One problem with Newton's method is that it can involve much storage space and processor time to calculate v^- or $\tilde{v}^-$ in models of high order. Another is that its behaviour may depend on where the iteration is started. If the starting value is far from the true maximum, it may not converge or approach the true maximum slowly. Often a combination of Newton's method and the method of iterative partial maximization (see Section A.4) is appropriate. When suitable functions of θ are held fixed, sections in Θ^* of (D.5) appear that satisfy the conditions necessary for Proposition A.3 to apply.

A typical variant of the algorithm of iterative partial maximization is as follows. Consider a full, regular and minimally represented exponential model of order k. Further, assume there is chosen a basis such that $V = \mathcal{R}^k$ and $\theta = (\theta^1, \ldots, \theta^k)$ and choose a fixed element θ_0 in the canonical parameter set Θ. Also let $\mathcal{A}$ denote a class of subsets of $\{1, \ldots, k\}$ that covers all coordinates, i.e. $\cup_{a \in \mathcal{A}} a = \{1, \ldots, k\}$. We next define sections for $\theta^* \in \Theta$:
$$\Theta_a(\theta^*) = \left\{ \theta \in \Theta \mid \theta^i = \theta^{*i}, i \notin a \right\}.$$

These are affine subspaces of Θ and therefore identify full and regular submodels of the original model with canonical statistics $t^a(x) = (t^i(x), i \in a)$. If we now cyclically maximize the likelihood function over the sections $\Theta_a(\theta^*)$, we obtain a convergent algorithm. If we let $\tau^a(\theta)$ denote the expected value of $t^a(X)$ under θ, the problem is at each stage to solve the equation $\tau^a(\theta) = t^a(x)$ for $\theta \in \Theta_a(\theta^*)$.

But this computation may itself be too involved. We can then instead do the following. First suppose that θ_0 can be chosen so that we can easily maximize the likelihood function over $\Theta_a(\theta_0)$, no matter what the observed value of the statistic t^a is. Let $\tilde{\theta}_a$ denote the value that maximizes the likelihood function over $\Theta_a(\theta_0)$ had the statistic been observed to be $\tau^a(\theta^*)$, and $\check{\theta}_a$ that which maximizes over $\Theta_a(\theta_0)$ for the actual observed value $t^a(x)$, i.e.
$$\tau^a(\check{\theta}_a) = t^a(x), \quad \tau^a(\tilde{\theta}_a) = \tau^a(\theta^*). \tag{D.6}$$

Further, define the sections
$$\tilde{\Theta}_a(\theta^*) = \Theta \cap \left\{ \theta^* + \lambda \left(\check{\theta}_a - \tilde{\theta}_a \right), \lambda \in \mathcal{R} \right\}.$$

If $\check{\theta}_a = \tilde{\theta}_a$, these sections consist of the single point θ^*. Otherwise the sections $\Theta_a(\theta^*)$ are affine hypotheses in Θ and form full and regular one-dimensional exponential models. These sections are called *line sections*. We can then maximize the likelihood function over $\tilde{\Theta}_a(\theta^*)$, for example using Newton's method applied to the reciprocal likelihood function as described earlier. This process of maximization is referred to as *line search* and we denote the maximizing value by $T_a(\theta^*)$, i.e.

$$L\{T_a(\theta^*)\} \geq L(\theta) \text{ for all } \theta \in \tilde{\Theta}_a(\theta^*).$$

By repeating these line searches cyclically over the sets $a \in \mathcal{A}$, we obtain another variant of the algorithm of iterative partial maximization. To show that the modification gives a convergent algorithm, we must ensure that if a value θ' maximizes the likelihood function $L(\theta)$ over all line sections, then θ' maximizes $L(\theta)$ globally, and the maximum over any section depends continuously on θ^*. But if we let $\theta^*_\lambda = \theta^* + \lambda(\check{\theta}_a - \tilde{\theta}_a)$, we get

$$\left.\frac{\partial}{\partial \lambda} \log L(\theta^*_\lambda)\right|_{\lambda=0} = (\check{\theta}_a - \tilde{\theta}_a)^\top \{t^a(x) - \tau^a(\theta^*)\}.$$

Since $\check{\theta}_a$ and $\tilde{\theta}_a$ solve the equations (D.6), this expression is by (D.3) strictly positive unless $\check{\theta}_a = \tilde{\theta}_a$, which again is equivalent to $t^a(x) = \tau^a(\theta^*)$. If this happens, the sections consist of a single point and consequently θ^* trivially maximizes L over the full section $\Theta_a(\theta^*)$. Otherwise $\log L(\theta^*_\lambda)$ is strictly increasing in λ at $\lambda = 0$ and its maximum is not at $\theta^*_0 = \theta^*$.

Hence θ^* maximizes L over all line sections $\tilde{\Theta}_a(\theta^*)$ if and only if it holds that $t^a(x) = \tau^a(\theta^*)$ for all $a \in \mathcal{A}$. This system of conditions is equivalent to having $\tau(\theta^*) = t(x)$, which implies that L is globally maximized at θ^*.

Finally, we must realize that the maximizing value $T_a(\theta^*)$ depends continuously on θ^*. We abstain from giving the details of this.

In the special case where $t^a(X)$ is a cut with $\lambda = (\theta_i, i \in a)$ (see Section A.5) we have from Corollary 10.4 of Barndorff-Nielsen (1978) that the difference $\check{\theta}_a - \tilde{\theta}_a$ does not depend on the chosen value of θ_0. Hence we could for a moment assume that $\theta_0 = \theta^*$, which would imply that $\tilde{\theta}_a = \theta^*$, and thus it must hold that

$$\check{\theta}_a - \tilde{\theta}_a = \check{\theta}_a - \theta^*,$$

whereby

$$\tau^a(\theta^* + \check{\theta}_a - \tilde{\theta}_a) = \tau^a(\check{\theta}_a) = t^a(x).$$

Hence $\theta^* + \check{\theta}_a - \tilde{\theta}_a$ maximizes L not only over the line section $\tilde{\Theta}_a(\theta^*)$ but also over the full section $\Theta_a(\theta^*)$.

When t^a is a cut, the iterative step for a fixed subset a changes the joint density as

$$f\{x, T_a(\theta^*)\} = f(x, \theta^*) \frac{f(x, \check{\theta}_a)}{f(x, \tilde{\theta}_a)} = f(x, \theta^*) \frac{f_a(t^a, \check{\theta}_a)}{f_a(t^a, \theta^*)},$$

where $f_a(t^a, \theta)$ denotes the marginal density of t^a. This only depends on θ through coordinates in θ_i with $i \in a$, because t^a is a cut. The algorithm therefore properly deserves the name of *general iterative proportional scaling* if all groups $a \in \mathcal{A}$ make t^a a cut.

When t^a is not a cut, it is sometimes favourable to let θ_0 vary with θ^* or $t(x)$. This can make t^a nearly a cut, which typically speeds up computation.

In practice, it is in general difficult and unnecessary to make a full line search over $\widetilde{\Theta}_a(\theta^*)$. Instead we just look for a new value of θ in the section that has higher likelihood. For example, inspired by the situation where we have a cut, one can choose

$$\lambda^* = \max\left\{\lambda \in \{1, 1/2, 1/4, 1/8, \ldots\} \mid \theta_\lambda^* \in \Theta \text{ and } L(\theta_\lambda^*) > L(\theta^*)\right\}.$$

If we have a cut as above, we have $\lambda^* = 1$ and this would lead to the full maximum over the large section. Otherwise the procedure is called *modified iterative proportional scaling* (Frydenberg and Edwards 1989). This variant of the algorithm is not theoretically known to be convergent, as the maximizing value at each stage is not a continuous function of θ^*. But, as implemented for mixed hierarchical models in the program MIM (Edwards 1995), it has so far appeared to converge in all examples.

D.2 Curved exponential models

There is no essential difference between the asymptotic theory of regular exponential models and that of curved models, so we might as well treat the curved case immediately. However, there are some technicalities in the singular case that could obscure the basic ideas. Therefore the non-singular case will be treated first.

D.2.1 The non-singular case

We consider a regular exponential base model of order k as described in the previous section and assume that the corresponding representation is minimal such that $\dim V = k$. Without loss of generality we can then assume a basis to be chosen such that $V = \mathcal{R}^k$.

Consider now a submodel $\Theta_0 \subseteq \Theta$ given as $\Theta_0 = \phi(B)$, where B is an open subset of $\mathcal{R}^m$ for some m with $1 \leq m \leq k$. Assume that the map ϕ satisfies

(i) ϕ is a homeomorphism onto Θ_0,

(ii) ϕ is twice continuously differentiable,

(iii) the matrix $\partial \phi / \partial \beta$ has full rank m for all $\beta \in B$,

where
$$\left(\frac{\partial \phi}{\partial \beta}\right)_{rs} = \frac{\partial}{\partial \beta_s}\phi_r(\beta), \quad r=1,\ldots,k, s=1,\ldots,m$$
is the matrix of partial derivatives. The condition (i) says that ϕ is continuous, globally one-to-one and has a continuous inverse. Condition (ii) is a smoothness condition that can be replaced by weaker conditions, and the last condition ensures that the hypothesis everywhere locally resembles $\mathcal{R}^m$, i.e. the local dimension is constant.

A hypothesis of this kind is called a *curved hypothesis* and the corresponding reduced exponential model a *curved exponential model*. Its *dimension* is the local dimension m of its parameter space.

Note in particular that if $\Theta_0 = \Theta$ and ϕ is the identity, the conditions are fulfilled, such that a regular exponential model is a special curved model. Also, affine hypotheses are special cases of curved hypotheses. The dimension is then simply the order of the corresponding regular model.

When we have observed $X = x$, the *observed Fisher information* about β under Θ_0 is the matrix function j_0 with entries

$$j_0(\beta;x)_{rs} = -\frac{\partial^2 \log f(x,\phi(\beta))}{\partial \beta_r \partial \beta_s}, \quad r,s = 1,\ldots,m.$$

Denoting as usual the corresponding random variable with the capital letter J_0, the *expected Fisher information* is its expected value $i_0 = \mathbf{E}J_0$ and can be expressed as

$$i_0(\beta) = \frac{\partial \phi}{\partial \beta}^\top v(\phi(\beta))\frac{\partial \phi}{\partial \beta}. \tag{D.7}$$

The condition (iii) is equivalent to $i_0(\beta)$ being positive definite for all $\beta \in B$.

Now let $X_1,\ldots,X_n$ be independent and identically distributed with distribution given by a member of a curved exponential model with true parameter $\theta_0 \in \Theta_0$. Multiplying the densities, we again obtain an exponential model with the same parameter space as before and $t(X_1)+\cdots+t(X_n)$ as canonical statistic. This holds in the base model Θ as well as in the reduced model Θ_0. Let

$$\bar{t}_n = \{t(X_1) + \cdots + t(X_n)\}/n$$

and let $\hat{\beta}_n$ denote the maximum likelihood estimate of β in the reduced model Θ_0 based upon these observations.

The basic asymptotic result is the following:

Proposition D.3 *As $n \to \infty$, the probability that $\hat{\beta}_n$ exists uniquely tends to 1. The estimate is then given as the solution to the equation*

$$\tau(\phi(\beta))^\top \frac{\partial \phi}{\partial \beta}(\beta) = \bar{t}_n^\top \frac{\partial \phi}{\partial \beta}(\beta) \tag{D.8}$$

and
$$\hat{\beta}_n \stackrel{a}{\sim} \mathcal{N}\left(\beta, i_0(\beta)^{-1}/n\right).$$

Here $X_n \stackrel{a}{\sim} \mathcal{N}(\xi_n, a_n \Sigma)$ means that the distribution of $a_n^{-1/2}(X_n - \xi_n)$ converges weakly to a normal distribution with mean zero and covariance matrix Σ.

The asymptotic distribution simplifies when the parameters and distribution decompose simultaneously. If Y is a *cut* in (X, Y) (see Section A.5) the joint density of (X, Y) factorizes as

$$f(x, y \,|\, \beta) = f(x \,|\, y, \kappa) f(y \,|\, \lambda).$$

In this case it follows directly from the definition of the Fisher information that this is block-diagonal such that $\hat{\kappa}$ and $\hat{\lambda}$ are asymptotically independent, since all mixed second derivatives vanish in the observed information and hence also in the expected. Similarly if Z is a *split* in (X, Y, Z), we have

$$f(x, y, z \,|\, \beta) = f(x \,|\, z, \kappa) f(y \,|\, z, \lambda) f(z \,|\, \mu).$$

When there is a split, the estimates $\hat{\kappa}$, $\hat{\lambda}$, and $\hat{\mu}$ are also asymptotically mutually independent, which is seen as in the case of a cut.

Consider next a further subhypothesis $\Theta_1 \subseteq \Theta_0$ with $\Theta_1 = \phi(b(A))$, where A is an open subset of $\mathcal{R}^l$ for some l with $1 \leq l \leq m$ and b satisfies similar conditions as did ϕ. Let us simultaneously consider the hypotheses

$$H : \theta \in \Theta, \quad H_0 : \theta \in \Theta_0, \quad H_1 : \theta \in \Theta_1$$

or, in an equivalent formulation,

$$H : \theta \in \Theta, \quad H_0 : \theta = \phi(\beta),\ \beta \in B, \quad H_1 : \theta = \phi(b(\alpha)),\ \alpha \in A.$$

A third equivalent formulation of H_1 is that H_0 holds and $\beta = b(\alpha)$.

We now omit the index n and let $\hat{\tau} = \bar{t}$ be the average statistic, $\hat{\theta}$ the estimate of θ under H, $\hat{\beta}$ the estimate of β under H_0 and $\hat{\alpha}$ the estimate of α under H_1. Further, we let

$$\hat{\theta}_0 = \phi(\hat{\beta}), \hat{\tau}_0 = \tau(\hat{\theta}_0), \quad \hat{\theta}_1 = \phi(b(\hat{\alpha})), \hat{\tau}_1 = \tau(\hat{\theta}_1).$$

Finally let i_1 denote the information about α under H_1.

Assume now that H_1 holds and fix α at the true value of the parameter. Then the matrices

$$\Pi_0 = \frac{\partial \phi}{\partial \beta} i_0^{-1} \frac{\partial \phi}{\partial \beta}^\top v, \quad \Pi_1 = \frac{\partial \phi}{\partial \beta} \frac{\partial b}{\partial \alpha} i_1^{-1} \frac{\partial b}{\partial \alpha}^\top \frac{\partial \phi}{\partial \beta}^\top v,$$

where expressions are evaluated at the value corresponding to α, are matrices for the orthogonal projections with respect to the inner product determined by v onto the tangent spaces of Θ_0 and Θ_1 at $\theta = \phi(b(\alpha))$.

If we further let $C_r = \tau(\Theta_r)$ for $r = 0, 1$, their transposes $\Pi_r^\top$ are projection matrices with respect to v^{-1} onto the tangent spaces of C_r at $\tau(b(\alpha))$. Alternatively, $\Pi_r^\top = \partial \hat{\tau}_r / \partial \hat{\tau}$.

Letting $X_n \stackrel{a}{=} Y_n$ denote that $X_n - Y_n$ converges to zero in probability, we moreover have the fundamental

Theorem D.4 *Under the assumptions above,*

$$\sqrt{n}(\hat{\theta} - \hat{\theta}_0, \hat{\theta}_0 - \hat{\theta}_1, \hat{\theta}_1 - \theta) \stackrel{a}{=} (I - \Pi_0, \Pi_0 - \Pi_1, \Pi_1)Y \quad \text{(D.9)}$$

$$\sqrt{n}(\hat{\tau} - \hat{\tau}_0, \hat{\tau}_0 - \hat{\tau}_1, \hat{\tau}_1 - \tau) \stackrel{a}{=} (I - \Pi_0^\top, \Pi_0^\top - \Pi_1^\top, \Pi_1^\top)Z, \quad \text{(D.10)}$$

where $Y \sim \mathcal{N}(0, v^{-1})$ and $Z \sim \mathcal{N}(0, v)$. Thus the three components on the left-hand side of the asymptotic equations are asymptotically independent and normally distributed.

It follows that if $\theta \in \Theta_0$ we have

$$\hat{\theta}_0 \stackrel{a}{\sim} \mathcal{N}(\theta, u/n), \quad \hat{\tau}_0 \stackrel{a}{\sim} \mathcal{N}(\tau, w/n), \quad \text{(D.11)}$$

where u and w are found by direct calculation using the expression for Π_0 and (D.7) for the Fisher information. We obtain

$$u = \Pi_0 v^{-1} \Pi_0^\top = \frac{\partial \phi}{\partial \beta} i_0^{-1} \frac{\partial \phi}{\partial \beta}^\top$$

and

$$w = \Pi_0^\top v \Pi_0 = v \frac{\partial \phi}{\partial \beta} i_0^{-1} \frac{\partial \phi}{\partial \beta}^\top v = vuv.$$

Note that u and w are mutual generalized inverses, i.e. we have that $wuw = w$ and $uwu = u$. Another consequence of Theorem D.4 is that if we let

$$\begin{aligned} K_0 &= n(\hat{\theta} - \hat{\theta}_0)^\top v(\theta)(\hat{\theta} - \hat{\theta}_0) \\ K_{01} &= n(\hat{\theta}_0 - \hat{\theta}_1)^\top v(\theta)(\hat{\theta}_0 - \hat{\theta}_1) \\ K &= n(\hat{\theta}_1 - \theta)^\top v(\theta)(\hat{\theta}_1 - \theta), \end{aligned}$$

we have

Proposition D.5 *The approximately quadratic forms K_0 and K_{01} defined above are asymptotically independent and χ^2-distributed with $k - m$ and $m - l$ degrees of freedom and independent of $\hat{\theta}_1$. Finally K is asymptotically distributed as χ^2 with l degrees of freedom.*

The importance of this result lies in the fact that the above quadratic forms approximate the deviance statistics. More precisely, let

$$D_0 = -2 \log \frac{L(\hat{\theta}_0)}{L(\hat{\theta})}, \quad D_{01} = -2 \log \frac{L(\hat{\theta}_1)}{L(\hat{\theta}_0)}, \quad D = -2 \log \frac{L(\theta)}{L(\hat{\theta}_1)}$$

and

$$\begin{aligned}
\tilde{D}_0 &= n(\hat{\tau} - \hat{\tau}_0)^\top v(\theta)^{-1}(\hat{\tau} - \hat{\tau}_0) \\
\tilde{D}_{01} &= n(\hat{\tau}_0 - \hat{\tau}_1)^\top v(\theta)^{-1}(\hat{\tau}_0 - \hat{\tau}_1) \\
\tilde{D} &= n(\hat{\tau}_1 - \tau)^\top v(\theta)^{-1}(\hat{\tau}_1 - \tau),
\end{aligned}$$

as well as

$$\tilde{K}_{01} = n\bigl(\hat{\beta} - b(\hat{\alpha})\bigr)^\top i_0\bigl(b(\hat{\alpha})\bigr)\bigl(\hat{\beta} - b(\hat{\alpha})\bigr), \quad \tilde{K} = (\hat{\alpha} - \alpha)^\top i_1(\hat{\alpha})(\hat{\alpha} - \alpha).$$

Then we have

Proposition D.6 *The following asymptotic equalities hold:*

$$D_0 \stackrel{a}{=} \tilde{D}_0 \stackrel{a}{=} K_0, \quad D_{01} \stackrel{a}{=} \tilde{D}_{01} \stackrel{a}{=} K_{01} \stackrel{a}{=} \tilde{K}_{01}, \quad D \stackrel{a}{=} \tilde{D} \stackrel{a}{=} K \stackrel{a}{=} \tilde{K}.$$

These statistics are therefore all asymptotically χ^2-distributed with appropriate degrees of freedom.

Thus the deviance test of a curved hypotheses assuming a larger one can asymptotically be performed by using any of the quadratic approximations and an appropriate χ^2-distribution. Note that the asymptotic independence implies that it does not matter asymptotically whether the test is conditional on the sufficient statistic under the hypothesis, should this be affine.

Note also that $v(\theta)$ in any of the quadratic expressions can be replaced with either of $v(\hat{\theta})$, $v(\hat{\theta}_0)$ or $v(\hat{\theta}_1)$ without altering the asymptotic distribution as long as the true value of θ is in Θ_1.

D.2.2 The singular case

In the case where the model is not minimally represented, most of the asymptotic results carry over with minor modifications. Essentially one only has to modify at the following points. Inverses have to be replaced with generalized inverses where appropriate and all expressions directly involving the canonical parameter – which is not uniquely defined – should be avoided or at least interpreted with care. Finally the formulation of the smoothness condition is replaced by

(a) $\tau \circ \phi$ is a homeomorphism onto C_0;

(b) $\tau \circ \phi$ is twice continuously differentiable;

(c)   the matrix $\partial(\tau \circ \phi)/\partial \beta = v(\theta)\partial \theta/\partial \beta$ has full rank m for all $\beta \in B$.

The main reason for this is that τ is always in a unique correspondence with the distribution.

The Fisher information matrices are defined exactly as in the non-singular case and condition (c) ensures that these are regular. The expected Fisher information can be calculated by either of the expressions

$$i_0(\beta) = \frac{\partial \tau \circ \phi}{\partial \beta}^{\mathsf{T}} v(\phi(\beta))^{-} \frac{\partial \tau \circ \phi}{\partial \beta} = \frac{\partial \phi}{\partial \beta}^{\mathsf{T}} v(\phi(\beta)) \frac{\partial \phi}{\partial \beta},$$

where $v(\theta)^-$ is any generalized inverse to $v(\theta)$.

Then the basic asymptotic result Proposition D.3 holds verbatim. In Theorem D.4 some care has to be taken to make sense out of (D.9), whereas (D.10) carries over directly as it stands. The same is true for the independence of the components.

Of the various approximate quadratic forms, those involving θ should be treated carefully. Also v^{-1} should be replaced by an arbitrary generalized inverse v^- in the $\tilde{D}$-versions. Otherwise both Propositions D.5 and D.6 hold. Just keep in mind that k is the order of the model rather than the dimension of V.

Bibliography

Abramowitz, M. and Stegun, I. (1965). *Handbook of mathematical functions.* Dover, New York.

Aigner, M. (1979). *Combinatorial theory.* Springer-Verlag, New York.

Akaike, H. (1974). A new look at the statistical model identification. *IEEE Transactions on Automatic Control*, **19**, 716–23.

Almond, R. (1995). *Graphical belief modeling.* Chapman and Hall, London.

Andersen, A. H. (1969). Asymptotic results for exponential families. *Bulletin of the International Statistical Institute*, **43**, 241–2.

Andersen, A. H. (1974). Multidimensional contingency tables. *Scandinavian Journal of Statistics*, **1**, 115–27.

Andersen, E. B. (1977). Multiplicative Poisson models with unequal cell rates. *Scandinavian Journal of Statistics*, **4**, 153–8.

Andersen, H. H., Højbjerre, M., Sørensen, D., and Eriksen, P. S. (1995). *Linear and graphical models for the multivariate complex normal distribution*, Lecture Notes in Statistics 101. Springer-Verlag, New York.

Andersen, S. K., Olesen, K. G., Jensen, F. V., and Jensen, F. (1989). HUGIN — a shell for building Bayesian belief universes for expert systems. In *Proceedings of the 11th international joint conference on artificial intelligence*, pp. 1080–5. Morgan Kaufmann, San Mateo. Also reprinted in Shafer and Pearl (1990).

Anderson, T. W. (1984). *An introduction to multivariate statistical analysis*, (2nd edn). John Wiley and Sons, New York.

Andersson, S. A. and Perlman, M. D. (1993). Lattice models for conditional independence in a multivariate normal distribution. *Annals of Statistics*, **21**, 1318–58.

Andersson, S. A., Madigan, D., and Perlman, M. D. (1995a). A characterization of Markov equivalence classes for acyclic digraphs. Technical Report 287, Department of Statistics, University of Washington, Seattle.

Andersson, S. A., Madigan, D., Perlman, M. D., and Triggs, C. M. (1995b). On the relation between conditional independence models determined by finite distributive lattices and by directed acyclic graphs. *Journal of Statistical Planning and Inference*, **48**, 25–46.

Andersson, S. A., Madigan, D., and Perlman, M. D. (1996). On the Markov equivalence of chain graphs, undirected graphs, and acyclic digraphs. *Scandinavian Journal of Statistics*, **26**, To appear.

Asmussen, S. and Edwards, D. (1983). Collapsibility and response variables in contingency tables. *Biometrika*, **70**, 567–78.

Badsberg, J. H. (1991). A guide to CoCo. Technical Report R-91-43, Department of Mathematics and Computer Science, Aalborg University.

Badsberg, J. H. (1992). Model search in contingency tables by CoCo. In *Computational statistics*, (ed. Y. Dodge and J. Whittaker), pp. 251–6. Physica Verlag, Heidelberg.

Badsberg, J. H. (1995). An environment for graphical models. Unpublished PhD thesis, Aalborg University.

Barndorff-Nielsen, O. E. (1978). *Information and exponential families in statistical theory*. John Wiley and Sons, New York.

Bartlett, M. S. (1935). Contingency table interactions. *Journal of the Royal Statistical Society, Supplement*, **2**, 248–52.

Bartlett, M. S. (1937). Properties of sufficiency and statistical tests. *Proceedings of the Royal Society of London, Series A*, **160**, 268–82.

Basu, D. (1955). On statistics independent of a complete sufficient statistic. *Sankhyā*, **15**, 377–80.

Beeri, C., Fagin, R., Maier, D., Mendelzon, A., Ullman, J., and Yannakakis, M. (1981). Properties of acyclic database schemes. In *Proceedings of 13th annual ACM symposium on the theory of computing, Milwaukee*. Association of Computing Machinery, New York.

Beeri, C., Fagin, R., Maier, D., and Yannakakis, M. (1983). On the desirability of acyclic database schemes. *Journal of the Association of Computing Machinery*, **30**, 479–513.

Berge, C. (1973). *Graphs and hypergraphs*. North-Holland, Amsterdam. Translated from French by E. Minieka.

Berk, R. H. (1972). Consistency and asymptotic normality of maximum likelihood estimates for exponential models. *Annals of Mathematical Statistics*, **43**, 193–204.

Bernardo, J. M. (1976). Psi (digamma) function. Algorithm AS 103. *Applied Statistics*, **25**, 315–17.

Besag, J. and Clifford, P. (1989). Generalized Monte Carlo significance tests. *Biometrika*, **76**, 633–42.

Besag, J. and Green, P. (1993). Spatial statistics and Bayesian computation. *Journal of the Royal Statistical Society, Series B*, **55**, 25–37.

Besag, J., York, J., and Mollié, A. (1991). Bayesian image restoration, with two applications in spatial statistics (with discussion). *Annals of the Institute of Statistical Mathematics*, **43**, 1–59.

Bhattacharjee, G. P. (1970). Algorithm AS 32. The incomplete gamma integral. *Applied Statistics*, **19**, 285–7.

Birch, M. W. (1963). Maximum likelihood in three way contingency tables. *Journal of the Royal Statistical Society, Series B*, **25**, 220–33.

Bishop, Y. M. M. (1971). Effects of collapsing multidimensional contingency tables. *Biometrics*, **27**, 545–62.

Bishop, Y. M. M., Fienberg, S. E., and Holland, P. W. (1975). *Discrete multivariate analysis: theory and practice*. MIT Press, Cambridge, Massachusetts.

Blalock, H. M., Jr (ed.) (1971). *Causal models in the social sciences*. Aldine-Atheston, Chicago.

Borosh, I. and Fraenkel, A. S. (1966). Exact solutions of linear equations with rational coefficients by congruence techniques. *Mathematics of Computation*, **20**, 107–12.

Bouckaert, R. R. and Studený, M. (1995). Chain graphs: Semantics and expressiveness. In *Symbolic and qualitative approaches to reasoning and uncertainty*, (ed. C. Froideveaux and J. Kohlas), pp. 69–76. Lecture Notes in Artificial Intelligence 946. Springer-Verlag, Berlin.

Box, G. E. P. (1949). A general distribution theory for a class of likelihood criteria. *Biometrika*, **36**, 317–46.

Brown, L. D. (1986). *Fundamentals of statistical exponential families*, IMS-monographs, Vol. IX. Institute of Mathematical Statistics, Hayward.

Buhl, S. L. (1993). On the existence of maximum likelihood estimators for graphical Gaussian models. *Scandinavian Journal of Statistics*, **20**, 263–70.

Chen, T. and Fienberg, S. E. (1974). Two-dimensional contingency tables with both completely and partially cross-classified data. *Biometrics*, **30**, 629–42.

Chen, T. and Fienberg, S. E. (1976). The analysis of contingency tables with incompletely classified data. *Biometrics*, **32**, 133–44.

Chentsov, N. N. (1972). *Statistical decision rules and optimal conclusions*. Nauka, Moscow. In Russian.

Christensen, E. S. (1989). Statistical properties of I-projections within exponential families. *Scandinavian Journal of Statistics*, **16**, 307–18.

Christensen, R. (1990). *Log-linear models*. Springer-Verlag, New York.

Clemmensen, J., Hansen, G., Nielsen, A., Røjel, J., Steensberg, J., Sørensen, S., and Toustrup, J. (1974). Lung cancer and air pollution in Fredericia. *Ugeskrift for Læger*, **136**, 2260–8.

Cooper, G. and Herskovits, E. (1992). A Bayesian method for the induction of probabilistic networks from data. *Machine Learning*, **9**, 309–47.

Cox, D. R. and Wermuth, N. (1993). Linear dependencies represented by chain graphs (with discussion). *Statistical Science*, **8**, 204–218; 247–277.

Cox, D. R. and Wermuth, N. (1996). *Multivariate dependencies*. Chapman and Hall, London.

Cramér, H. (1946). *Mathematical methods of statistics*. Princeton University Press, Princeton, N.J.

Csiszár, I. (1975). I-divergence geometry of probability distributions and minimization problems. *Annals of Probability*, **3**, 146–58.

Darroch, J. N. and Ratcliff, D. (1972). Generalized iterative scaling for log-linear models. *Annals of Mathematical Statistics*, **43**, 1470–80.

Darroch, J. N. and Speed, T. P. (1983). Additive and multiplicative models and interactions. *Annals of Statistics*, **11**, 724–38.

Darroch, J. N., Lauritzen, S. L., and Speed, T. P. (1980). Markov fields and log-linear interaction models for contingency tables. *Annals of Statistics*, **8**, 522–39.

Dawid, A. P. (1979). Conditional independence in statistical theory (with discussion). *Journal of the Royal Statistical Society, Series B*, **41**, 1–31.

Dawid, A. P. (1980). Conditional independence for statistical operations. *Annals of Statistics*, **8**, 598–617.

Dawid, A. P. (1984). Statistical theory. The prequential approach (with discussion). *Journal of the Royal Statistical Society, Series A*, **147**, 278–92.

Dawid, A. P. (1990). Contribution to the discussion of David Edwards: Hierarchical interaction models. *Journal of the Royal Statistical Society, Series B*, **52**, 55.

Dawid, A. P. (1992). Applications of a general propagation algorithm for probabilistic expert systems. *Statistics and Computing*, **2**, 25–36.

Dawid, A. P. and Lauritzen, S. L. (1993). Hyper Markov laws in the statistical analysis of decomposable graphical models. *Annals of Statistics*, **21**, 1272–317.

Deming, W. E. and Stephan, F. F. (1940). On a least squares adjustment of a sampled frequency table when the expected marginal totals are known. *Annals of Mathematical Statistics*, **11**, 427–44.

Dempster, A. P. (1972). Covariance selection. *Biometrics*, **28**, 157–75.

Dempster, A. P. (1973). Aspects of the multinomial logit model. In *Proceedings of the 3rd symposium on multivariate analysis, Dayton, Ohio*, (ed. P. R. Krishnaiah), pp. 129–42. Academic Press, New York.

Dempster, A. P., Laird, N., and Rubin, D. B. (1977). Maximum likelihood from incomplete data via the EM algorithm (with discussion). *Journal of the Royal Statistical Society, Series B*, **39**, 1–38.

Diaconis, P. and Sturmfels, B. (1994). Algebraic algorithms for generating from conditional distributions. Unpublished manuscript.

Diestel, R. (1987). Simplicial decompositions of graphs – some uniqueness results. *Journal of Combinatorial Theory, Series B*, **42**, 133–45.

Diestel, R. (1990). *Graph decompositions*. Clarendon Press, Oxford.

Dirac, G. A. (1961). On rigid circuit graphs. *Abhandlungen Mathematisches Seminar Hamburg*, **25**, 71–6.

Eaton, M. L. (1983). *Multivariate statistics. A vector space approach*. John Wiley and Sons, New York.

Edwards, D. (1990). Hierarchical interaction models (with discussion). *Journal of the Royal Statistical Society, Series B*, **52**, 3–20 and 51–72.

Edwards, D. (1995). *Introduction to graphical modelling*. Springer-Verlag, New York.

Edwards, D. and Havránek, T. (1985). A fast procedure for model search in multidimensional contingency tables. *Biometrika*, **72**, 339–51.

Edwards, D. and Havránek, T. (1987). A fast model selection procedure for large families of models. *Journal of the American Statistical Association*, **82**, 205–11.

Edwards, D. and Kreiner, S. (1983). The analysis of contingency tables by graphical models. *Biometrika*, **70**, 553–62.

Eriksen, P. S. (1996). Tests in covariance selection models. *Scandinavian Journal of Statistics*, **23**, To appear.

Fisher, R. A. (1925). *Statistical methods for research workers*. Harner Press, New York.

Frydenberg, M. (1990a). The chain graph Markov property. *Scandinavian Journal of Statistics*, **17**, 333–53.

Frydenberg, M. (1990b). Marginalization and collapsibility in graphical interaction models. *Annals of Statistics*, **18**, 790–805.

Frydenberg, M. and Edwards, D. (1989). A modified iterative proportional scaling algorithm for estimation in regular exponential families. *Computational Statistics and Data Analysis*, **8**, 143–53.

Frydenberg, M. and Jensen, J. L. (1989). Is the 'improved likelihood ratio statistic' really improved in the discrete case. *Biometrika*, **76**, 655–61.

Frydenberg, M. and Lauritzen, S. L. (1989). Decomposition of maximum likelihood in mixed interaction models. *Biometrika*, **76**, 539–55.

Fuchs, C. (1982). Maximum likelihood estimation and model selection in contingency tables with missing data. *Journal of the American Statistical Association*, **77**, 270–8.

Gavril, T. (1972). Algorithms for minimum coloring, maximum clique, minimum coloring by cliques and maximum independent set of a graph. *SIAM Journal of Computing*, **1**, 180–7.

Geiger, D. and Pearl, J. (1990). On the logic of causal models. In *Uncertainty in artificial intelligence IV*, (ed. R. D. Schachter, T. S. Levitt, L. N. Kanal, and J. F. Lemmer), pp. 136–47. North-Holland, Amsterdam.

Geiger, D. and Pearl, J. (1993). Logical and algorithmic properties of conditional independence and graphical models. *Annals of Statistics*, **21**, 2001–21.

Gelfand, A. E. and Smith, A. F. M. (1990). Sampling-based approaches to calculating marginal densities. *Journal of the American Statistical Association*, **85**, 398–409.

Geman, S. and Geman, D. (1984). Stochastic relaxation, Gibbs distributions, and the Bayesian restoration of images. *IEEE Transactions on Pattern Analysis and Machine Intelligence*, **6**, 721–41.

Geng, Z. (1989). Algorithm AS 244. Decomposability and collapsibility for log-linear models. *Applied Statistics*, **38**, 189–97.

Geng, Z. (1992). Collapsibility of relative risk in contingency tables with a response variable. *Journal of the Royal Statistical Society, Series B*, **54**, 585–93.

Geng, Z. and Asano, C. (1988). Recursive procedures for hierarchical log-linear models on high-dimensional contingency tables. *Journal of the Japanese Society for Computational Statistics*, **1**, 17–26.

Geng, Z. and Asano, C. (1993). Strong collapsibility of association measures in linear models. *Journal of the Royal Statistical Society, Series B*, **55**, 741–7.

Geyer, C. J. and Thompson, E. A. (1992). Constrained Monte Carlo maximum likelihood for dependent data (with discussion). *Journal of the Royal Statistical Society, Series B*, **54**, 657–99.

Gibbs, W. (1902). *Elementary principles of statistical mechanics.* Yale University Press, NewHaven, Connecticut.

Gilks, W. R., Clayton, D. G., Spiegelhalter, D. J., Best, N. G., McNeil, A. J., Sharples, L. D., and Kirby, A. J. (1993). Modelling complexity: application of the Gibbs sampler in medicine. *Journal of the Royal Statistical Society, Series B*, **55**, 39–52.

Gilks, W. R., Thomas, A., and Spiegelhalter, D. J. (1994). A language and a program for complex Bayesian modelling. *The Statistician*, **43**, 169–78.

Glonek, G. (1987). Some aspects of log-linear models. Unpublished PhD thesis, Flinders University.

Goldberger, A. S. and Duncan, O. D. (ed.) (1973). *Structural equation models in the social sciences.* Seminar Press, New York.

Golumbic, M. C. (1980). *Algorithmic graph theory and perfect graphs*. Academic Press, London.

Goodman, L. A. (1970). The multivariate analysis of qualitative data: interaction among multiple classifications. *Journal of the American Statistical Association*, **65**, 226–56.

Goodman, L. A. (1971). Partitioning of chi-square, analysis of marginal contingency tables, and estimation of expected frequencies in multidimensional contingency tables. *Journal of the American Statistical Association*, **66**, 339–44.

Goodman, L. A. (1973). The analysis of multidimensional contingency tables when some variables are posterior to others. A modified path analysis approach. *Biometrika*, **60**, 179–92.

Green, P. J. (1990). On use of the EM algorithm for penalized likelihood estimation. *Journal of the Royal Statistical Society, Series B*, **52**, 443–52.

Haberman, S. J. (1974). *The analysis of frequency data*. University of Chicago Press, Chicago.

Hammersley, J. M. and Clifford, P. E. (1971). Markov fields on finite graphs and lattices. Unpublished manuscript.

Hastings, W. K. (1970). Monte Carlo sampling methods using Markov chains and their applications. *Biometrika*, **57**, 97–109.

Heckerman, D., Geiger, D., and Chickering, D. M. (1994). Learning Bayesian networks: the combination of knowledge and statistical data. In *Proceedings of the 10th conference on uncertainty in artificial intelligence*, (ed. R. L. de Mantaras and D. Poole), pp. 293–301. Morgan Kaufmann, San Mateo.

Hocking, R. R. and Oxspring, H. H. (1974). The analysis of partially categorized contingency data. *Biometrics*, **30**, 469–83.

Højsgaard, S. and Thiesson, B. (1995). BIFROST — Block recursive models Induced From Relevant knowledge, Observations, and Statistical Techniques. *Computational Statistics & Data Analysis*, **19**, 155–75.

Ireland, C. T., Ku, H. H., and Kullback, S. (1969). Symmetry and marginal homogeneity of an $r \times r$ contingency table. *Journal of the American Statistical Association*, **64**, 1323–41.

Jensen, C. S., Kong, A., and Kjærulff, U. (1995). Blocking–Gibbs sampling in very large probabilistic expert systems. *International Journal of Human–Computer Studies*, **42**, 647–66.

Jensen, F., Jensen, F. V., and Dittmer, S. L. (1994). From influence diagrams to junction trees. In *Proceedings of the 10th conference on uncertainty in artificial intelligence*, (ed. R. L. de Mantaras and D. Poole), pp. 367–73. Morgan Kaufmann, San Mateo.

Jensen, F. V. (1996). *An introduction to Bayesian networks*. University College London Press, London.

Jensen, F. V. and Jensen, F. (1994). Optimal junction trees. In *Proceedings of the 10th conference on uncertainty in artificial intelligence*, (ed. R. L. de Mantaras and D. Poole), pp. 360–6. Morgan Kaufmann, San Mateo.

Jensen, F. V., Lauritzen, S. L., and Olesen, K. G. (1990). Bayesian updating in causal probabilistic networks by local computation. *Computational Statistics Quarterly*, **4**, 269–82.

Jensen, J. L. (1991). A large deviation type approximation for the Box class of likelihood ratio criteria. *Journal of the American Statistical Association*, **86**, 437–40.

Jensen, J. L. (1995). Correction to: A large deviation type approximation for the Box class of likelihood ratio criteria. *Journal of the American Statistical Association*, **90**, 812.

Jensen, S. T. (1989). *Exponentielle familier*, Lecture notes. Institute of Mathematical Statistics, University of Copenhagen. In Danish.

Jensen, S. T., Johansen, S., and Lauritzen, S. L. (1991). Globally convergent algorithms for maximizing a likelihood function. *Biometrika*, **78**, 867–77.

Johansen, S. (1979). *Introduction to the theory of regular exponential families*, Lecture notes. Institute of Mathematical Statistics, University of Copenhagen.

Jöreskog, K. G. (1973). Analysis of covariance structures. In *Proceedings of the 3rd symposium on multivariate analysis, Dayton, Ohio*, (ed. P. R. Krishnaiah), pp. 263–85. Academic Press, New York.

Jöreskog, K. G. (1977). Structural equation models in the social sciences: specification, estimation and testing. In *Applications of statistics*, (ed. P. R. Krishnaiah), pp. 267–87. North-Holland, Amsterdam.

Jöreskog, K. G. (1981). Analysis of covariance structures. *Scandinavian Journal of Statistics*, **8**, 65–92.

Kellerer, H. G. (1964a). Maßtheoretische Marginalprobleme. *Mathematische Annalen*, **153**, 168–98.

Kellerer, H. G. (1964b). Verteilungsfunktionen mit gegebenen Marginalverteilungen. *Zeitschrift für Wahrscheinlichkeitstheorie und verwandte Gebiete*, **3**, 247–70.

Kiiveri, H. and Speed, T. P. (1982). Structural analysis of multivariate data: a review. In *Sociological methodology*, (ed. S. Leinhardt). Jossey-Bass, San Francisco.

Kiiveri, H., Speed, T. P., and Carlin, J. B. (1984). Recursive causal models. *Journal of the Australian Mathematical Society, Series A*, **36**, 30–52.

Kjærulff, U. (1990). Graph triangulation — algorithms giving small total state space. Technical Report R 90-09, Aalborg University.

Kjærulff, U. (1992). Optimal decomposition of probabilistic networks by simulated annealing. *Statistics and Computing*, **2**, 19–24.

Koster, J. T. A. (1994). Gibbs–factorization and the Markov property. Unpublished manuscript.

Koster, J. T. A. (1996). Markov properties for non-recursive causal models. *Annals of Statistics*, **24**, To appear.

Kreiner, S. (1987). Analysis of multidimensional contingency tables by exact conditional tests: techniques and strategies. *Scandinavian Journal of Statistics*, **14**, 97–112.

Kreiner, S. (1989). User's guide to DIGRAM - a program for discrete graphical modelling. Technical Report 89-10, Statistical Research Unit, University of Copenhagen.

Krzanowski, W. J. (1975). Discrimination and classification using both binary and continuous variables. *Journal of the American Statistical Association*, **70**, 782–90.

Krzanowski, W. J. (1976). Canonical representation of the location model for discrimination and classification. *Journal of the American Statistical Association*, **71**, 845–8.

Kullback, S. (1959). *Information theory and statistics*. John Wiley and Sons, New York.

Kullback, S. and Leibler, R. A. (1951). On information and sufficiency. *Annals of Mathematical Statistics*, **22**, 79–86.

Lancaster, H. O. (1949). The derivation and partition of χ^2 in certain discrete distributions. *Biometrika*, **36**, 117–29.

Lauritzen, S. L. (1985). Test of hypotheses in decomposable mixed interaction models. Technical Report R 85-11, Institute for Electronic Systems, Aalborg University.

Lauritzen, S. L. (1989a). Lectures on contingency tables, (3rd edn). Technical Report R-89-29, Institute for Electronic Systems, Aalborg University.

Lauritzen, S. L. (1989b). Mixed graphical association models (with discussion). *Scandinavian Journal of Statistics*, **16**, 273–306.

Lauritzen, S. L. (1992). Propagation of probabilities, means and variances in mixed graphical association models. *Journal of the American Statistical Association*, **87**, 1098–108.

Lauritzen, S. L. (1995). The EM algorithm for graphical association models with missing data. *Computational Statistics & Data Analysis*, **19**, 191–201.

Lauritzen, S. L. and Spiegelhalter, D. J. (1988). Local computations with probabilities on graphical structures and their application to expert systems (with discussion). *Journal of the Royal Statistical Society, Series B*, **50**, 157–224.

Lauritzen, S. L. and Wermuth, N. (1984). Mixed interaction models. Technical Report R 84-8, Institute for Electronic Systems, Aalborg University.

Lauritzen, S. L. and Wermuth, N. (1989). Graphical models for associations between variables, some of which are qualitative and some quantitative. *Annals of Statistics*, **17**, 31–57.

Lauritzen, S. L., Speed, T. P., and Vijayan, K. (1984). Decomposable graphs and hypergraphs. *Journal of the Australian Mathematical Society, Series A*, **36**, 12–29.

Lauritzen, S. L., Dawid, A. P., Larsen, B. N., and Leimer, H.-G. (1990). Independence properties of directed Markov fields. *Networks*, **20**, 491–505.

Lauritzen, S. L., Thiesson, B., and Spiegelhalter, D. J. (1994). Diagnostic systems created by model selection methods: a case study. In *Selecting models from data: AI and statistics IV*, (ed. P. Cheeseman and R. Oldford), pp. 143–52. Lecture Notes in Statistics 89. Springer-Verlag, New York.

Leimer, H.-G. (1985). Strongly decomposable graphs and hypergraphs, PhD thesis. Berichte zur Stochastik und verwandte Gebiete 85-1, University of Mainz.

Leimer, H.-G. (1989). Triangulated graphs with marked vertices. In *Graph theory in memory of G. A. Dirac*, (ed. L. D. Andersen, C. Thomassen, B. Toft, and P. D. Vestergaard), pp. 311–24. Annals of Discrete Mathematics 41. Elsevier Science Publishers B.V. (North-Holland), Amsterdam.

Leimer, H.-G. (1993). Optimal decomposition by clique separators. *Discrete Mathematics*, **113**, 99–123.

Madigan, D. and Mosurski, K. (1990). An extension of the results of Asmussen and Edwards on collapsibility in contingency tables. *Biometrika*, **77**, 315–19.

Madigan, D. and Raftery, A. (1994). Model selection and accounting for model uncertainty using Occam's window. *Journal of the American Statistical Association*, **89**, 1535–46.

Madigan, D. and York, J. (1995). Bayesian graphical models for discrete data. *International Statistical Review*, **63**, 215–32.

Martin-Löf, P. (1970). *Statistiska modeller*, Lecture notes by Rolf Sundberg. University of Stockholm. In Swedish.

Matúš, F. (1992a). Ascending and descending conditional independence relations. In *Transactions of the 11th Prague conference on information theory, statistical decision functions and random processes*, pp. 189–200. Academia, Prague.

Matúš, F. (1992b). On equivalence of Markov properties over undirected graphs. *Journal of Applied Probability*, **29**, 745–9.

Matúš, F. and Studený, M. (1995). Conditional independences among four random variables I. *Combinatorics, Probability and Computing*, **4**, 269–78.

Metropolis, N., Rosenbluth, A. W., Rosenbluth, M. N., Teller, A. H., and Teller, E. (1953). Equations of state calculations by fast computing machines. *Journal of Chemical Physics*, **21**, 1087–92.

Morgan, W. M. and Blumenstein, B. A. (1991). Exact conditional tests for hierarchical models in multidimensional contingency tables. *Applied Statistics*, **40**, 435–42.

Moussouris, J. (1974). Gibbs and Markov random systems with constraints. *Journal of Statistical Physics*, **10**, 11–33.

Neapolitan, E. (1990). *Probabilistic reasoning in expert systems*. John Wiley and Sons, New York.

Oliver, R. M. and Smith, J. Q. (1990). *Influence diagrams, belief nets and decision analysis*. John Wiley and Sons, Chichester.

Olkin, I. and Tate, R. F. (1961). Multivariate correlation models with mixed discrete and continuous variables. *Annals of Mathematical Statistics*, **32**, 448–65.

Parter, S. (1961). The use of linear graphs in Gauss elimination. *SIAM Review*, **3**, 119–30.

Patefield, W. M. (1981). Algorithm AS 159. An efficient method of generating random $r \times c$ tables with given row and column totals. *Applied Statistics*, **30**, 91–7.

Paz, A. and Geva, R. (1996). Representation of irrelevance relations by annotated graphs. *Fundamenta Informaticae*. To appear.

Pearl, J. (1986a). A constraint propagation approach to probabilistic reasoning. In *Uncertainty in artificial intelligence*, (ed. L. M. Kanal and J. Lemmer), pp. 357–70. North-Holland, Amsterdam.

Pearl, J. (1986b). Fusion, propagation and structuring in belief networks. *Artificial Intelligence*, **29**, 241–88.

Pearl, J. (1988). *Probabilistic reasoning in intelligent systems*. Morgan Kaufmann, San Mateo.

Pearl, J. (1995). Causal diagrams for empirical research (with discussion). *Biometrika*, **82**, 669–710.

Pearl, J. and Paz, A. (1987). Graphoids: a graph based logic for reasoning about relevancy relations. In *Advances in Artificial Intelligence—II*, (ed. B. D. Boulay, D. Hogg, and L. Steel), pp. 357–63. North-Holland, Amsterdam.

Pearl, J. and Verma, T. (1987). The logic of representing dependencies by directed graphs. In *Proceedings of the 6th conference of American Association of Artificial Intelligence*, pp. 374–9. American Association of Artificial Intelligence.

Porteous, B. T. (1985a). Improved likelihood ratio statistics for covariance selection models. *Biometrika*, **72**, 473–5.

Porteous, B. T. (1985b). Properties of log-linear and covariance selection models. Unpublished PhD thesis, University of Cambridge.

Porteous, B. T. (1989). Stochastic inequalities relating a class of log-likelihood ratio statistics to their asymptotic χ^2 distribution. *Annals of Statistics*, **17**, 1723–34.

Range, P. R. (1979). Will he be the first? *The New York Times Magazine*. 11th March, 72–82.

Read, T. R. C. and Cressie, N. A. C. (1988). *Goodness-of-fit statistics for discrete multivariate data*. Springer-Verlag, New York.

Ripley, B. (1987). *Stochastic simulation*. John Wiley and Sons, New York.

Rissanen, J. (1987). Stochastic complexity (with discussion). *Journal of the Royal Statistical Society, Series B*, **49**, 223–239 and 253–265.

Rose, D. J. (1970). Triangulated graphs and the elimination process. *Journal of Mathematical Analysis and its Applications*, **32**, 597–609.

Rose, D. J., Tarjan, R. E., and Lueker, G. S. (1976). Algorithmic aspects of vertex elimination on graphs. *SIAM Journal of Computing*, **5**, 266–83.

Sakamoto, Y. and Akaike, H. (1978). Analysis of cross-classified data by AIC. *Annals of the Institute of Statistical Mathematics*, **30**, 185–97.

Schneider, B. E. (1978). The trigamma function. Algorithm AS 121. *Applied Statistics*, **27**, 97–9.

Schwarz, G. (1978). Estimating the dimension of a model. *Annals of Statistics*, **6**, 461–4.

Shafer, G. R. (1976). *A mathematical theory of evidence*. Princeton University Press, Princeton, N.J.

Shafer, G. R. (1996). *The art of causal conjecture*. MIT Press, Cambridge, Massachusetts.

Shafer, G. R. and Pearl, J. (ed.) (1990). *Readings in uncertain reasoning*. Morgan Kaufmann, San Mateo.

Sheehan, N. and Thomas, A. (1993). On the irreducibility of a Markov chain defined on a space of genotype configurations by a sampling scheme. *Biometrics*, **49**, 163–75.

Shenoy, P. P. and Shafer, G. R. (1990). Axioms for probability and belief-function propagation. In *Uncertainty in artificial intelligence IV*, (ed. R. D. Shachter, T. S. Levitt, L. N. Kanal, and J. F. Lemmer), pp. 169–98. North-Holland, Amsterdam.

Simpson, E. H. (1951). The interpretation of interaction in contingency tables. *Journal of the Royal Statistical Society, Series B*, **13**, 238–41.

Smith, A. F. M. and Roberts, G. O. (1993). Bayesian computation via the Gibbs sampler and related Markov chain Monte Carlo methods. *Journal of the Royal Statistical Society, Series B*, **55**, 3–23.

Smith, J. Q. (1989). Influence diagrams for statistical modelling. *Annals of Statistics*, **17**, 654–72.

Smith, P. W. F. (1992). Assessing the power of model selection procedures used when graphical modelling. In *Computational statistics*, (ed. Y. Dodge and J. Whittaker), pp. 275–80. Physica Verlag, Heidelberg.

Speed, T. P. (1979). A note on nearest-neighbour Gibbs and Markov probabilities. *Sankhyā, Series A*, **41**, 184–97.

Speed, T. P. and Kiiveri, H. (1986). Gaussian Markov distributions over finite graphs. *Annals of Statistics*, **14**, 138–50.

Spiegelhalter, D. J. and Lauritzen, S. L. (1990). Sequential updating of conditional probabilities on directed graphical structures. *Networks*, **20**, 579–605.

Spiegelhalter, D. J., Dawid, A. P., Lauritzen, S. L., and Cowell, R. G. (1993). Bayesian analysis in expert systems (with discussion). *Statistical Science*, **8**, 219–247 and 204–283.

Spirtes, P., Glymour, C., and Scheines, R. (1993). *Causation, prediction and search*. Lecture Notes in Statistics 81. Springer-Verlag, New York.

Stone, M. (1974a). Cross-validation and multinomial prediction. *Biometrika*, **61**, 509–15.

Stone, M. (1974b). Cross-validatory choice and assessment of statistical predictions. *Journal of the Royal Statistical Society, Series B*, **36**, 111–33.

Studený, M. (1989). Multiinformation and the problem of characterization of conditional independence relations. *Problems of Control and Information Theory*, **18**, 3–16.

Studený, M. (1992). Conditional independence relations have no finite complete characterization. In *Transactions of the 11th Prague conference on information theory, statistical decision functions and random processes*, pp. 377–96. Academia, Prague.

Studený, M. (1993). Structural semigraphoids. *International Journal of General Systems*, **22**, 207–17.

Sundberg, R. (1975). Some results about decomposable (or Markov-type) models for multidimensional contingency tables: Distribution of marginals and partitioning of tests. *Scandinavian Journal of Statistics*, **2**, 71–9.

Tarjan, R. E. (1985). Decomposition by clique separators. *Discrete Mathematics*, **55**, 221–32.

Tarjan, R. E. and Yannakakis, M. (1984). Simple linear-time algorithms to test chordality of graphs, test acyclicity of hypergraphs, and selectively reduce acyclic hypergraphs. *SIAM Journal of Computing*, **13**, 566–79.

Thiesson, B. (1991). (G)EM algorithms for maximum likelihood in recursive graphical association models. Unpublished Master's thesis, Aalborg University.

Thompson, E. A. and Guo, S. W. (1991). Evaluation of likelihood ratios in complex genetic models. *IMA Journal of Mathematics Applied in Medicine and Biology*, **8**, 149–69.

Verma, T. and Pearl, J. (1990a). Causal networks: semantics and expressiveness. In *Uncertainty in artificial intelligence IV*, (ed. R. D. Schachter, T. S. Levitt, L. N. Kanal, and J. F. Lemmer), pp. 69–76. North-Holland, Amsterdam.

Verma, T. and Pearl, J. (1990b). Equivalence and synthesis of causal models. In *Proceedings of the 6th conference on uncertainty in artificial intelligence*, (ed. P. Bonissone, M. Henrion, L. N. Kanal, and J. F. Lemmer), pp. 255–70. North-Holland, Amsterdam.

Vorobev, N. N. (1962). Consistent families of measures and their extensions. *Theory of Probability and its Applications*, **7**, 147–63.

Vorobev, N. N. (1963). Markov measures and Markov extensions. *Theory of Probability and its Applications*, **8**, 420–9.

Vorobev, N. N. (1967). Coalition games. *Theory of Probability and its Applications*, **12**, 250–66.

Wagner, K. (1937). Über eine Eigenschaft der ebenen Komplexe. *Mathematische Annalen*, **114**, 570–90.

Wedelin, D. (1993). Efficient algorithms for probabilistic inference, combinatorial optimization, and the discovery of causal structure from data. Unpublished PhD thesis, University of Gothenburg.

Wermuth, N. (1976a). Analogies between multiplicative models in contingency tables and covariance selection. *Biometrics*, **32**, 95–108.

Wermuth, N. (1976b). Model search among multiplicative models. *Biometrics*, **32**, 253–63.

Wermuth, N. (1980). Linear recursive equations, covariance selection and path analysis. *Journal of the American Statistical Association*, **75**, 963–72.

Wermuth, N. (1987). Parametric collapsibility and the lack of moderating effects in contingency tables. *Journal of the Royal Statistical Society, Series B*, **49**, 353–64.

Wermuth, N. (1989a). Moderating effects in multivariate normal distributions. *Methodika*, **3**, 74–93.

Wermuth, N. (1989b). Moderating effects of subgroups in linear models. *Biometrika*, **76**, 81–92.

Wermuth, N. (1992). Block-recursive regression equations (with discussion). *Revista Brasileira de Probabilidade e Estatistica*, **6**, 1–56.

Wermuth, N. and Lauritzen, S. L. (1983). Graphical and recursive models for contingency tables. *Biometrika*, **70**, 537–52.

Wermuth, N. and Lauritzen, S. L. (1990). On substantive research hypotheses, conditional independence graphs and graphical chain models (with discussion). *Journal of the Royal Statistical Society, Series B*, **52**, 21–72.

Whittaker, J. (1984). Fitting all possible decomposable and graphical models to multiway contingency tables. In *COMPSTAT 84*, (ed. T. Havránek), pp. 98–108. Physica Verlag, Wien.

Whittaker, J. (1990a). Contribution to the discussion of David Edwards: Hierarchical interaction models. *Journal of the Royal Statistical Society, Series B*, **52**, 54.

Whittaker, J. (1990b). *Graphical models in applied multivariate statistics*. John Wiley and Sons, Chichester.

Whittemore, A. (1978). Collapsibility of multidimensional contingency tables. *Journal of the Royal Statistical Society, Series B*, **40**, 328–40.

Wilks, S. S. (1932). Certain generalizations in the analysis of variance. *Biometrika*, **24**, 471–94.

Wilks, S. S. (1935). On the independence of k sets of normally distributed statistical variables. *Econometrica*, **3**, 309–26.

Wold, H. D. A. (1954). Causality and econometrics. *Econometrica*, **22**, 162–77.

Wold, H. D. A. (1960). A generalization of causal chain models. *Econometrica*, **28**, 443–63.

Wright, S. (1921). Correlation and causation. *Journal of Agricultural Research*, **20**, 557–85.

Wright, S. (1923). The theory of path coefficients: a reply to Niles' criticism. *Genetics*, **8**, 239–55.

Wright, S. (1934). The method of path coefficients. *Annals of Mathematical Statistics*, **5**, 161–215.

Yule, G. U. (1903). Notes on the theory of association of attributes in statistics. *Biometrika*, **2**, 121–34.

Zadeh, L. A. (1965). Fuzzy sets. *Information and Control*, **8**, 338–53.

Zadeh, L. A. (1978). Fuzzy sets as a basis for a theory of possibility. *Fuzzy Sets and Systems*, **1**, 3–28.

Index

adjacent 5
affine hypothesis 132, 268
ancestor 6
ancestral set 6

Basu's theorem 241
beta distribution 262
BIFROST 230
block-recursive
    Markov property 54
block-recursive model 113
    decomposable 119
    graphical 114
    hierarchical 118
boundary 5
Box type 171, 192, 262
BUGS 233

canonical parameter 132, 266
cells 67, 158, 246
    marginal 68, 246
CG distribution 158, 159
    canonical characteristics 159
    conditional 164
        canonical characteristics 164
        moment characteristics 165
    homogeneous 160
    marginal 160, 161
        moment characteristics 161, 162
        weak 162, 165
    mixed characteristics 160
    moment characteristics 159
CG regression 165, 217
    homogeneous 165
chain 6
    active 48
    blocked 48
    dependence 7, 113
chain component 7
    terminal 7

chain graph 7
    factorization 53
    Markov property 54
chain graph model 114
    estimation 115, 154, 218
characteristic function 253
child 5
chord 9
classification criteria 67
clique 5
closure 6
CoCo 84, 101, 105, 230
collapsibility 121, 156, 219
complete
    graph 5
    statistic 268, 269
    subset 5
component
    chain 7
    connectivity 6
concentration 251
    matrix 129, 251
    operator 251
conditional independence 28, 32
    normal distribution 129
contingency table 67
    asymptotic distributions 76
    cells 67
    counts 68
    dimension 68
    levels 67
    log–affine model 72
        deviance 79
        estimation 74
        extended 72
    log–linear model 72
    marginal 68
        cells 68
        counts 68
    sampling schemes 69, 70

contingency table (*cont.*)
   saturated model 70
      estimation 71
   slice 68
   structural zeros 72
counts 68
covariance 250
   matrix 251
   operator 251
covariance selection model 131
   decomposable 144
      estimation 145
      test 149, 151, 152
   decomposition 136
   direct join 136
   estimation 133, 135, 138
   likelihood equations 133
   likelihood function 132
   test 142
cross-product ratio
   partial 37
curved exponential model 273
cut 204, 242, 274
   mixed model 183
cycle 6
   directed 6

d-separation 48
decomposable
   graph 8
   hypergraph 23
decomposable model 90
   asymptotics 95
   estimation 91
      sampling distribution 93
   test 98
decomposition 7
   proper 8
   strong 7
   weak 8
dependence chain 7, 53, 113
descendant 6
DIGRAM 101
direct join 22
   covariance selection model 136
   hierarchical model 84
divergence 162, 238

edge 4
   directed 4
   undirected 4
EM algorithm 233
exact test 269
expectation 250
expert system 221
exponential model 266
   base measure 266
   canonical parameter 266
   canonical statistic 266
   convex support 267
   curved 272, 273
   order 267
   regular 267

factor subspace 74, 246
factorization
   chain graph 53
   recursive 46
Fisher information 273
forest 7
   junction 24

gamma distribution 263
Gaussian graphical model 131
generalized inverse 245, 251
generating class 82, 249
   graphical 89
   mixed 202
      graphical 203
Gibbs sampler 231
graph 4
   chain 7
   complete 5
   decomposable 8
   directed 5
   directed version 5
   dual 33
   interaction 88
   marked 4
   moral 7
   perfect 7
   simple 4
   star 10
   triangulated 9
   undirected 5

INDEX

undirected version 5
weakly decomposable 9
graphical hypergraph 22, 89
graphical model 64, 89
   decomposable 90
   recursive 64
graphoid 30

hierarchical model 81
   dimension 86
   direct join 84
   estimation 82
   graphical 89
   subspace 247
      mixed 213, 215
   test 85, 87
hierarchical models
   likelihood equations 82
history 15
HUGIN 227, 236
hyper Markov 230
   directed 109
   property 85, 94, 148
hyperedge 21
hypergraph 21
   clique 21
   decomposable 23
   graphical 22, 89
   reduced 22
   simple 21

information inequality 238
interaction 1, 82, 131, 248
   CG 173
   discrete 174
   linear 174
   mixed 174
   quadratic 174
interaction graph 88, 203
IPS-algorithm 83, 134
iterative partial maximization 135, 239, 271
iterative proportional scaling 82
   covariance selection model 134
   general 272
   modified 206, 272

join
   direct 22
junction
   forest 24
   property 24
   tree 24, 227

Kronecker product 244
Kullback–Leibler divergence 162, 238

Lancaster components 101
learning 228
levels 67
line search 271
log–affine model 67, 72
log–linear model 64, 72
logarithm 237

Möbius inversion 36, 239
main effect 174, 248
   discrete 174
   linear 174
   quadratic 174
marginal
   cells 246
   table 68, 246
Markov chain Monte Carlo 231
Markov equivalence 60
Markov property
   chain graph
      block-recursive 54
      global 55
      local 55
      pairwise 55
   directed
      global 47
      local 50
      pairwise 50
   hyper 85, 94, 148
   undirected 131
      decomposition 42–44
      extended 40
      factorization 34, 36
      global 32
      local 32
      pairwise 32, 36
      positive 36

MCMC 231
mean 250
MIM 101, 201, 230
MIM model 201, 203
mixed interaction model
    decomposable 187
        estimation 188
        test 191
    graphical 173
        decomposition 179
        estimation 175
        likelihood equations 178
    hierarchical 199
        estimation 205
        likelihood equations 206
mixed model
    cut 183, 204
    dimension 172
    generating class 202
    graphical 203
    likelihood function 172
    sampling distributions 168
    saturated 168
        estimation 169
        homogeneous 168
        test 170

neighbour 5
non-adjacent 5
non-descendant 6
normal distribution 254
    concentration 255
    conditional 256, 258
    conditional independence 129
    covariance 254
    density 254, 258
    marginal 256

outer product 246, 259

parent 5
partial correlation coefficient 130
partial regression coefficient 130
path 6
perfect
    directed version 18
    graph 7

numbering 15
sequence 14
probability propagation 227, 235

recursive graphical model
    estimation 108
recursive hierarchical model
    estimation 112
residual 15
running intersection property 15, 25

saturated model
    contingency table 70
    Gaussian 124
        estimation 125
        likelihood function 124
    mixed 168
        estimation 169
    recursive 107
semi-graphoid 30
separate 6
separator 6, 15
simplicial 13
split 242, 274
statistic
    complete 241, 268, 269
    sufficient 241
sufficiency 241

tree 6
    junction 24, 227

vertex 4
    continuous 4
    discrete 4
    simplicial 13
    terminal 7

Wilks's distribution 263
Wishart distribution 258
    constant 261
    covariance 258
    density 261
    mean 258

Yule–Simpson paradox 65